D0758796

ASTEROIDS:
Their Nature and Utilization

THE ELLIS HORWOOD LIBRARY OF SPACE SCIENCE AND SPACE TECHNOLOGY
SERIES IN ASTRONOMY

Series Editor: JOHN MASON
Consultant Editor: PATRICK MOORE

In establishing this new series, we aim to co-ordinate a team of international authors of the highest reputation, integrity and expertise in all aspects of astronomy. This series will make a valuable contribution to the existing literature encompassing all areas of astronomical research. The titles in this series will be illustrated with both black and white and colour photographs, and include many line drawings and diagrams, with tabular data and extensive bibliographies. Aimed at a wide readership, the books will appeal to the professional astronomer, undergraduate students, the high-flying 'A' level student, and the non-scientist with a keen interest in astronomy.

PLANETARY VOLCANISM*
PETER CATTERMOLE, Department of Geology, Sheffield University, UK
SATELLITE ASTRONOMY: The Principles and Practice of Astronomy from Space
JOHN K. DAVIES, Royal Observatory, Edinburgh, UK
THE DUSTY UNIVERSE*
ANEURIN EVANS, Department of Physics, University of Keele, UK
SPACE-TIME AND THEORETICAL COSMOLOGY*
MICHEL HELLER, Department of Philosophy, University of Cracow, Poland
ASTEROIDS: Their Nature and Utilization
CHARLES T. KOWAL, Space Telescope Institute, Baltimore, Maryland, USA
ELECTRONIC AND COMPUTER-AIDED ASTRONOMY*
IAN S. McLEAN, United Kingdom Infrared Telescope Unit, Hilo, Hawaii, USA
URANUS: The Planet, Rings and Satellites*
ELLIS D. MINER, Jet Propulsion Laboratory, Pasadena, California, USA
THE PLANET NEPTUNE*
PATRICK MOORE
ACTIVE GALACTIC NUCLEI*
IAN ROBSON, Director of Observatories, Lancashire Polytechnic, Preston, UK
ASTRONOMICAL OBSERVATIONS FROM THE ANCIENT ORIENT*
RICHARD F. STEPHENSON, Department of Physics, Durham University, Durham, UK
THE TERRESTRIAL PLANETS FROM SPACECRAFT*
YURI A. SURKHOV, Chief of the Laboratory of Geochemistry of the Planets, USSR Academy of Sciences, Moscow, USSR
THE HIDDEN UNIVERSE*
ROGER J. TAYLER, Astronomy Centre, University of Sussex, UK
THE ORIGIN OF THE SOLAR SYSTEM*
MICHAEL M. WOOLFSON, Department of Physics, University of York, UK and
JOHN R. DORMAND, Department of Mathematics and Statistics, Teeside Polytechnic, Middlesborough, UK
TO THE EDGE OF THE UNIVERSE*
ALAN and HILARY WRIGHT, Australian National Radio Astronomy Observatory, Parkes, New South Wales

* *In preparation*

ASTEROIDS: Their Nature and Utilization

CHARLES T. KOWAL
Computer Sciences Corporation
Space Science Telescope Institute
Balitmore, Maryland, USA

ELLIS HORWOOD LIMITED
Publishers · Chichester

Halsted Press: a division of
JOHN WILEY & SONS
New York · Chichester · Brisbane · Toronto

First published in 1988 by
ELLIS HORWOOD LIMITED
Market Cross House, Cooper Street,
Chichester, West Sussex, PO19 1EB, England
The publisher's colophon is reproduced from James Gillison's drawing of the ancient Market Cross, Chichester.

Distributors:
Australia and New Zealand:
JACARANDA WILEY LIMITED
GPO Box 859, Brisbane, Queensland 4001, Australia
Canada:
JOHN WILEY & SONS CANADA LIMITED
22 Worcester Road, Rexdale, Ontario, Canada
Europe and Africa:
JOHN WILEY & SONS LIMITED
Baffins Lane, Chichester, West Sussex, England
North and South America and the rest of the world:
Halsted Press: a division of
JOHN WILEY & SONS
605 Third Avenue, New York, NY 10158, USA
South-East Asia
JOHN WILEY & SONS (SEA) PTE LIMITED
37 Jalan Pemimpin # 05–04
Block B, Union Industrial Building, Singapore 2057
Indian Subcontinent
WILEY EASTERN LIMITED
4835/24 Ansari Road
Daryaganj, New Delhi 110002, India

British Library Cataloguing in Publication Data
Kowal, Charles T.
Asteroids
1. Asteroids
I. Title
523.4'4

Library of Congress Card No. 88–13454

ISBN 0–7458–0136–6 (Ellis Horwood Limited)
ISBN 0–470–21230–6 (Halsted Press)

Typeset in Times by Ellis Horwood Limited
Printed in Great Britain by Hartnolls, Bodmin

Table of contents

Introduction page 7

Astronomical and geological significance of the asteroids. Reasons for current activity in this field. What we can learn about the asteroids from ground-based observations. Source of meteorites. Future utilization. Basic facts and definitions — what is an asteroid? Asteroids and the amateur astronomer.

Chapter 1 — History and cataloging page 11

The Titius–Bode Law. The 'missing planet'. Piazzi's discovery of Ceres. Gauss's invention of a method for the computation of orbits. More discoveries — the 'missing planet' is a whole swarm of planetoids. Visual searches for asteroids. Photographic searches. Unusual asteroids. Trojans. The Lagrangian points. Hidalgo and Chiron. The Golden Age of asteroid discovery. Apollos, Amors, and Atens. Collision with the Earth. Cataloging. The Minor Planet Center. Naming and numbering the asteroids. Keeping track of the asteroids. Appendix: A chronology of asteroid studies.

Chapter 2 — The Main Belt asteroids page 21

Distribution of the asteroids in space. Kirkwood gaps, resonances, and commensurabilities. Asteroid families. Collisions between asteroids. Almost all asteroids are fragments from collisions. A plot of asteroid orbital parameters. What we can learn from such plots. Groups. Secular resonances. The McDonald Survey. The Palomar–Leiden Survey. Size distribution. IRAS. Factors affecting the brightness of an asteroid. Rotation, shape, and light curves. Appendix: The 30 largest asteroids.

Chapter 3 — Observational techniques page 35

New techniques allow us to determine the composition of tiny rocks, billions of miles away. Determining surface composition through photometry and spectrophotometry. Types of detectors. Compositional types of the asteroids. Obtaining sizes and temperatures through infrared radiometry. Polarimetry. Surface texture. Comparison with meteorites. Types of meteorites, and corresponding asteroid types. Occultations. Mass and density of the asteroids.

Chapter 4 — The nature of the asteroids page 49

A trip through the asteroid belt. What the asteroids look like, and the changes in their appearance as a function of location. How the distribution of asteroid types supports current theories of the origin of the solar system. Thermal history. Methods of heating small bodies. Evolution. Alternative theory of asteroid origin — the exploded planet. Descriptions of some representative asteroids and families. Captured asteroids.

Chapter 5 — Asteroids beyond the Main Belt page 63

Physical and orbital characteristics of the Hildas, Thule, and the Trojans. Organic compounds on the asteroids. The shape and origin of Hektor, Hidalgo, and Chiron. Possible relation to comets. A personal account of the discovery of Chiron. Chiron's past and future. Theories of origin.

Chapter 6 — Asteroids near the Earth page 69

Orbital characteristics of the Apollos, Amors, and Atens. Number of objects. Collisions with planets. Cratering rates. Asteroids and the dinosaurs. Origin of the Apollos. Relation to comets and meteors. Composition of the Apollos. Oljato. 1983TB. Radar studies of 1986DA.

Chapter 7 — Utilization of the asteroids page 77

Earth's limitations. Reasons for mining the asteroids. Relative advantages over mining the Moon. Retrieving an asteroid. Solar power satellites. Space habitats. Colonizing the solar system. Candidates for a space mission.

Chapter 8 — The future of asteroid research page 83

The Hubble Space Telescope. The Shuttle Infrared Telescope Facility. Being there. Galileo. The Comet-Rendezvous Asteroid-Flyby mission. Exploration of the Main Belt. Near-Earth asteroids. Remote sensing. Ground-based research. The Spacewatch Camera.

Appendix A — Asteroids and the amateur astronomer page 89

Appendix B — The numbered asteroids page 93

The plates page 137

Glossary page 145

Bibliography page 149

Index page 151

Introduction

The study of the asteroids, or 'minor planets' has had a checkered history among astronomers. When the first asteroid was discovered, in 1801, it was a major event in the history of science, apparently confirming speculation about the existence of a 'missing planet' between the orbits of Mars and Jupiter. As more asteroids were found in the next few years, the situation became even more intriguing. Were these objects the remnants of a planet that had disintegrated? Or were they fragments which had never coalesced into a single planet? These questions are unanswered even today.

Within a few years, *dozens* of asteroids were discovered. Then, after the advent of astronomical photography, asteroids were discovered by the *hundreds*. Today, over 3000 asteroids have been observed well enough for their orbits to be determined accurately, and many thousands more have been observed only briefly.

By the middle of the twentieth century, many astronomers had come to regard asteroids as common pieces of 'junk'. The little trails they made on astronomical photographs were merely nuisances. After all, once you find an asteroid, and determine its orbit, what more can you do with it?

The situation is changing once again. Since the 1960s, advances in astronomical techniques have made it possible to actually determine the mineralogical composition of these small objects. Using our instruments here on the Earth, we can determine the nature of the asteroid surfaces, and the sizes of these objects, which are tens of millions of kilometers away. Now, in answer to the question, 'What can you do with an asteroid?', we can reply: 'You can use photometry, spectrophotometry, radiometry, and polarimetry, to determine the size, shape, temperature, and composition of the asteroid!' Furthermore, the information thus gained tells us much about the origin and history of the solar system as a whole. These new capabilities, combined with the exciting plans to send space probes to the asteroids, have again made the asteroids the center of attention of a large number of scientists.

Space probe photographs of the cratered surfaces of Mars, Mercury, and the planetary satellites have shown that meteoritic and asteroidal bombardment has been an important process in the formation and development of the planets, including the Earth. For this reason, geologists, as well as astronomers, are now studying the asteroids.

Aside from the scientific importance of the asteroids, these objects may one day affect the lives of everyone on the Earth. The asteroids are a rich storehouse of valuable materials such as iron, nickel, carbonaceous compounds, and water. The utilization of this mineral supply could put an end to the 'limits to growth' imposed by the dwindling natural resources of the Earth. By using the vast resources of the asteroids, we can reach out to colonize space, and can enrich our lives here on Earth.

When we finally venture into space in a big way — with gigantic satellites, and space colonies housing 10 000 people — it will no longer be feasible to launch the millions of tons of needed materials from the surface of the Earth. The construction of these huge facilities will have to be done in space, and most of the raw materials for this construction will have to come from an object which is already in space — i.e., the Moon, or the asteroids. While the Moon can be used as a source of raw materials, its surface is not rich in some of the important minerals. Water seems to be completely absent on the Moon. The Moon's gravity is also a substantial barrier to the transport of materials from the Moon's surface. On the other hand, we believe that the asteroids are rich in metals, other important minerals, and water; and they have no significant gravity. Some asteroids can be reached almost as easily as the Moon. These little planets are therefore the logical sources of materials for building large space facilities.

Before we examine these various aspects of asteroid science in detail, let us review a few basic facts and definitions.

First of all, the word 'asteroid' means 'star-like'. This describes the visual appearance of these objects in a telescope, but is totally inappropriate to their physical nature. The term 'planetoid' is much better, but for some reason this word is seldom used nowadays. Technically, the asteroids are called 'minor planets'. This is a good term, but it is generally used only in technical publications. Since the word 'asteroid' is familiar to almost everyone, it is the term we will use throughout most of this book.

An asteroid, or minor planet, is any natural object orbiting the Sun, other than a planet, comet, or satellite. As we shall see later in this book, there are so many different kinds of asteroids that the term is excessively broad. We have 'Main Belt' asteroids. Trojan asteroids, Apollo asteroids, etc. Some of these different types of asteroids may be related to each other, some may not be. Some 'asteroids' may, in fact, be inactive comets. The reader should therefore be aware that the terminology is inexact, and that one object called an asteroid may be totally different from another object called an asteroid.

The largest known asteroid is **Ceres**, which is less than 1000 km in diameter. The smallest planet, Pluto, is just over 2000 km in diameter. What will happen if an object is found with a diameter of 1500 km? Will it be called an asteroid or a planet? You can be sure that astronomers will not answer this question until they are forced to!

Asteroid-watching is becoming a popular activity among amateur astronomers, as well as professionals. Since the introduction of inexpensive programmable calculators, and personal computers, some amateurs are even computing the orbits of the asteroids they observe. When you can watch an object through your own telescope, compute its orbit, predict where it will be weeks or months later, and then find it at the predicted position, it adds a whole new dimension to 'star-gazing'!

Another possibility for amateurs is to measure the changes in brightness of the asteroids, either photographically, or with a photoelectric photometer. In this way, the rate of rotation of the asteroids can be determined. The shape of the asteroid can be derived from the amplitude of the light-variations. Three-color photometry can also be used to obtain crude mineralogical classifications for the brighter asteroids.

For those amateurs who want to do more than just look through their telescopes, or take pretty pictures, the asteroids provide an enjoyable opportunity to do 'real' astronomy. For the professional scientist, the asteroids provide clues to the origin and development of the solar system. For the interested layman, the asteroids demonstrate how much we can learn from some large rocks millions of kilometers away.

Here is the story of these exciting objects.

1

History and cataloging

THE MISSING PLANET

In 1766, Johannes Titius, of Wittenberg, Germany, tried to find a mathematical formula which would describe the distances of the planets from the Sun. He succeeded rather well. A few years later, Johann Elert Bode hit upon the same formula, and publicized it. Now known as the Titius–Bode Law, or simply Bode's Law, this formula is not really a 'law' of nature, but it does give a fair approximation to the distances of most of the planets. The formula is:

$$\text{distance (in A.U.)} = 0.4+(0.3\times N)$$

where $N = 0, 1, 2, 4, 8$, etc., doubling for each successive planet. (An A.U. is one 'astronomical unit', or 150 000 000 kilometers, or 93 000 000 miles).

The results of the computation are shown in Table 1.1. You can see that the

Table 1.1 — The Titius–Bode Law

Planet	N	Predicted distance (A.U.)	Actual distance (A.U)
Mercury	0	0.4	0.39
Venus	1	0.7	0.72
Earth	2	1.0	1.00
Mars	4	1.6	1.52
'gap'	8	2.8	2.77 (Ceres)
Jupiter	16	5.2	5.20
Saturn	32	10.0	9.54
Uranus	64	19.6	19.19
Neptune	128	38.8	30.07
Pluto	256	77.2	39.53

Note: If Pluto is regarded as an 'interloper', rather than a true planet, then the next 'real' planet beyond Neptune would be at 77.2 A.U.

computed values are approximately correct, up to Uranus, but the law fails for Neptune and Pluto. The latter two planets were not discovered until long after the Titius–Bode Law was formulated, however, so astronomers of the late eighteenth century had little reason to doubt the Law's validity. To this day, a few people still try to formulate 'improved' versions of the Titius–Bode Law, although most astronomers feel it has little physical significance.

The important thing about this Law is that is shows a 'gap' at 2.8 astronomical units from the Sun. No planet exists in this gap. Instead of regarding this as a failure of the Law, many astronomers of the eighteenth century were convinced that a small, undiscovered planet must exist within that gap. Toward the end of the year 1800, a group of astronomers decided to organize a search for the 'missing planet'. Before this search could get started, however, Giuseppe Piazzi, in Sicily, discovered the planet accidentally! He named the little planet **Ceres**, in honor of the patron goddess of Sicily.

After the discovery of Ceres, it became necessary to compute the **orbit** of the new object, so that it could be observed in future months and years. Piazzi was not able to observe Ceres long enough for a reliable orbit to be computed with the methods then available. There was a danger that the object might become lost. The great mathematician, Carl Friedrich Gauss, quickly came to the rescue by inventing a powerful new method for the calculation of orbits. Gauss's invention was one of the great milestones of mathematical astronomy. With some modifications, the 'Method of Gauss' is still used today for the computation of orbits.

Gauss computed the orbit of Ceres, and found that it was, indeed, in the region expected for the 'missing planet'. He then predicted where the new planet would be during the winter of 1801–02. (Ceres had meanwhile become unobservable, because it had moved into the daytime sky).

Wilhelm Olbers used Gauss's prediction to search for Ceres, toward the end of 1801. On December 31 he found it, thus confirming Piazzi's discovery, and showing the accuracy of Gauss's computations.

MORE DISCOVERIES

Olbers continued to observe the new object for several more months. During this follow-up work, in March 1802, Olbers found another asteroid, which he named **Pallas**. This second discovery raised the possibility that there might be *many* minor planets, instead of the single, larger planet which was expected in the Titius–Bode gap. Olbers therefore started to search in earnest. In 1804, his colleague, Karl Harding, found the third asteroid — **Juno**. Olbers found the fourth — **Vesta** — in 1807.

In the following decades, many more of these minor planets were found. It should be mentioned that all of these objects were found by the laborious method of looking through a telescope and comparing each star-like image with a chart which had been drawn previously. Any object which was not on the chart then had to be observed for several hours, to see if it moved. The job of charting every star visible with even a small telescope is incredibly tedious, and astronomers usually had to concentrate on relatively small parts of the sky.

When astronomical photography became practical, in the late 1800s, the situation was radically altered. Asteroids could be found by the hundreds. During a long-exposure photograph, the telescope is slowly moved, to follow the stars and compensate for the Earth's rotation. The stars appear as little dots on the photograph, while the moving asteroids make short trails. To find the asteroids, the astronomer simply looks for the trails on the photographs.

The pioneer in this work was Max Wolf, of Heidelberg. Beginning in 1891, Wolf used the photographic technique to hunt for asteroids, and eventually made 228 confirmed discoveries. In addition, he found hundreds of others which were not observed long enough for orbits to be computed. Today, instruments such as the 48-inch Schmidt telescopes in California and Australia can easily record 200 to 300 asteroids on a single photograph. There are so many asteroids, in fact, that the vast majority of them are ignored.

UNUSUAL ASTEROIDS

There are a few types of asteroids, however, which are still actively hunted. These are the asteroids which have unusual orbits.

All of the hundreds of asteroids discovered before the end of the nineteenth century traveled in orbits within the asteroid 'belt', between Mars and Jupiter. In 1898, Dr G. Witt, of Berlin, discovered an object which appeared to be moving at an exceptionally fast rate of speed. When the orbit of this object was computed, it was found that this asteroid actually crossed the orbit of Mars. In fact, the asteroid spent most of its time inside Mars' orbit. (By 'inside', we mean it is closer to the Sun than Mars). This asteroid was named **Eros**.

The next unusual asteroid to be discovered was **Achilles**, found in 1906, by Max Wolf. Achilles is far outside of the asteroid belt. It travels in the same orbit as Jupiter, but about 60 degrees ahead of that planet. At first, it seemed impossible for two objects to occupy the same orbit. Why didn't Achilles eventually collide with Jupiter? Just a slight difference in their motions would cause Jupiter to catch up with Achilles, or Achilles could go around its orbit and overtake Jupiter. The answer was actually found by the French mathematician, Joseph Louis Lagrange, in 1772 — one hundred and thirty-four years before Achilles was discovered! Lagrange found that a small object could travel in the same orbit as a larger object, such as a planet, if the small object formed an equilateral triangle with the planet and the Sun. In other words, the angle between the planet and the object must be about 60 degrees, as seen from the Sun. The small object could be 60 degrees behind, or ahead of, the planet. (See Fig. 1). These two stable points are called 'libration points', or 'Lagrangian points'. Achilles travels near the Lagrangian point which *precedes* Jupiter. Later, other asteroids were found near this point, a well as at the point which *follows* Jupiter in its orbit.

Since Achilles was a hero of the Trojan war, the other asteroids later found at Jupiter's Lagrangian points were named after other heroes of that war. These asteroids, as a group, are therefore called the **Trojan** asteroids. Hundreds of Trojan asteroids are now known, traveling on both sides of Jupiter.

An even more distant asteroid was found by Walter Baade in 1920. This is **Hidalgo**, which travels from the asteroid belt to the orbit of Saturn — an incredibly eccentric orbit, which resembles the orbit of a comet more than an asteroid's orbit.

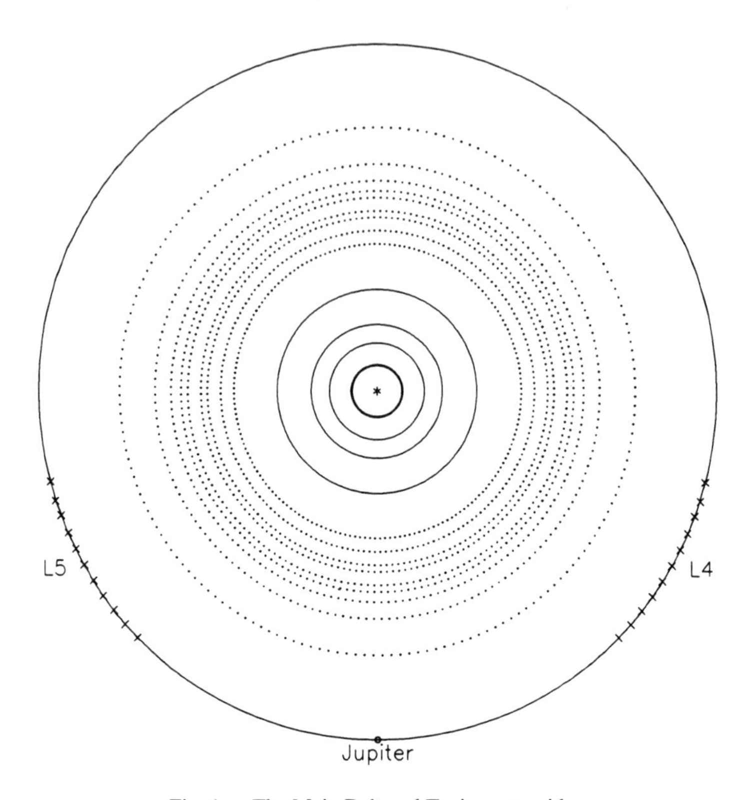

Fig. 1 — The Main Belt and Trojan asteroids.
The four inner circles represent the orbits of Mercury, Venus, Earth, and Mars. The outermost circle is the orbit of Jupiter. Between Mars and Jupiter lies the Main Belt of asteroids. Note that the Main Belt is not a continuous band, but is broken up into discrete regions separated by gaps. These are the Kirkwood gaps.
The areas marked L4 and L5 are near the Lagrangian points of Jupiter's orbit. The Trojan asteroids lie within these regions.
All orbits drawn to scale.

Hidalgo held the record for distance until 1977, when Charles Kowal discovered **Chiron**. Chiron travels between the orbits of Saturn and Uranus. It never goes anywhere near the asteroid belt at all. Both Chiron and Hidalgo are so unusual that they may not be related to the other asteroids. They could be extinct comets, escaped satellites, or something completely different.

A kind of 'Golden Age' of asteroid discoveries occurred in the 1930s, when Karl Reinmuth, in Heidelberg, and E. Delporte, in Belgium, competed in discovering ever more unusual asteroids.

Delporte began the race in March 1932, when he discovered **Amor**, an asteroid which not only crossed the orbit of Mars, but almost reached the Earth's orbit. One month later, Reinmuth discovered **Apollo**, which actually crossed the orbit of the Earth. These asteroids became the prototypes of two classes of asteroids. 'Amor asteroids' are those asteroids which come between 1.0 and 1.3 astronomical units from the Sun, at perihelion. 'Apollo asteroids' are those which actually cross the

Earth's orbit, getting closer than 1.0 A.U. from the Sun. (See Fig. 2.) Some Apollos even cross the orbit of Mercury. In 1932, Apollo passed within 11 million kilometers of the Earth — a near miss, by astronomical standards.

Delporte ran ahead of the race, by finding **Adonis**, in 1936. Adonis passed within 2 million kilometers of the Earth. Finally, Reinmuth, in 1937, discovered **Hermes**, which sometimes comes closer to the Earth than the Moon does! It came within 800 000 kilometers of the Earth in 1937, and it sometimes can come much closer.

After the Reinmuth–Delporte 'race', the asteroids were pretty well neglected by astronomers. Of course, World War II interrupted astronomy in general. But even after the war, astronomers felt that little could be learned about asteroids, and, perhaps, the asteroids were not as interesting as the new discoveries which changed our ideas about the universe as a whole. It was learned that our galaxy is only one of many. Astronomers suddenly had enlarged the universe manyfold. It is not really surprising that they were not very interested in the nearby 'rocks'.

In 1957, Sputnik was launched into orbit around the Earth, and scientists again realized that our own neighborhood is, indeed, relevant to us. By the late 1960s, observational techniques had advanced to the point where we could, in fact, learn something about those little rocks in our neighborhood. Interest in the asteroids gradually increased. Young astronomers wanted to learn about the asteroids (perhaps to the chagrin of their professors). The young astronomers came into their own, and applied the new techniques to determine the chemical compositions and shapes of the asteroids. Most of all, they learned that the asteroids were quite exciting!

In 1971, a landmark meeting was held in Tucson, Arizona. Organized by Dr Tom Gehrels, this meeting brought the asteroids to the attention of the rest of the astronomical community. Once again, astronomers were stimulated to search for more asteroids. In the forefront of the new searches was Dr Eugene Shoemaker. From the work of Dr Shoemaker and many others, we have come to understand something about the role asteroids play in the solar system.

In 1976, Dr Shoemaker's colleague, Mrs E. Helin, found an asteroid whose orbit lies mostly *inside* the orbit of the Earth. Mrs Helin named this asteroid **Aten**. Aten became the prototype of yet another class of asteroids. The Aten asteroids all have average distances from the Sun of *less* than 1 A.U. The Amor, Apollo, and Aten asteroids are often referred to collectively as the 'Earth-approaching asteroids'. Any asteroid that crosses the orbit of a planet is called a 'planet-crossing asteroid'.

It is only a question of time before some 'Earth-approaching' asteroid collides with the Earth. It is now estimated that a typical Apollo asteroid collides with the Earth every 250 000 years or so. That seems like a long time between 'hits', but that quarter of a million years might expire a week from next Tuesday! Imagine the effect of a kilometer-wide chunk of rock striking the Earth's surface as a speed of 30 kilometers per *second*! It has happened in the past, and is sure to happen again. It is for this reason that geologists and planetary scientists are interested in planet-crossing asteroids. The cratered surfaces of the Moon, Mars, and Mercury, show the effects of asteroid collisions. The number of craters on a planetary surface can even be used as an indication of the age of that surface. For example, Jupiter's satellite **Io** has very few craters. From this we infer that Io's volcanoes are constantly renewing the surface of that satellite. **Callisto**, on the other hand, is completely covered with

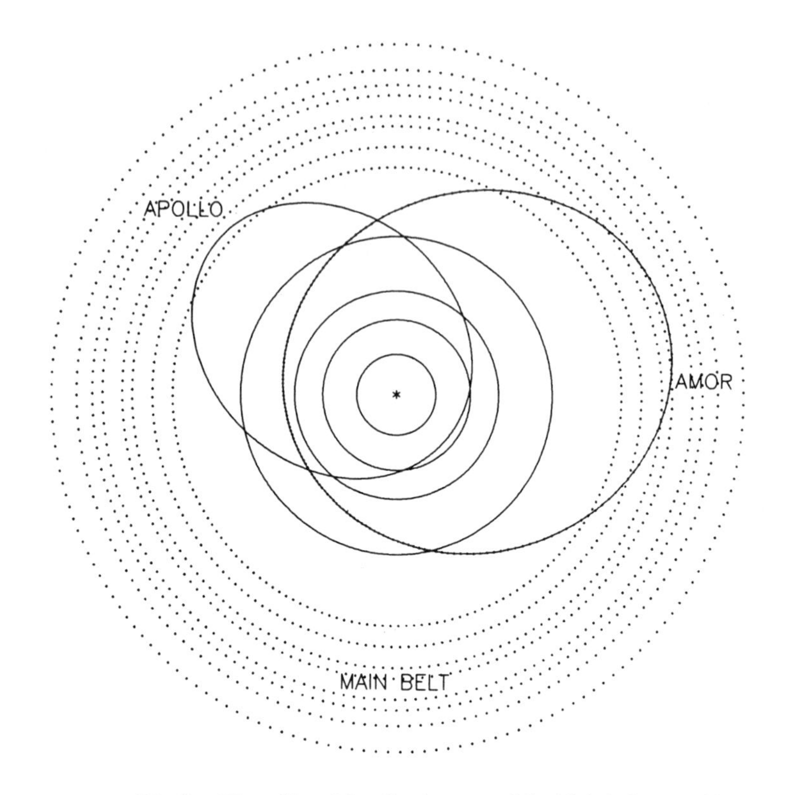

Fig. 2 — The orbits of Apollo, Amor, and the Main belt asteroids.
Note that Amor crosses the orbit of Mars, and almost reaches the orbit of Earth. Apollo crosses the orbits of Mars, Earth, and Venus. Aten could not be drawn on this diagram because its orbit is too close to the Earth's orbit.

craters. Its surface has not been altered at all, except by the influx of asteroidal objects. The Earth's weathering processes have obliterated most of the evidence of past asteroid collisions, but those collisions certainly do occur. As we shall see later in this book, an asteroid collision may have wiped out the dinosaurs from the face of the Earth, and made it possible for the mammals to attain dominance.

For the immediate future, we are not really worried about the effects of collisions between the Earth and the Apollo asteroids; instead, we are interested in exploiting these asteroids, in order to obtain their raw materials for our benefit. This topic will be discussed in Chapter 7 of this book. The unusual asteroids will be discussed in Chapters 5 and 6.

CATALOGING

The job of cataloging and keeping track of all the thousand of asteroids, of all types, is shared by the Minor Planet Center in Cambridge, Massachusetts, and the Institute for Theoretical Astronomy in Leningrad, USSR. The Minor Planet Center maintains a computerized file of every asteroid observation ever made, and publishes the

Minor Planet Circulars. The Russian institute publishes an annual ephemeris of all the known asteroids observable in that particular year. This publication is simply called *Ephemerides of Minor Planets*.

The Minor Planet Center was founded in 1947, with Paul Herget as its first director, at the University of Cincinnati. Dr Herget pioneered the use of modern data-processing systems to handle the vast number of asteroid observations, and to compute orbits and ephemerides. Upon Dr Herget's retirement in 1978, the Minor Planet Center was transferred to the Harvard-Smithsonian Center for Astrophysics in Cambridge, Mass. Now under the direction of Dr Brian G. Marsden, the Center continues to publish the *Minor Planet Circulars*, or *MPCs*. These circulars contain all reported observations of asteroids (and also comets), orbital elements, and ephemerides.

When an asteroid is discovered, it receives a temporary designation. This designation consists of the year of discovery, followed by two letters. The first letter indicates the half-month during which the asteroid was discovered. (The letters 'I' and 'Z' are not used). The second letter shows the order of discovery within that half-month. Thus, the asteroid 1976 AA was the first asteroid to be discovered in the first half of January, 1976. Asteroid 1976 AB was the second object to be found in that half of the month, while 1976 BA was found in the second half of January. If there are so many discoveries that all of the letters are used, then the letters are followed by a subscript number. For example, asteroid 1977 HZ would be followed by 1977 HA_1, if both objects were found during the same two-week interval.

After an asteroid has been observed well enough for an accurate orbit to be determined, the object is given a permanent number, and usually, a name. The numbers are simply assigned in chronological order, whenever the orbit is considered to be accurate enough. This usually requires that the asteroid be observed for several years. The permanent asteroid numbers have no relation to the order of discovery.

The discoverer of an asteroid has the right to name it, after it has received a permanent number. In the past, all asteroids were given feminine names, usually from Classical mythology. The only exceptions were the unusual asteroids, such as the Trojans and the Apollos, which received masculine names. Although some astronomers continue to adhere to this tradition, it is no longer followed rigorously. In fact, it is safe to say that the situation has degenerated to the point of absurdity. Asteroids have been named after girlfriends, financial supporters, cats, and computers. For a traditionalist like myself, it seems a pity that the naming of asteroids has become so trivialized.

When an asteroid is mentioned in a book, or a journal article, it is usually referred to by its name, or the complete name-and-number designation. For example: (2063) Bacchus, (1) Ceres, etc. Of course, if the asteroid does not yet have a permanent number, it is identified by its temporary year-and-letters designation.

When an astronomer finds an interesting asteroid on his photographs, he first must check to see if it is a known object, or a new discovery. This is done by checking the *Ephemerides of Minor Planets*, published in Leningrad. Recent discoveries are listed in the *Minor Planet Circulars*, and in the *International Astronomical Union Circulars*.

If the asteroid is a new discovery, or if it is a known object in need of further observations, the astronomer must measure its position as accurately as possible. He then notifies the Minor Planet Center, or the Central Bureau for Astronomical Telegrams. The latter Bureau is responsible for transmitting information about new discoveries to the astronomical community. At this point, the asteroid receives a temporary designation.

After several observations of an asteroid have been obtained, specialists will compute its orbit and an ephemeris. Other astronomers may then observe the object. If enough follow-up observations are made, the asteroid will eventually receive a permanent number. This usually requires observations to be made over a span of several years. The discoverer may then name it. If the object is not followed up, it can become lost, and may never be seen again until someone else 'rediscovers' it, years later. This is especially likely to happen with the fast-moving Apollo asteroids, which are only visible for a few weeks, while they are near the Earth.

Of the three 'Classical' Apollo asteroids — Apollo, Adonis, and Hermes — *all* became lost shortly after they were discovered! Thanks largely to the efforts of Brian G. Marsden, who used high-speed computers to redetermine their orbits, Apollo and Adonis have been found again. Apollo was recovered in 1973 by McCrosky and Shao, at Harvard. Adonis was recovered in 1977 by C. Kowal, at Palomar. Hermes remains lost.

APPENDIX: A CHRONOLOGY OF ASTEROID STUDIES

1766 Johannes Titius develops a mathematical formula which describes the distances of all the known planets from the Sun. Later popularized by Johann Bode, the Titius–Bode Law suggests that there should be an unknown planet between the orbits of Mars and Jupiter.

1801 Giuseppe Piazzi discovers the first asteroid, Ceres.

1867 Kirkwood shows that there are gaps in the asteroid belt. These gaps occur at simple fractions of Jupiter's orbital period.

1891 Max Wolf begins a photographic survey of the asteroids.

1898 Witt discovers Eros, the first asteroid known to cross the orbit of Mars.

1906 Wolf discovers Achilles, the first Trojan asteroid.

1918 Hirayama shows that some asteroids belong to families, whose members have similar orbital characteristics.

1920 Baade discovers Hidalgo, the first asteroid known to travel beyond the orbit of Saturn.

1932 Delporte discovers Amor.

1932 Reinmuth discovers Apollo, the first asteroid known to cross the Earth's orbit.

1937 Reinmuth discovers Hermes, which passes less than a million kilometers from the Earth.

1949 Kuiper begins the Yerkes–McDonald asteroid survey.

1960 Gehrels, *et al.*, begin the Palomar–Leiden survey.

1968 Icarus is observed by radar.

1971 The first Tucson asteroid conference is held.

1976 Helin discovers Aten, the first asteroid whose orbital radius is less than the Earth's.

1977 Kowal discovers Chiron, whose orbit lies between Saturn and Uranus.

1983 IRAS observes thousands of asteroids at infrared wavelengths.

2

The Main Belt asteroids

DISTRIBUTION

The vast majority of asteroids are found within a 'belt' between 2.0 and 3.3 A.U. from the Sun, with a few stragglers around 4 A.U. The asteroids which lie outside of this belt will be discussed in Chapters 5 and 6. Let us examine the structure of this belt, and see what we can learn from it. (The reader is advised to check the Glossary for the definitions of unfamiliar terms. Note that some terms, such as **eccentricity**, do not have the same meanings that they have in everyday life!)

The histogram in Fig. 3 shows the number of asteroids found at various distances from the Sun. About 1800 asteroids are plotted in this diagram. Note that the distribution is distinctly non-uniform. In particular, there are several gaps in which relatively few asteroids are found. These gaps, called the **Kirkwood gaps**, occur in locations which are 'commensurable' with the period of revolution of Jupiter. That is, their periods of revolution are simple fractions of Jupiter's period (one-half, one-third, etc.).

(The period of revolution of an object is related to its distance from the Sun by Kepler's formula: $\boldsymbol{P}^2 = \boldsymbol{a}^3$, where $\boldsymbol{P}$ is the period of revolution, in years, and $\boldsymbol{a}$ is the distance from the Sun, in A.U.)

Jupiter revolves around the Sun with a period of about 12 years. There are relatively few asteroids with periods of 4 years (the '1:3 commensurability'), or 4.8 years (the '2:5 commensurability'). The inner edge of the asteroid belt is near the 1:4 commensurability, having a period of 3 years, but Mars also has an influence here. The outer edge of the belt occurs at the 1:2 commensurability, with a period of 6 years. Notice, however, that there is a small group of asteroids clustered around the 2:3 commensurability (a period of 8 years). These are the **Hilda** asteroids, which will be discussed in Chapter 5.

It is logical to assume that the Kirkwood gaps are somehow caused by Jupiter. The details of the mechanism which creates the gaps is not well understood, but an over-simplified explanation goes something like this: An asteroid at the 1:3 commensurability, with a period of revolution of 4 years, will complete exactly three revolutions for every one revolution of Jupiter. At the time of closest approach to

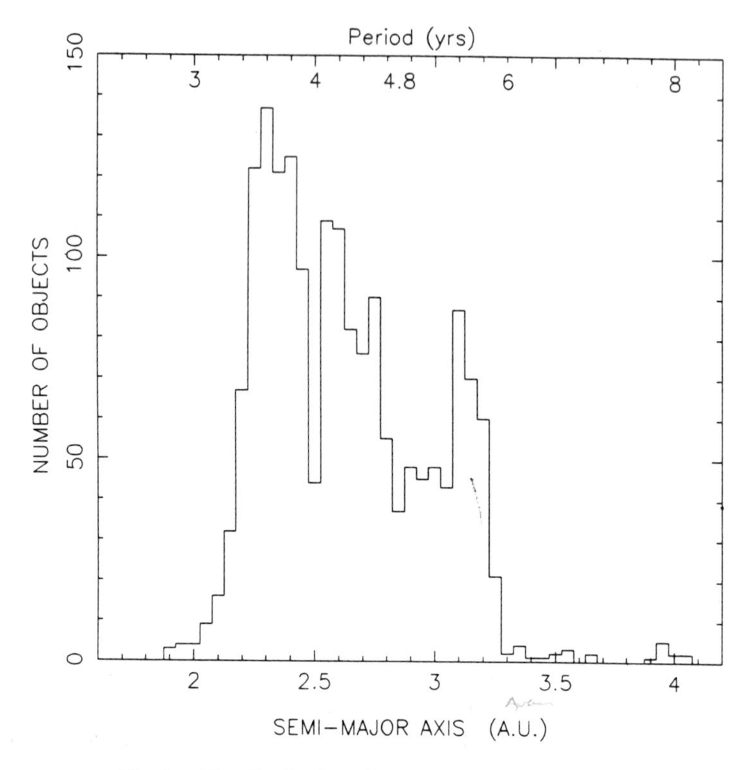

Fig. 3 — The distribution of the asteroids in the Main Belt.
Note that the distribution of asteroids is quite irregular. In particular, there are very few asteroids at 2.5 and 2.85 A.U.

Jupiter, the asteroid receives its maximum pull from Jupiter's gravity. This gravitational pull occurs at the same part of the asteroid's orbit every 12 years. This rhythmic disturbance, or 'perturbation', repeated every 12 years, will gradually pull the asteroid into a somewhat different orbit, until it is no longer in 'resonance' with Jupiter. In this way, regions around commensurabilities are cleared out, forming the Kirkwood gaps. Asteroids which are *not* in a resonance are also affected by Jupiter, of course, but at completely random time intervals. The perturbations therefore tend to cancel each other out, over long time intervals, and the asteroid orbits are not drastically changed.

ASTEROID FAMILIES

Since there are so many asteroids, much can be learned from statistical studies, and from plots of various asteroid characteristics. The histogram in Fig. 3 is such a plot. Another interesting effect can be seen by plotting the asteroids in a graph with two orbital elements as axes; for example, the semi-major axis and the inclination, (or, sometimes, the **Sine** of the inclination). Such a plot is shown in Fig. 4. Note that there are several concentrations of asteroids, forming 'clumps' in the diagram. This clumping would be even more evident if we could also plot the orbital eccentricity as a third dimension in the diagram. This clumping was first noticed by K. Hirayama in 1918. The groups of asteroids having similar elements, which form the concentrations in the diagram, are often called the **Hirayama families**. It is thought that each family is the result of the break-up of a single, larger asteroid at some time in the distant past. Studies of individual members of these families provide a way of examining a 'cross section' of the original asteroid parent bodies. In a way, nature has dissected the asteroids for us. Some of the families have members which are quite homogeneous in composition; other families are composed of members of differing types. The latter families must have come from 'differentiated' parent bodies. That is, the parent bodies must have been large enough to have had metallic cores, and lighter surface layers. Since almost half of all the asteroids occur in families, it is clear that the asteroid belt once had far fewer, though larger, members, and that fragmentation into smaller pieces is a relatively common occurrence.

COLLISIONS

We sometimes think of the asteroid belt as a dense accumulation of rocks. Actually, even the myriad of small asteroids are separated by thousands of kilometers. Nevertheless, the solar system is about 4 500 000 000 years old, and in this enormous span of time the asteroids have collided with each other many times. A typical velocity of collision is about 5 kilometers per second. When a small asteroid hits a larger one, it will make a crater in the larger body. If the two objects are of comparable size, they will completely destroy each other, leaving hundreds or thousands of fragments. This collisional destruction has occurred so often during the lifetime of the solar system, that practically all of the asteroids we now see are mere fragments of the original parent bodies. Only the very largest asteroids seem immune from catastrophic collisions, because of their size and strength. We can expect these largest asteroids to be covered with craters and collisional debris, however. The asteroids which are some tens of miles in diameter have life expectancies of only a billion years, or less. The smaller the asteroid, the shorter is its life span.

Now, let us go back to Fig. 4, and see how much information we can derive from this plot. The diagram is simply a plot of the semi-major axis (a) of each asteroid's orbit, versus the inclination (i), of those orbits. Over 3000 asteroids are plotted in this diagram.

The Kirkwood gaps are again obvious. The clearest one is at 2.5 A.U., which corresponds to a period of 4 years; (1:3 commensurability). There is a sharp break at 3.3 A.U., which corresponds to a period of 6 years, or one-half of Jupiter's period. Another gap is visible at 2.83 A.U. (period = 4.8 years, or two-fifths of Jupiter's period).

It is also obvious that the vast majority of asteroids have inclinations of less than 20 degrees. The statistics of asteroid inclinations and eccentricities will be discussed later in this chapter.

Three Hirayama families are conspicuous at: a=2.88, i=2; a=3.02, i=10; and a=

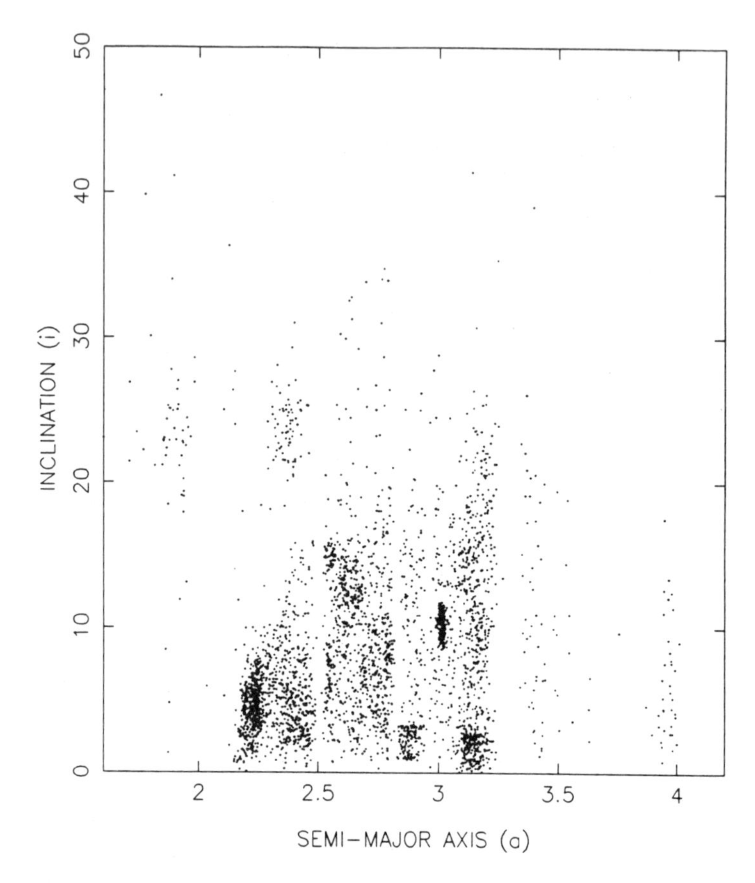

Fig. 4 — The semi-major axes of the asteroids *vs.* their inclinations.
The distribution of these orbital parameters is decidedly 'clumpy'. The regions of high density constitute the asteroid families. The Kirkwood gaps are also noticeable.

3.13, $i = 1.5$. These are the Koronis, Eos, and Themis families, respectively. The families are named for their most prominent members.

The concentration at $a = 2.2$ A.U., $i = 5$, is the complex **Flora region**. This region is sometimes subdivided into several separate families.

Many more families could be found by plotting more asteroids, and by using the eccentricities as a third dimension, to separate the families from the 'background'.

Notice the groups of asteroids at high inclinations. At $a = 1.95$ and $i = 23$ we have the **Hungaria group**, and at $a = 2.36$ and $i = 24$ we have the **Phocaea group**. These groups are not true families, but are merely groups of asteroids separated from the main belt by 'secular resonances'. Secular resonances occur when the **precession** of an asteroid's orbit is at some critical value related to other planetary motions. The effect of these resonances is to clear asteroids out of certain locations, just as the Kirkwood gaps were cleared out.

An asteroid which enters a region of secular resonance, for example by a collision, will experience an oscillation of the eccentricity of its orbit. The eccentricity may become so large that the asteroid will actually cross the orbit of Mars. Perturbations from Mars can then change the orbit of the asteroid drastically, and bring it into the inner parts of the solar system. Thus, the regions of secular resonance become depleted, and some asteroids are brought into Earth-crossing orbits. This mechanism may be an important source of the meteorites which hit the Earth. The meteorites which we see in our museums are almost certainly fragments of asteroids which were once in the Main Belt.

SURVEYS

Although many hundreds of asteroids have been discovered accidentally, or in various surveys, this material generally has too many 'selection effects' to permit detailed statistical studies of asteroid distribution. For example, bright, nearby asteroids are much more likely to be discovered than faint, distant asteroids. If we merely counted all the asteroids we can photograph, our counts would show an excess of large, nearby asteroids. To learn how the asteroids are *truly* distributed in space, it is necessary to have a relatively 'unbiased' sample of asteroids, and to understand any remaining biases and their effects on the completeness of the sample.

In the 1950s and 1960s, two surveys were conducted specifically for the purpose of obaining a statistically meaningful sample of asteroids.

Dr G. P. Kuiper directed a survey of asteroids at the McDonald Observatory from 1950 to 1952. The entire ecliptic region was photographed twice, down to a limiting magnitude of about 16. (**Magnitude**, in this context, is a measure of the **brightness** of an asteroid. See the Glossary.) Accurate magnitudes and distances were determined for all of the asteroids found. In this way, a consistent sample of asteroids was obtained. Corrections could then be made for incompleteness, and good statistics were obtained for the number, intrinsic brightness, and distances of the asteroids.

A rather different survey was conducted in 1960 by van Houten, van Houten-Groeneveld, Herget, and Gehrels. This so-called 'Palomar–Leiden' survey examined a relatively small part of the sky in great detail. Orbital information and magnitudes were obtained for about 1800 asteroids down to the 20th magnitude.

As a result of these detailed surveys, we now have a fairly clear picture of the over-all distribution of the asteroids down to very faint magnitudes.

Figs 5 through 7 show plots of the distribution of eccentricities, inclinations, and magnitudes of the asteroids from the Palomar–Leiden Survey report, and from other lists of asteroids. It is important to note that most asteroids do *not* have circular, zero-inclination orbits. Instead, they have moderate eccentricities and inclinations. If all asteroids traveled in circular orbits, in the plane of the ecliptic, asteroid collisions would be relatively gentle. However, since the orbits criss-cross each other, collisions are quite violent. It is this fact which prevents the asteroids from accumulating into a single planet.

Note also, how the number of asteroids increases as we go to fainter magnitudes. For example, there are about two-and-a-half times as many 19th-magnitude asteroids as 18th-magnitude asteroids. If we can assume that the brightness of an asteroid is proportional to its size (that is, that bright asteroids are bigger than faint asteroids), then it follows that there are many more small asteroids than large ones.

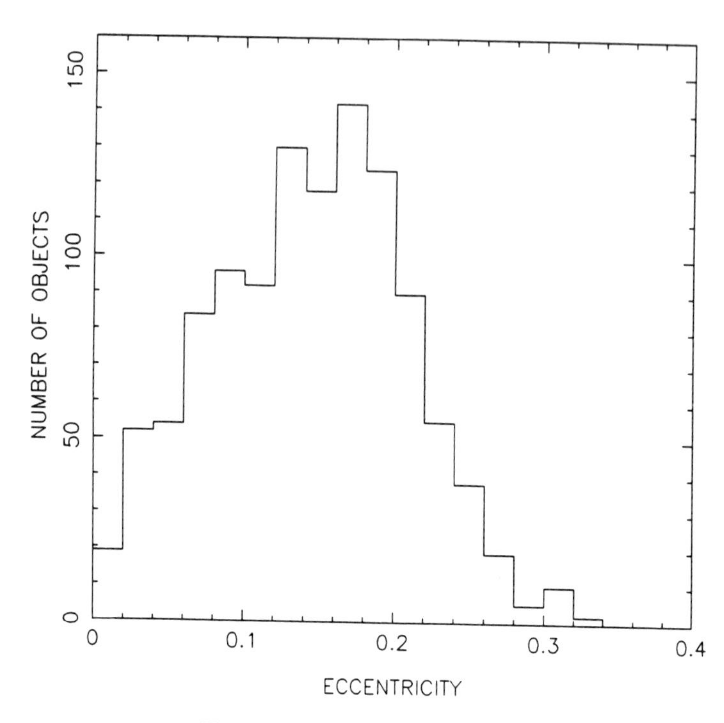

Fig. 5 — Distribution of eccentricities.
Most asteroids do not travel in circular orbits. The mean eccentricity is about 0.17.

THE INFRARED ASTRONOMICAL SATELLITE

A far different kind of survey was conducted in 1983, by the InfraRed Astronomical Satellite (IRAS). This satellite surveyed the entire sky at infrared wavelengths. During the survey, the satellite 'incidentally' observed many thousands of asteroids. Analysis of these observations is still underway, but one of the first things noticed was that there are several bands of *dust* in the solar system. The most prominent of these bands are inclined to the ecliptic by the same amount as the asteroids of the Eos and Themis families. It seems very likely that the dust we see is the results of collisions within those families.

If you smash two bricks together, you will end up with a few large fragments, many smaller fragments, and countless numbers of tiny dust particles. The same statistical relation holds true for the asteroids, which supports the theory that the asteroids we see today are largely the result of collisional fragmentation.

The asteroids range in size from almost 1000 km in diameter, down to small rocks or dust particles. Gehrels has estimated that there are about half-a-million asteroids larger than 1.6 km in diameter. Since the vast majority of asteroids are so small, however, the total mass of the asteroid belt is also quite small. If all of the asteroids were lumped together, their total mass would probably be less than one-thousandth the mass of the Earth.

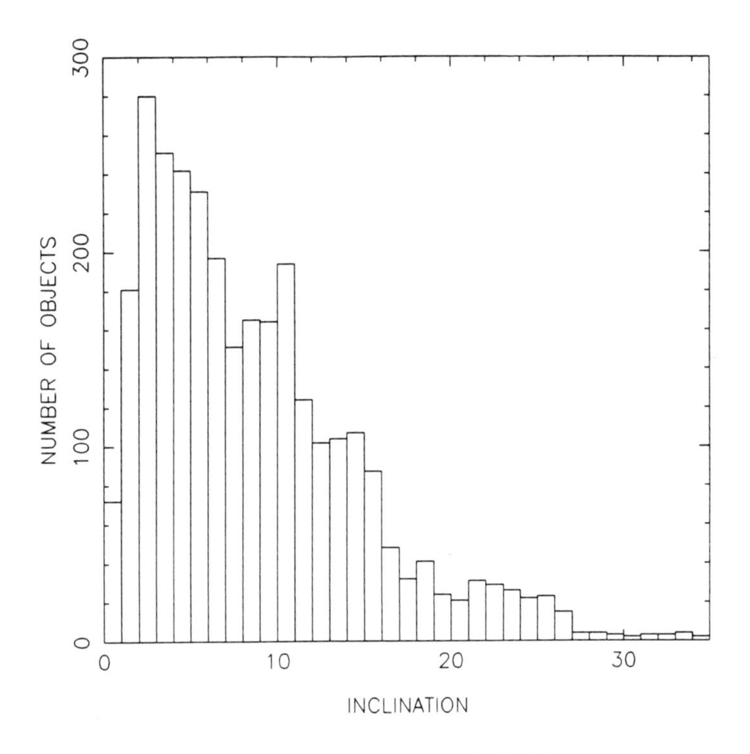

Fig. 6 — Distribution of asteroid inclinations.
The peak of this distribution is not at zero, but at about two degrees.

BRIGHTNESS

The apparent brightness of an asteroid depends partly on the object's size, and partly on the object's distance from us. Since the asteroids, and the Earth, are constantly moving, it is convenient to define an **absolute magnitude** for the asteroids, to indicate their *intrinsic* brightness, independent of their distance. We say that the absolute magnitude of an asteroid is the magnitude it would have if it were 1 A.U. from the Earth, 1 A.U. from the Sun, and fully illuminated. This is obviously a geometrically impossible, 'artificial' definition, but it is quite useful. The absolute magnitude is usually called '*g*' in the older literature, but nowadays we use the symbols: B(1,0) and V(1,0) for the absolute magnitudes in the blue, and yellow (visual), standard photometric systems, respectively.

The apparent brightness of an asteroid depends not only on its size and distance, but also on its degree of illumination, that is, its **phase**. Just as the full moon is much brighter than the quarter moon, so too, the asteroids are brightest at full phase. In fact, an asteroid at full phase is about 0.3 magnitudes brighter than would be expected from simple extrapolation of its partial phases. An asteroid starts to brighten by this amount when it comes within a few degrees of opposition from the Sun. Fig. 8 shows how the brightness of an asteroid increases as it approaches 180

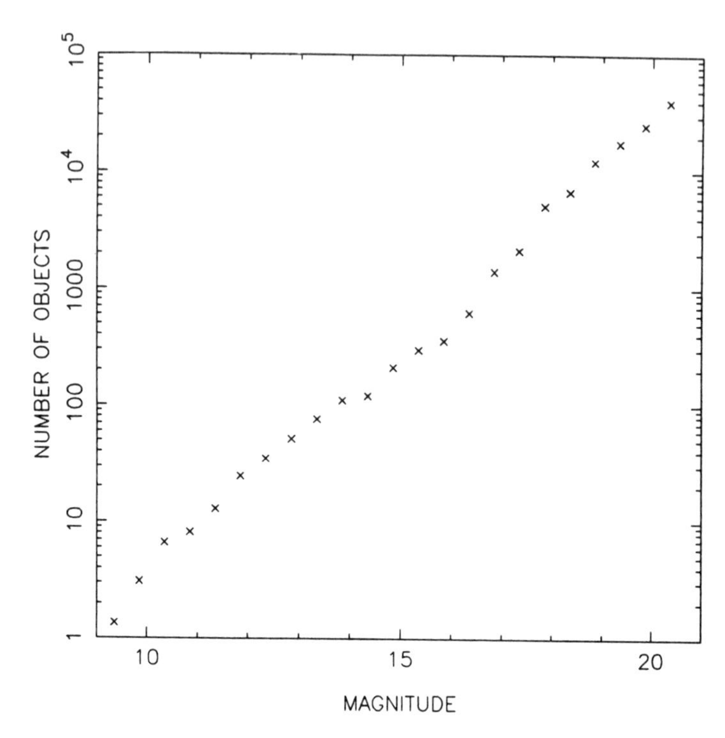

Fig. 7 — Distribution of asteroid magnitudes.
Faint objects have higher-numbered magnitudes. The number of asteroids almost doubles with each magnitude.

degrees from the Sun, or zero degrees from opposition, that is, full phase. The extra brightening at full phase is called the **opposition effect**.

Asteroids are usually irregular in shape. This is partly the result of collisional fragmentation, and partly because most asteroids are too small to 'pull' themselves into a spherical shape. In general, only the biggest asteroids are spherical. As an irregular asteroid rotates, the area of the face which is turned toward us will change. As a result, the apparent brightness of the asteroid will change. By measuring the brightness of an asteroid continuously for several hours, its rate of rotation can be determined. Furthermore, some idea of the shape of the asteroid can be deduced from the shape of its **light curve**. For example, if an asteroid is observed throughout one rotation, and its maximum brightness is twice as great as its minimum brightness, we can infer that the area of the largest side of the asteroid is twice as great as the area of its smallest side. Some asteroids, in fact, are quite 'cigar-shaped'. When we see the *end* of the 'cigar', the asteroid looks relatively faint. When we see the *long side* of the cigar, it looks much brighter. Of course, the problem is not really this simple. The shape and amplitude of the light curve also depend on the orientation of the rotational axis of the asteroid. If, for example, we happen to be looking at the 'north pole' of the asteroid, there will be *no* light variations because we always see the same side of the asteroid. It is therefore necessary to observe the light curves at different

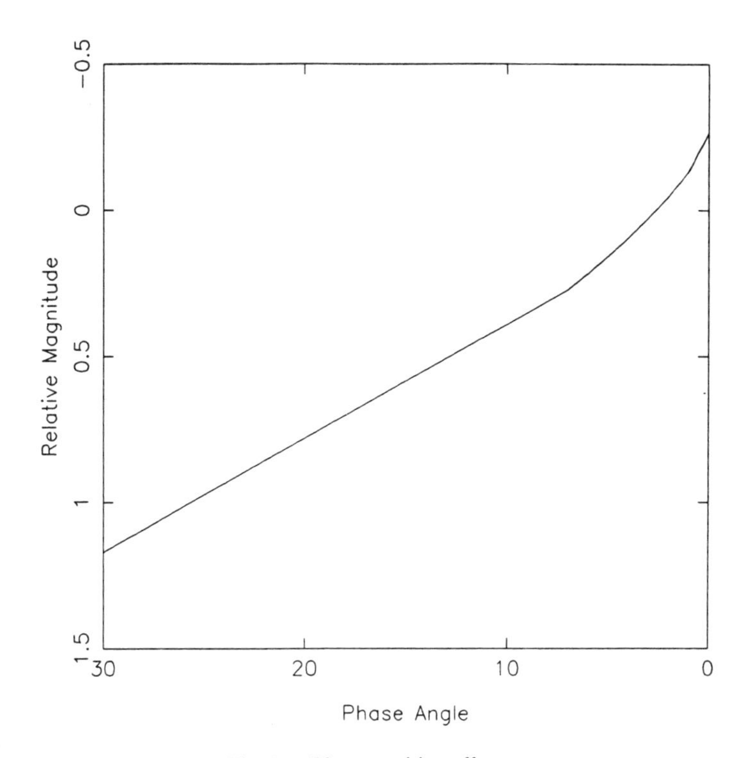

Fig. 8 — The opposition effect.
As an asteroid approaches full phase (phase angle zero), it steadily becomes brighter. When the phase angle is less than 10 degrees, however, the brightness increases more sharply. This sharp increase is called the opposition effect.

parts of the asteroid's orbit. From this sequence of light curves, it is possible to derive the orientation of the rotational axis and, finally, the shape of the asteroid.

The apparent brightness of an asteroid therefore depends on the distance of the asteroid, the percentage of the illuminated surface that we can see (the phase), and the area of that surface. The brightness also depends on the **reflectivity**, or **Albedo**, of the surface. A shiny, white asteroid will look brighter than a dull, black asteroid of the same size, shape, and distance. Some asteroids reflect only 3 or 4 per cent of the sunlight that strikes them. Other asteroids reflect as much as 40 per cent of the incident sunlight. The sunlight which is not reflected is absorbed by the asteroid. This absorbed radiation causes the asteroid to heat up, until it reaches thermal equilibrium. Then, the excess heat is radiated by the asteroid as infrared radiation. This is a very useful fact, which we will discuss in greater detail in the next chapter.

The information presented so far, shows what could be done with the asteroids up to about 1970. Surveys of the asteroid population, statistical studies of asteroid distribution, and photometric studies of asteroid magnitudes and light curves, have all contributed greatly to our understanding of the asteroid belt. But all of this tells us little about the composition of the individual asteroids, or about their origin and relation to other bodies in the solar system. Information about the physical

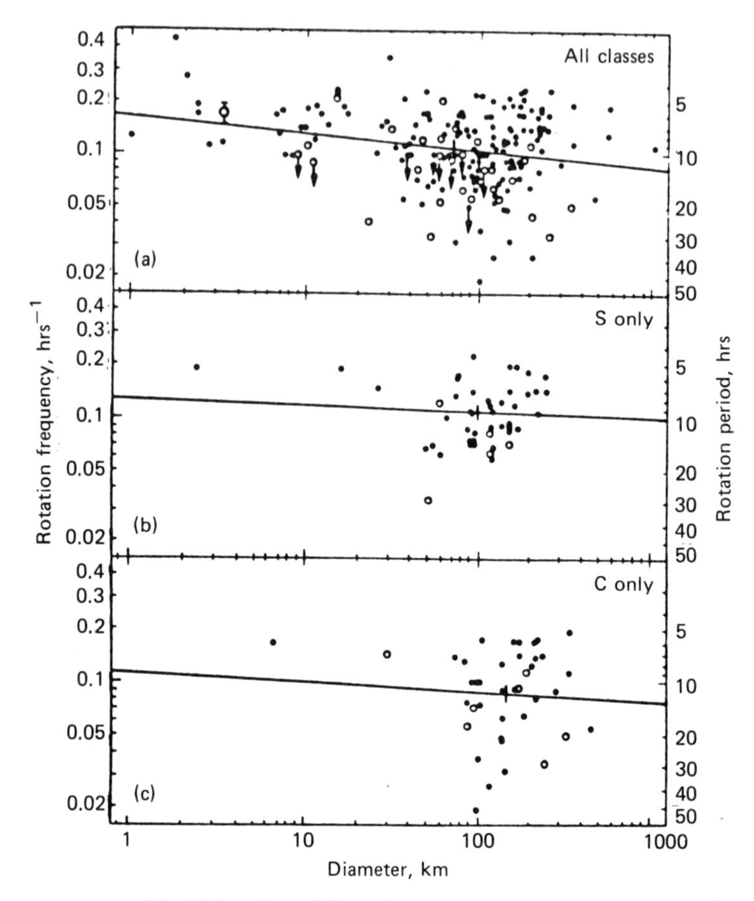

Fig. 8(a) — Asteroid rotation rates *vs.* diameter.
The period of rotation is shown on the right side of the graphs, while rotation frequency, in rotations per hour, is shown on the left. The open circle on the left side of the upper diagram is the average for ten Mars-crossing asteroids. There is a tendency for small asteroids to rotate faster than larger asteroids. (From Burns and Tedesco, in *Asteroids*, Tom Gehrels, ed. (1979) p. 513.)

composition of individual asteroids could not be obtained until new techniques were developed in the 1970s. (Although inferences could be obtained from studies of the meteorites). The new techniques have created a revolution in asteroid science. We can now study the asteroids in such detail that these objects are no longer in the exclusive domain of astronomers. Geologists study the mineralogical composition of the asteroids and their relation to meteorites. Physical chemists study the chemical composition and thermal histories of the asteroids and meteorites for clues to the origin and evolution of the entire solar system. These new techniques will be described in the next chapter.

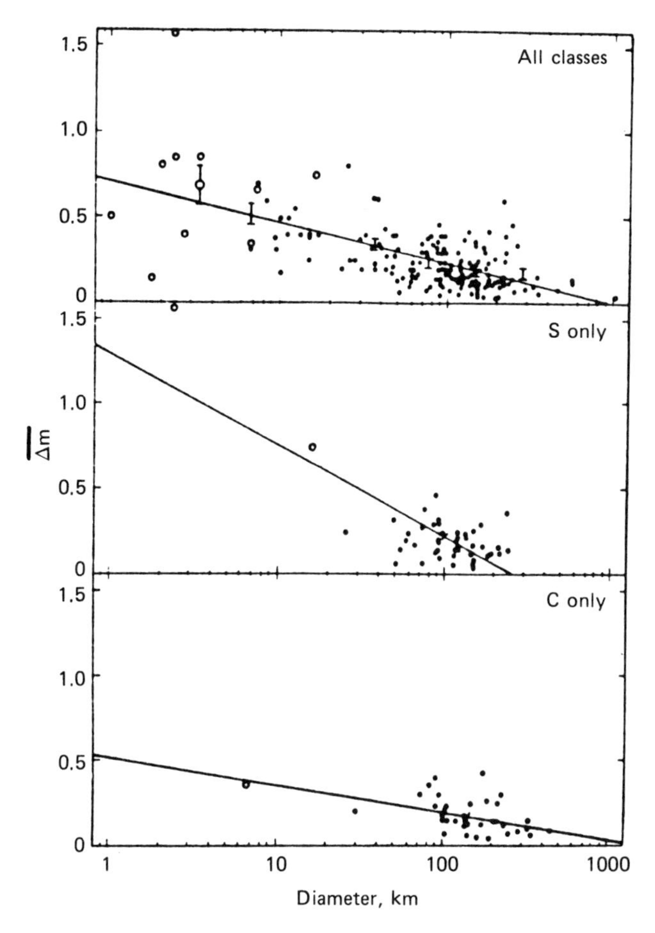

Fig. 8(b) — Light curve amplitudes *vs.* diameter.
In general, light curve amplitude is an indicator of the shape of an asteroid — the greater the amplitude, the more irregular the shape. Small asteroids tend to have the most irregular shapes (i.e., greatest amplitudes). The open circles are Earth- or Mars-crossing asteroids. (From Burns and Tedesco, in *Asteroids*, Tom Gehrels, ed. (1979) p. 519.)

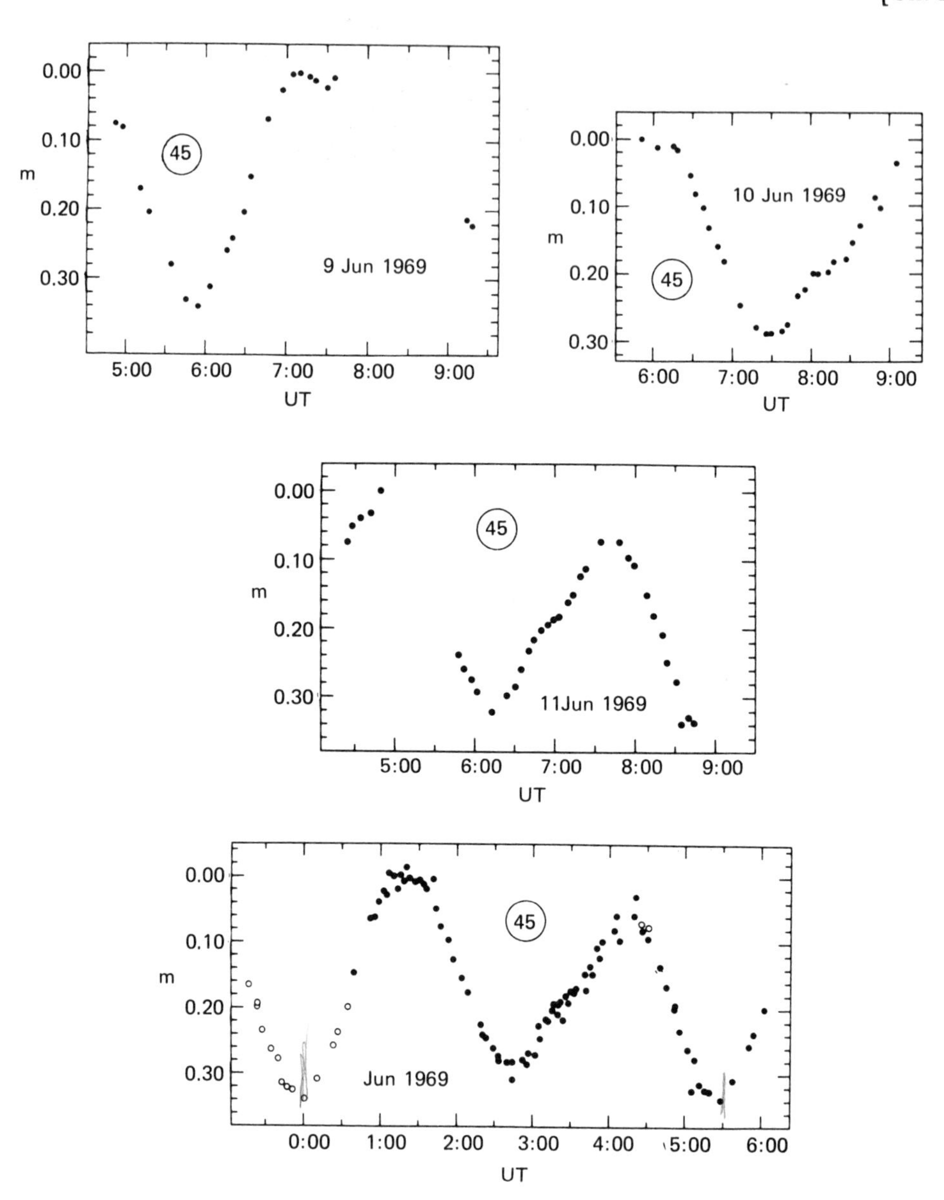

Fig. 8(c) — Rotational light curves of (45)Eugenia.
The light curve shows that the period of rotation of this asteroid is 5 hours 41 minutes 57 seconds. The double maximum during each rotation can be caused either by an elongated shape, or by two spots on Eugenia's surface. (Taylor *et al.* (1968) *Icarus*, **73** 314.)

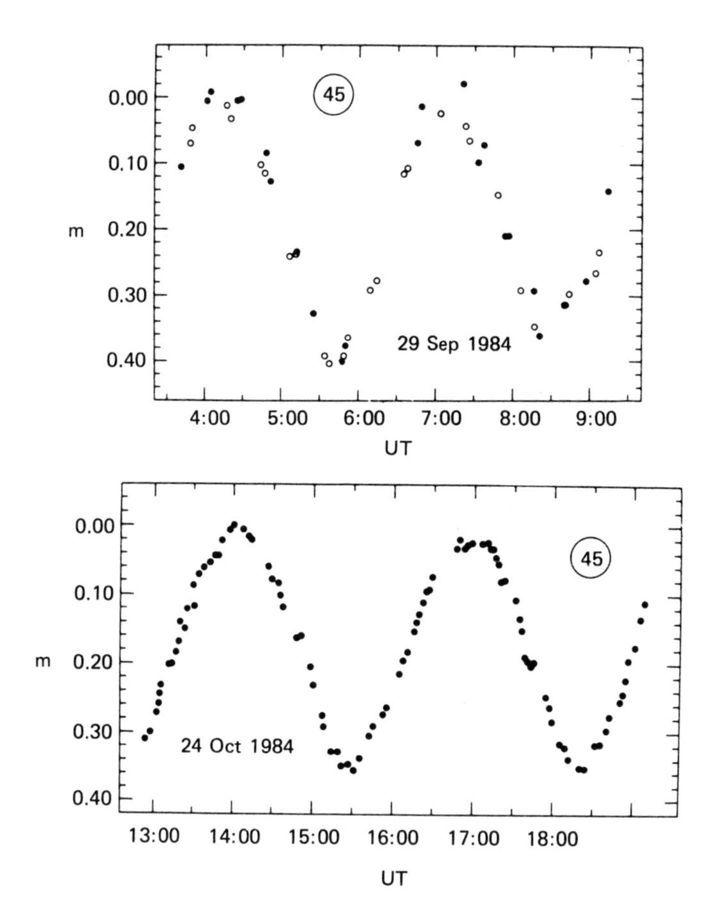

Fig. 8(c) continued — Rotational light curves of (45)Eugenia.
The light curve shows that the period of rotation of this asteroid is 5 hours 41 minutes 57 seconds. The double maximum during each rotation can be caused either by an elongated shape, or by two spots on Eugenia's surface. (Taylor *et al.* (1968) *Icarus*, **73** 314.)

APPENDIX: THE 30 LARGEST ASTEROIDS

Designation	Type	Diameter (km)
(1)Ceres	C	940
(2)Pallas	CU	588
(4)Vesta	U	576
(10)Hygeia	C	430
(704)Interamnia		338
(511)Davida	C	324
(65)Cybele	C?	308
(52)Europa	C	292
(87)Sylvia	P	282
(451)Patientia	C	280
(31)Euphrosyne	C	270
(15)Eunomia	S	260
(324)Bamberga	C	252
(3)Juno	S	248
(16)Psyche	M	248
(48)Doris	C	246
(13)Egeria	C	244
(45)Eugenia	C	244
(624)Hector	D	232
(24)Themis	C	228
(95)Arethusa	C	228
(165)Loreley	C	228
(153)Hilda	P	222
(532)Herculina	S	220
(2060)Chiron		220
(702)Alauda	C	216
(107)Camilla	C	212
(88)Thisbe	C	210
(7)Iris	S	208
(423)Diotima	C	208

3

Observational techniques

It is quite true that the asteroids were neglected by astronomers for many years, but this was for a very good reason. The asteroids were simply too small for detailed investigations, and their importance to the general study of the solar system was not fully realized. As mentioned in the last chapter, it was not until almost 1970 that observational techniques became sufficiently advanced to enable us to study the size, composition, and other characteristics of individual asteroids. These new techniques have made the asteroids a rich source of information about the origin and evolution of the solar system.

PHOTOMETRY

As we have seen in the previous chapter, much can be learned about the asteroids by simply measuring their brightness. This technique is called **photometry**. Photometry is particularly valuable when used in conjunction with **radiometry**, which will be described later. Photometry consists of measuring the brightness of the asteroids visually, photographically, or electronically. Often, one or more colored filters are used, to measure the brightness in a particular color. If many filters are used, this technique is equivalent to **spectrophotometry**.

Photometry is often used to measure the brightness variations of asteroids over a period of time. Variations in brightness are caused by a number of things — the position of the asteroid in its orbit, spottiness of the asteroid's surface, the shape of the asteroid, and the rotation of the asteroid. By measuring the brightness of an asteroid, we can obtain information about all of these things. This, indeed, is the principle behind much of observational astronomy. If some property of an object produces an observable effect, we can learn about that property by measuring the effect. Conversely, by measuring just about everything that we can, we often see effects which lead us to uncover previously-unknown characteristics of the bodies we study. Knowing the characteristics of individual objects, and of groups of those objects, we can then perform statistical analyses which lead to even more information about the objects, and even about their origins.

Through observations of their light curves, the rotation rates and shapes of many asteroids have been determined. In general, we find that small asteroids are more irregular, and spin more rapidly, than larger asteroids. This is particularly pronounced in the case of the Apollo asteroids. Many of these small objects are rapidly-rotating 'slivers' of rock. This is precisely what we would expect to see, if the Apollos are fragments produced by collisions between asteroids. Thus, by simply measuring the brightness of these tiny objects, we can obtain clues to their origin. We may never obtain final answers, but at least we can find clues.

If we can learn this much through simple photometry, it should not be surprising that we can learn even more by measuring the **color**, as well as the brightness, of the asteroids.

SPECTROPHOTOMETRY

How can we determine the composition of a rock which is a quarter-of-a-billion kilometers away? If you look at various rocks in a field, or in a stream bed, you can see that each type of rock *looks* different. Some rocks are light, some are dark, some are shiny, some are dull. There are white rocks, black rocks, gray rocks, and colored rocks. A geologist, or a rock hound, can often identify certain types of rocks just by looking at them. The same thing can be done in a quantitative, objective way, by *measuring* the light reflected by different types of rocks. The rocks do not need to be in a laboratory, they can be on the moon, in the asteroid belt, or anywhere in space. The light from an asteroid can be measured at many wavelengths, and compared with the color of sunlight. This tells you exactly how the asteroid *reflects* sunlight, and makes it possible to determine the surface composition of the asteroids. This technique is called **spectrophotometry**.

There are several ways of doing spectrophotometry. It is not sufficient to simply photograph the spectrum of an object. Quantitative information about the light output, or **flux** of the object is needed, and it must be compared with the light flux reaching the asteroid from the Sun. It is not possible to calibrate a photograph with sufficient accuracy to do this. The spectrum must be recorded with a linear device such as a photomultiplier tube or a solid-state detector.

In a 'spectrum scanner', the spectrum of an object is slowly moved across the cathode of a photomultiplier tube. In this way, the intensity at each point of the spectrum is measured. This method has some disadvantages. Since it takes several minutes to complete a scan, changes in the opacity of the sky can occur during the observation, making the results unreliable. Thus, spectrum scanners can only be used when the sky is perfectly uniform and unchanging. Furthermore, since only one spectral region is observed at a time, this method makes very inefficient use of telescope time.

A 'multi-channel spectrophotometer' is similar to a spectrum scanner, but uses several photomultiplier tubes instead of just one. Each tube records the intensity of a small part of the spectrum. Together, the whole spectrum is observed at once. This method can be used even through thin clouds. Clouds are very nearly 'neutral'. They absorb all colors equally. So, even though the sky opacity might change during an observation, that change will affect the entire spectrum equally. With a single-tube scanner, however, a cloud might block the object's light while the instrument is measuring one part of the spectrum, and then clear up while the rest of the spectrum is being measured. Observing the whole spectrum at once also increases the speed and efficiency of this method.

Solid-state detectors, such as the 'CCD', (Charge-Coupled device), or the 'Reticon', are less cumbersome than a multi-channel device. Instead of a bank of photomultiplier tubes, the solid-state devices are small electronic 'chips' which contain hundreds or thousands of tiny light-detectors. When the spectrum of an object is focused on one of these devices, each little light detector measures the intensity of a small part of the spectrum. These devices are extremely sensitive — often 100 times more sensitive than photographic materials.

A cruder way of measuring a spectrum is simply to measure the brightness of the object through several narrow-band filters in succession. In this case, the technique is comparable in many ways to the spectrum-scanner method, but requires much simpler equipment.

When measuring very faint objects, broad-band filters are often used. The most common are the U, B, and V filters, which stand for ultraviolet, blue, and visual (yellow), respectively. The color of an asteroid can be defined as the *difference* between its magnitudes in two of these bandpasses. For example, the 'B–V' color of an asteroid is its magnitude in the B-system, *minus* its magnitude in the V-system. The smaller the value of B–V, the bluer is the asteroid. A high value of B–V means the object is reddish. For comparison, the Sun has a B–V of 0.63, and a U–B of 0.10. All asteroids are redder than the Sun.

CLASSIFICATION

Asteroids of various colors have been divided into several taxonomic classes, as shown in Fig. 9. This diagram is called a 'two-color plot'. An attempt has been made to relate these classes to similar-appearing classes of meteorites. (The meteorites will be discussed in greater detail later). The **C-type** asteroids have relatively blue colors, and fairly flat, featureless spectra, similar to the colors and spectra of carbonaceous chondrite meteorites. (The 'C' stands for 'carbonaceous'). The **S-type** asteroids have a reddish color, and spectra similar to the stony-iron meteorites. (The 'S' stands for 'silicaceous').

The C and S categories include most of the asteroids, but there are a few smaller classes, including **E-type** (for 'enstatite'), **M-type** (for 'metallic'), and **R-type** (for 'red'). There is also the ever-popular **U-type** (for 'unclassified'). As we obtain more information about asteroid spectra, we tend to introduce more and more new types. This is what happens with every classification scheme, whether it be for asteroids, galaxies, or birds. If you learn enough about the objects you are studying, you will find that each object is a unique individual. This should not prevent us from recognizing that the objects do, in fact, fall into certain general categories.

Although it is often useful to obtain broad-band colors of asteroids, it is naturally more informative to observe the whole spectrum of an asteroid in detail. Since the asteroids are illuminated by the Sun, the spectrum of an asteroid is just a reflection of the solar spectrum — but with subtle differences. Those small differences are what we need to measure. We want to know how the asteroid reflects sunlight at each wavelength. We need to derive the **reflection spectrum**. Various minerals reflect

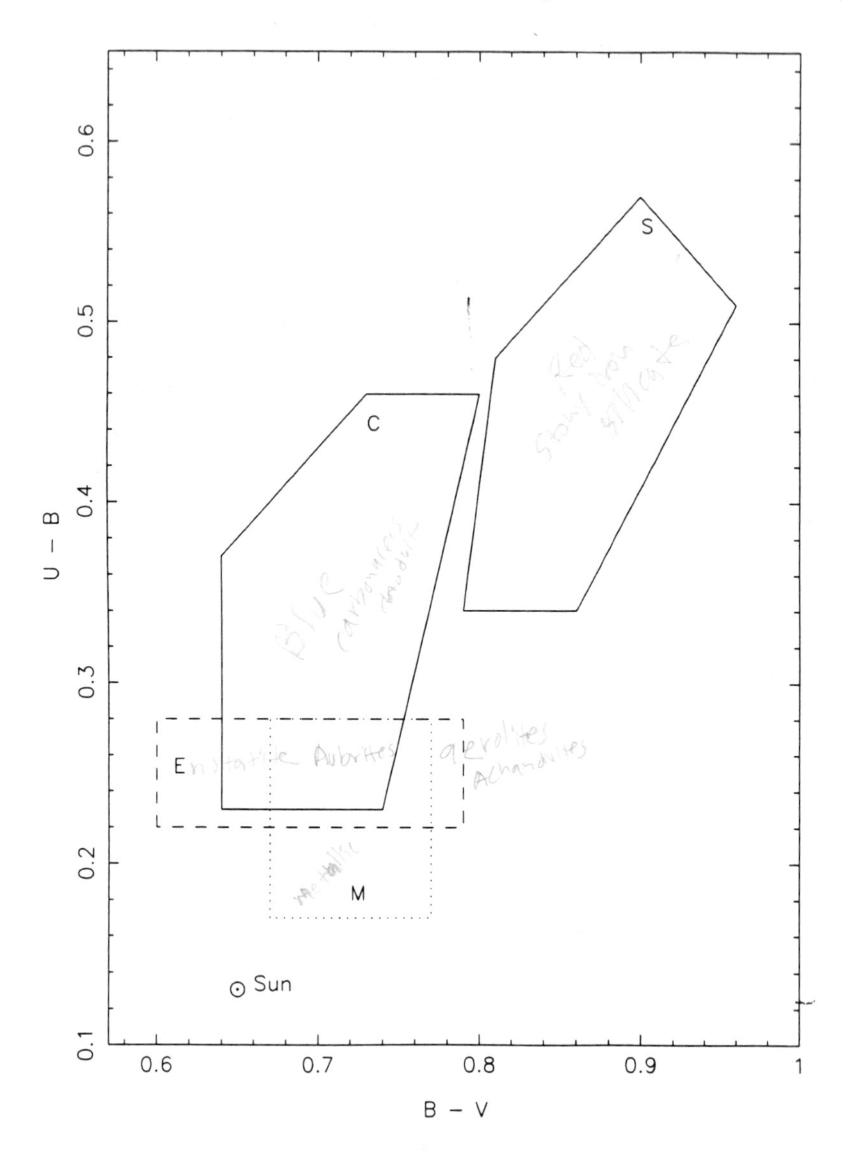

Fig. 9 — Asteroid color classes.
The color domains of the C. S. M. and E-type asteroids. The reddest asteroids are at the top right of the diagram. There is some overlap in the colors of the various classes; asteroids in the overlap regions can only be distinguished by their spectra or their albedoes.

sunlight in characteristic ways, so the reflection spectrum of the asteroid tells us something about the mineralogical composition of the asteroid. To obtain a reflection spectrum, we simply *divide* the flux from the asteroid at each wavelength, by the solar flux at those same wavelengths. Your intuition may tell you that we should subtract the solar spectrum from the asteroid spectrum, but that is not the case. We are interested in the *ratio* between incident light and reflected light, so we must *divide* the spectra. The resulting 'ratio spectrum' does not tell us how much of the sunlight is actually reflected. It merely tells us how much red light is reflected, compared to the amount of blue light. For example, the asteroid might reflect 10 per cent of the blue sunlight, and 20 per cent of the red light. Or, it might reflect 1 per cent of blue light and 2 per cent of red light. In both cases the reflection spectrum will look the same — twice as much red light as blue light. To compensate for this uncertainty, and to enable us to compare the reflectance spectra of different asteroids, we must adjust the spectra by some factor to normalize them at some arbitrary wavelength. Usually, the spectrum is adjusted so that the relative flux at 5600 ångstroms will have a value of 1.0. The reflectance spectrum of **Vesta** is shown in Fig. 10. Note that Vesta

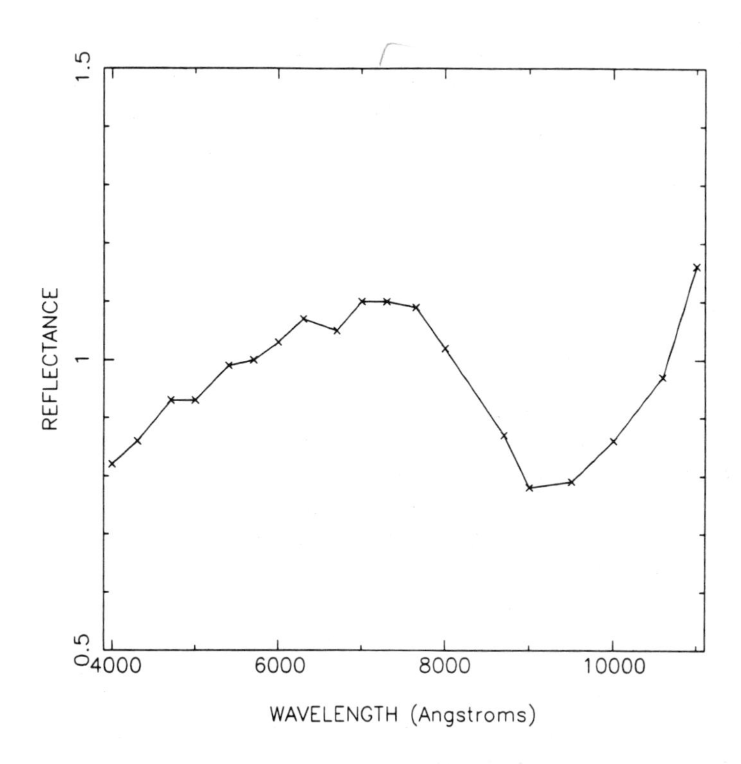

Fig. 10 — The reflectance spectrum of Vesta.
Notice how the reflectance drops at about 9000 ångstroms. This dip is caused by the minerals pyroxene and olivine. Compare this with the spectrum of Ceres in Fig. 13.

is a relatively poor reflector of blue light, and has a pronounced dip in its reflectivity at 9000–9500 ångstroms. This dip is characteristic of the silicate mineral pyroxene.

RADIOMETRY

Another powerful technique for studying asteroids is **radiometry**, that is, measuring the **infrared radiation** from an asteroid. When sunlight strikes an asteroid, some of the light is reflected, but most of it is absorbed. The absorbed radiation causes the asteroid to heat up. This heat can be measured by measuring the infrared light emitted by the asteroid. The amount of light reflected by an asteroid depends on its size and albedo. Neither of these quantities is known. But, by measuring the infrared radiation from the asteroid, and comparing it to the amount of reflected visible light, we can determine its albedo. Knowing the apparent brightness, distance, and albedo of the asteroid, we can easily compute its size.

As you might guess, this procedure is not really as simple as it sounds. Actually, some assumptions must be made about the nature of the asteroid material before this technique can be applied. For example, a metallic object would tend to heat up uniformly through its whole depth, while an insulating material will be hottest at its surface. Asteroids rotate, and the rate of rotation determines how much time the dark side has to cool off. The shape of the asteroid also has an effect on the way radiation is reflected or emitted. These complications are fairly well understood, however, and the radiometric method can now be used to obtain the diameters of the asteroids to an accuracy of 10 to 15 per cent. This can be checked in various ways for some of the asteroids.

POLARIMETRY

The third 'new' technique for the study of individual asteroids is **polarimetry**. Actually, the first studies of light polarization on asteroids were done in 1934, by Bernard Lyot. This technique reached maturity only in the 1970s, however.

In the polarimetric technique, a sheet of polarizing material is placed in front of a photometer. The light from the asteroid is measured by the photometer while the polarizer is rotated. In this way, the amount and direction of polarization can be measured. (Try looking at something through polarizing sunglasses. Then, rotate the glasses. The appearance of the object you are looking at will change. Try this with different kinds of objects — metals, pavement, water.) The polarization depends on the mineralogy and texture of the asteroid surface, and on the geometry of the situation. A bare metallic surface has little or no polarization, while a surface covered with powdery dust shows strong polarization. Typically, we measure polarizations of a few per cent. The polarization must be measured at different **phase angles**, that is, at different Sun–asteroid–Earth angles. (The phase angle is *zero* degrees when the asteroid is directly *opposite* the Sun.) The polarimetric technique therefore requires that the asteroid be observed for several weeks or months. When the amount of polarization is plotted against the phase angle, we get a **curve of polarization**, such as the one shown in Fig. 11.. Negative polarizations mean that the 'electric vector' is maximum in the plane of the Sun–asteroid–Earth, while in positive polarization the electric vector is perpendicular to that plane.

It was found empirically, that the shape of the polarization curve is related to the albedo of the material being measured. In particular, the slope of the ascending branch of the curve shows a strong correlation with albedo. This correlation breaks down for albedos of less than 6 per cent, however, so the polarimetric method cannot be used for the darkest asteroids. For the other asteroids, the diameters determined

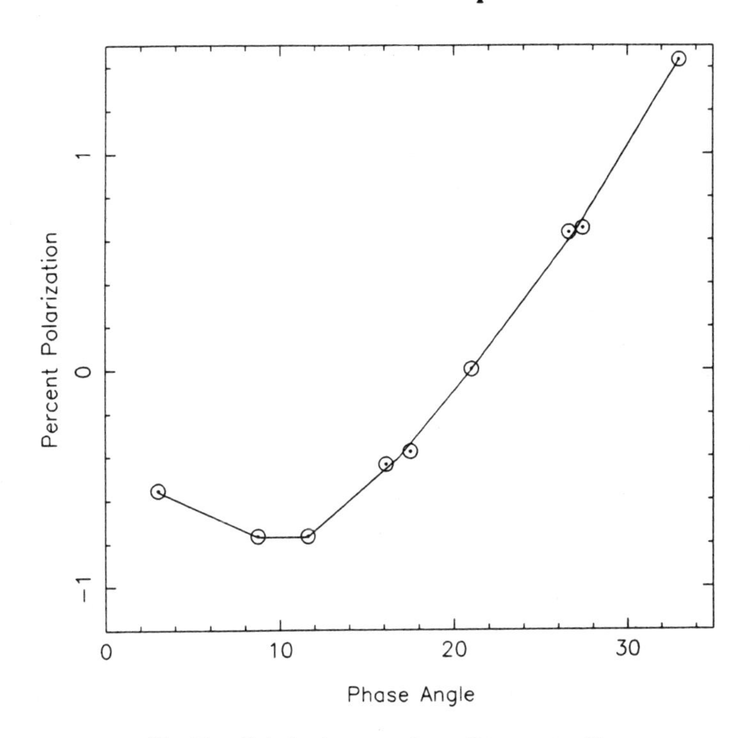

Fig. 11 — Polarization curve for an S-type asteroid.
The parameters of interest in these curves are: the depth of the negative portion, the slope of the positive branch, and the phase angle at which the polarization changes from negative to positive. (The data for this graph were taken from: Zellner and Gradie (1976) *Astronomical Journal* **81** 262.)

by polarimetry generally agree with the diameters derived from radiometry to within 10 per cent. New theoretical and laboratory studies are continually improving the accuracy of both of these observational techniques.

METEORITES

The techniques of spectrophotometry and polarimetry require that the observations of asteroids be compared with measurements of known materials. It is not usually suitable to use terrestrial rocks for these comparisons. The obvious materials to use are the meteorites, since we believe that the meteorites came from the asteroid belt.

At this point, we must get into a rather detailed discussion of meteorites and meteoritics. Until lunar rocks were returned to Earth during the American and Russian space programs, the meteorites provided our only samples of extraterrestrial material. Meteorites were therefore studied intensively in our laboratories, and these studies are continuing, with methods of ever-greater sophistication.

First of all, let us review the terminology, **Meteors** are the bright streaks of light which we can see in the sky every night. These 'shooting stars' are caused by

meteoroids striking the Earth's atmosphere and burning up. A chunk of rock floating in space is called a meteoroid until it enters the Earth's atmosphere. Most of the meteors we see are caused by particles no bigger than grains of sand, but those which are large enough to survive their trip through the atmosphere, become **meteorites**. Thus, it is the meteorites that we can pick up and study.

Several times each year, we can see showers of meteors, all emanating from one area of the sky. It was found in the nineteenth century, that many meteor showers travel in the same orbits as some of the known comets. We therefore believe that meteor showers are the debris of disintegrating comets. Cometary particles are too fragile to survive their passage through the Earth's atmosphere, so meteor showers rarely, if ever, produce meteorites. The great Tunguska meteoroid, which exploded over Siberia in 1908, may have been a large fragment of a comet. It is noteworthy that this object did not produce a crater, but disintegrated completely in the atmosphere, in spite of its size. Although this object did not reach the ground, its explosion in the atmosphere knocked down trees over a radius of 50 kilometers. Its explosive force was the equivalent of a 12-megaton bomb.

It should be noted that *any* object is subjected to tremendous stresses when it enters the Earth's atmosphere at a speed of tens of kilometers per second. 'Fragility', under these conditions, is a very relative term. There is some controversy about whether the Tunguska meteoroid was necessarily a comet.

At any rate, it is only the very brightest meteors, called **fireballs**, or **bolides**, which produce recoverable meteorites. Since bolides do not appear to be associated with meteor showers, they are probably not associated with comets. Furthermore, although our knowledge of cometary materials is extremely limited, we do not think that comets contain the wide diversity of materials that are present in meteorites. The asteroids do exhibit such diversity.

Meteorites can be broadly classified into three categories — irons, stones, and stony-irons. The iron meteorites are solid chunks of metal — usually 90–95 per cent iron, and 5–10 per cent nickel. The M-type asteroids have spectra that resemble the spectra of iron meteorites, and we believe that such meteorites came from M-type asteroids. A large fraction of the meteorites we find are iron meteorites, but there are only a few M-type asteroids. The reason for this is simply that iron meteorites are relatively easy to find. They have a distinctive appearance, and are quickly noticed by anyone who sees them. Stony meteorites are not so easy to spot, and are easily mistaken for 'ordinary' Earth rocks. So, iron meteorites are over-represented in our collections.

The stony meteorites are composed of silicates, with about a 10 per cent metal content. (Stony-iron meteorites are midway between the other two categories, in terms of metal content.) The stony meteorites are sub-divided into two broad categories — **chondrites** and **achondrites**. Many of the stony meteorites contain small inclusions called chondrules. Chondrules are typically 1 mm in diameter, or less, and can be composed of metal, silicates, or sulfides. They show evidence of having been rapidly cooled from a molten state, probably at the very birth of the solar system. In general, the meteorites which contain chondrules are called chondrites, those which do not contain chondrules are achondrites. The spectra of achondritic meteorites, and some chondrites, resemble the S-type asteroid spectra.

Of special interest are the **carbonaceous chondrites**. These resemble the C-type asteroids — the most numerous asteroids of all. The carbonaceous chondrites were never in a molten state; their mineralogy shows that these objects were never hotter than 130°C. Although these objects were not subjected to large-scale heating, they did not escape other types of physical alteration. The carbonaceous meteorites show evidence of violent collisions in the past. There is even evidence of *aqueous* activity. **Water of hydration** is still present in the carbonaceous meteorites and in asteroids such as Ceres.

Later in this book we will discuss the practical utilization of the asteroids. Obviously, an asteroid that contains water, carbon, and metals, would be a prime target for exploitation. These asteroids contain vast quantities of almost everything we need to build habitations in space.

The mineralogical classification of meteorites is shown in Table 3.1. A detailed

Table 3.1 — Classification of meteorites

Class	Sub-class	Type	Remarks
Irons (siderites)		Hexahedrites	Nickel 4–6% Iron ~95%
		Octahedrites	Nickel 6–14%
		Nickel-rich ataxites	Nickel >10% to 66%
Stony-irons (siderolites)		Pallasites Siderophytes Lodranites Mesosiderites	Contain silicates, with much free metal
Stones (aerolites)	Achondrites	Aubrites (enstatite) Diogenites (hypersthene) Chassignites (olivine) Ureilites (olivine-pigeonite)	Calcium-poor
		Angrites Nakhalites Howardites Eucrites	Calcium-rich (Howardites, Eucrites: Basaltic achondrites)
	Chondrites	Enstatite (E) Olivine-bronzite (H)[a] Olivine-hypersthene (L,L)[a]	'Ordinary' chondrites
		Carbonaceous (C1, C2, C3) or (CI, CM, CO, CV)	Contain water, carbon, silicates, and metals

[a]The letters H, L, and LL, refer to high, low, and very low iron content, respectively.

discussion of meteorite mineralogy is beyond the scope of this book. Suffice it to say that different minerals are formed at different temperatures and pressures. The mineral composition of a meteorite therefore tells us something about the environment in which the rock was formed. We can determine the rate at which the material cooled from an earlier molten state, the depth at which it formed, and the length of time the rock was exposed to space before striking the Earth.

The techniques of spectrophotometry, radiometry, and polarimetry, as well as laboratory studies of meteorites, are providing most of the evidence we are obtaining about the nature of the asteroids. These techniques can be used at any time. There is another technique which can only be used under favorable circumstances, in certain parts of the Earth. This is the observation of asteroid occulations.

OCCULTATIONS

Observing **occultations** of stars by asteroids is a very powerful and precise technique, which has produced some unexpected results. When an asteroid passes between the Earth and a star, an occultation, or eclipse, of the star will be observable from some parts of the Earth. The asteroid, in effect, casts a shadow on the Earth. This shadow moves across the Earth's surface as the Earth rotates and the asteroid moves in its orbit. The width of the shadow is practically equal to the diameter of the asteroid. Just before an occultation, an observer on the Earth sees the combined light of the star and the asteroid. When the asteroid passes in front of the star, the light received at the telescope will decrease, because the star's light is no longer visible. After the star emerges from behind the asteroid, the total amount of light returns to its original value. By measuring the length of time that the star is invisible, we can obtain very accurate measurements of the diameter of the asteroid. To obtain true diameters, it is necessary to observe the occultation from several stations spaced across the width of the occultation path. In this way, diameters have been accurately measured for several asteroids. These measurements generally confirm the accuracy of the radiometric and polarimetric methods, but there have occasionally been some sizable discrepancies.

Occasionally, observers see multiple disappearances of the occulted star. This implies that the asteroid has one or more companions, or satellites. A few years ago, there was a flurry of reports of such multiple disappearances, and much speculation about how many asteroids have satellites. Nowadays there are far fewer such reports, probably because observers are becoming more experienced and more careful. There is still a possibility that some asteroids have satellites, but no undisputed observations of satellites have yet been obtained.

The observation of asteroid occultations is an ideal project for well-equipped amateur astronomers. It requires cooperation between a team of observers who have highly portable equipment. Such mobility and flexibility are not often available to the professional astronomer.

SPECKLE INTERFEROMETRY

Since the asteroids are so small, we cannot see details on their surfaces even with the largest telescopes. This is not really because of limitations of the telescopes themselves. The problem is caused by turbulence in the Earth's atmosphere.

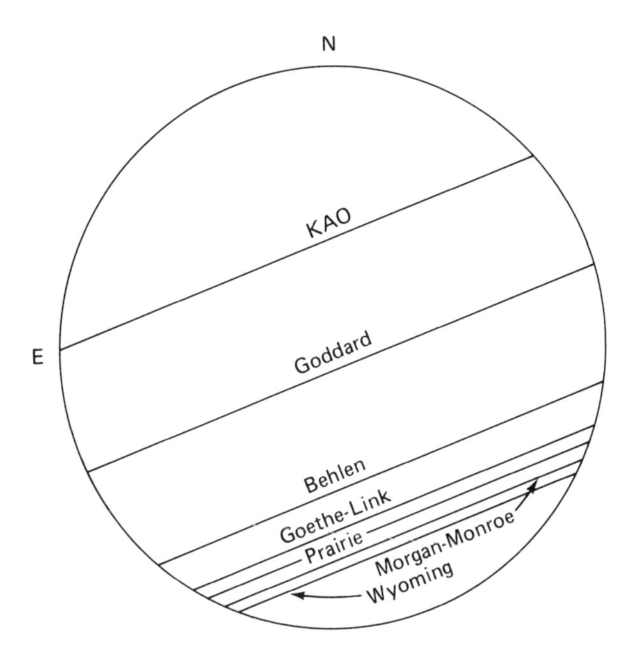

Fig. 11a — The profile of Pallas, as determined from the occultation of May 9, 1978. Each chord represents the time and duration of the occultation, as measured at the specified observatories. (From Millis and Elliot, in *Asteroids*, Tom Gehrels, ed. (1979), p. 109.)

The atmosphere causes every object to look somewhat 'blurred', no matter how large the telescope may be. This blurring is called 'seeing' by astronomers. Because of bad 'seeing', the asteroids appear indistinguishable from stars. The stars, too, are blurred. One solution, of course, is to place a telescope in orbit above the Earth's atmosphere. This will be accomplished when the Hubble Space Telescope is launched, in 1989. However, there are other ways to overcome the limitations imposed by the atmosphere. **Interferometry** is one such method.

Interferometry was used successfully several decades ago, by Michelson and Pease, to measure the diameters of a few of the larger stars. There are several different techniques which use the principles of interferometry. Today, the most popular technique is called **speckle interferometry**.

Atmospheric turbulence causes the air to be broken up into separate 'cells'. Each cell is typically several centimeters in diameter, and the cells are all in constant motion. The resolving power of a telescope is limited by the size of these cells. If the cells are 10 centimeters in diameter, a 10-meter telescope will not resolve details any better than a 10-centimeter telescope! (Of course, the larger telescope collects far more light, so it can see fainter objects, but its ability to resolve fine details is no better than the small telescope's.)

If you take a short-enough exposure with a lage telescope, you can effectively 'freeze' the motion of the atmosphere. The exposures must be only a few milliseconds in length. The resulting photograph will have a 'speckled' appearance. The image of every object will be composed of many separate small images. Each of the small images is produced by an individual atmospheric cell. In principal, all we have

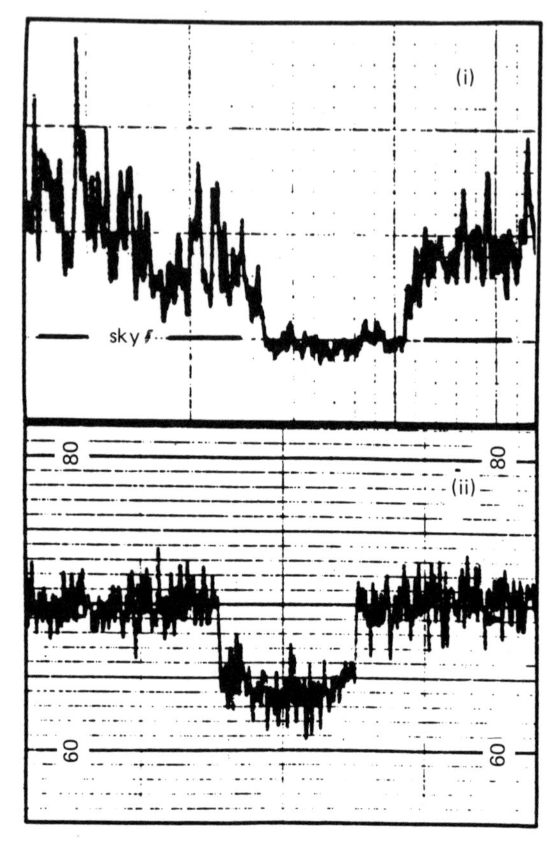

Fig. 11b — Light curves of two possible secondary occultations.
The dip in (i) occurred a few minutes prior to the occultation of the star SAO 120774 by (532) Herculina. Herculina was only 3 degrees above the horizon at this time. (ii) shows a dip which was observed during an appulse of the star SAO 114159 by (18) Melpomene. This observation was made from a location outside of the path of the occultation. Clouds interrupted the observations shortly after this dip was observed, on December 11, 1978.
Although these observations are sometimes cited as evidence for the existence of satellites around some asteroids, both observations were made under difficult circumstances. The existence of such satellites remains unconfirmed. From Millis and Elliot, in *Asteroids,* Tom Gehrels, ed. (1979), p. 113.)

to do is combine each of the speckles into a single image, which will then show almost as much detail as the telescope would show if there were no atmosphere.

Though relatively simple in principle, speckle interferometry is actually quite complex in practice. Nevertheless, this technique has been used quite successfully to measure the diameters of several asteroids. Recently, speckle interferometry has been used to obtain images of Vesta's surface. Although these images are quite blurry, they do show some markings on the surface of this asteroid.

MASS AND DENSITY

Among the goals of asteroid researchers are the determination of the gross physical properties of the asteroids — size, mass, and density. We have already described several ways of measuring the diameters of the asteroids, but the determination of mass and density are much more difficult. To determine the density, one must know the volume and the mass of the object. An error of 10 per cent in the measured diameter of an object results in an error of 33 per cent in its volume. The only way to determine the mass is to measure the gravitational effect of the object on some other body — such as another asteroid. Asteroids rarely come close enough to each other to produce measurable perturbations. The best way to obtain asteroid masses and densities will be to send a space probe to those objects. Until then, we must be content with observing the occasional close approaches between two asteroids.

Our best present estimates of the masses and densities of Ceres, Pallas, and Vesta, are shown in Table 3.2. For comparison the density of the Earth is 5.5 gm/cm^3. The mass of the Moon is 7.35×10^{25} gms.

Table 3.2 — Physical parameters of a few asteroids

Object	Diameter (km)	Mass (gm)	Density (gm/cm^3)
1 Ceres	940	1.17×10^{24}	2.6 ± 1.1
2 Pallas	538	2.26×10^{23}	2.6 ± 0.9
4 Vesta	576	2.75×10^{23}	3.1 ± 1.5

RADAR

Astronomical research suffers from one huge handicap. We cannot usually interact directly with the objects we study. A geologist can go into the field with his hammer, grab some samples, and return them to his laboratory for further study. But an astronomer must passively watch the objects of his interest from a great distance. In the future, space probes will help to alleviate this problem, but so far, the only extraterrestrial materials we have been able to study closely are the meteorites and the moon rocks. We wish that we could 'poke and prod' the asteroids in a more direct way.

There is one ground-based technique which allows us to overcome our near-total passivity. We can probe the asteroids with radar beams.

Studying the asteroids with radar provides a number of advantages over simply observing the reflected sunlight from these objects. Most of all, we can control the radar beam. The frequency, duration, and strength of the beam are known precisely. The beam can be pulsed, polarized, or varied in frequency. By comparing the known characteristics of the transmitted beam with the characteristics of the returned 'echo' we can derive a considerable amount of information about the composition, surface texture, and rotation of the asteroids.

A radar beam is somewhat like a beam of laser light. The radar beam is emitted at a single frequency, and the beam is **coherent**. That is, the phase of the beam is the same across the entire wavefront. When the radar beam is reflected by an asteroid, the returning beam will be shifted in frequency, depending on whether the asteroid is moving toward or away from us. (This is the 'Doppler shift'.) In addition, the rotation of the asteroid causes the returning beam to be spread out over a range of frequencies. The faster the asteroid rotates, the broader is the bandwidth of the reflected signal. The size of the asteroid also affects the spectrum of the radar echo.

The roughness of the asteroid's surface affects the way in which the radar beam is scattered. A smooth surface tends to preserve the coherency of the reflected beam, while a rough surface produces a completely **incoherent** echo. Furthermore, the *composition* of the asteroid affects the strength of the returned signal. Bare metallic objects are strong reflectors of radar, while dusty, insulating surfaces reflect radar poorly. Thus from an analysis of the characteristics of the radar echo, we can learn quite a lot about the surface texture and composition of the asteroids.

Finally, by simply measuring the time it takes for the radar beam to reach the asteroid, and return, we obtain an accurate measure of the distance to the asteroid. In many cases, such measurements can greatly improve our knowledge of the asteroid's orbit.

Radar study of the asteroids obviously has great potential. The major limitation of this technique is that the returned radar echoes are always very weak. Only objects which are very large, or very close to us, can be studied. Thus, only the largest main-belt asteroids can be probed with radar. The main contributions of radar techniques come from the study of the near-by Apollo asteroids. In Chapter 6, we will see how radar has produced some exciting results in the case of at least one such object.

We still have much to learn about the asteroids, but we have already learned a surprising amount. It is truly amazing that we can sit at our telescopes here on the Earth, and discover so much about the nature of some rocks hundreds of millions of kilometers away. We have discussed the techniques used to study the asteroids. Now, let us see what we have learned from this study.

4

The nature of the asteroids

We have seen how modern observational techniques have enabled us to determine the composition and physical parameters of individual asteroids. It is now time to discuss the results of these observations, and derive a general picture of the nature, origin, and evolution of the asteroids.

Let us take an imaginary trip through the asteroid belt. As we pass through the belt, we will see nothing like the dense cloud of rocks which is often portrayed in science fiction movies. In fact, we will probably not notice that we are within the belt at all. The asteroids are actually separated from each other by thousands of kilometers. On any random path through the asteroid belt, the chances of colliding with any substantial-size rock are close to zero. The Pioneer and Voyager space probes have also shown that the density of tiny dust particles within the belt is not a hazard. So we can proceed on our journey in complete safety. To make our trip more interesting, we will land on several asteroids along the way. We will find that almost all of the asteroids have powdery surfaces. This dusty surface layer is called a **regolith**. On the largest asteroids, the regolith is a few kilometers deep, while the smallest asteroids have only a thin coating of dust. We will find considerable diversity in the composition of this surface layer, as we go from one asteroid to another. As we travel from the inner parts of the asteroid belt to the more distant asteroids, however, we will notice a systematic trend in the surface compositions. At the inner edge of the belt, the asteroids tend to be grayish, perhaps with a tinge of red or brown. Some asteroids along the way will show a metallic surface of iron and nickel chunks. As we travel outwards, however, we will quickly see that more and more of the asteroids are covered with a dull black material resembling coal dust. These black asteroids, in fact, are by far the most numerous of all asteroid types. The percentage of the gray S-type asteroids decreases from about 60 per cent at 2.2 A.U. from the Sun, to about 15 per cent at 3.0 A.U. The C-type, black, carbonaceous asteroids increase from about 10 per cent of the total population at 2.2 A.U., to 80 per cent at 3.0 A.U. (See Fig. 12, and the discussion of asteroid photometric types in the preceding chapter.) Overall, about 75 per cent of all asteroids are type C, 15 per cent are type S, and the remainder are a variety of other types. The gradual, fairly uniform change in

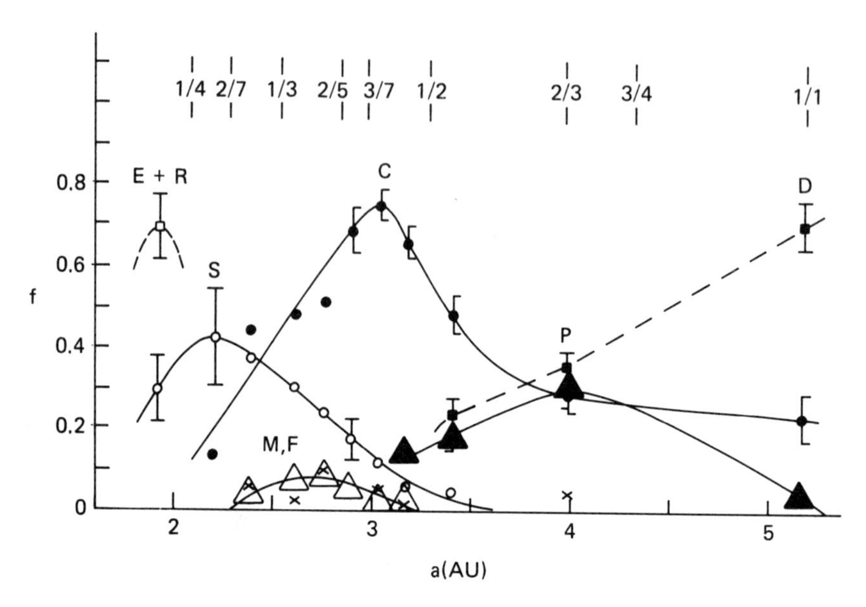

Fig. 12 — The percentage of asteroids of each compositional class at different parts of the asteroid belt.
Note that C-types constitute 80 per cent of all asteroids at 3.0 A.U., while S-types constitute 40 — 60 per cent of all asteroids at 2.2 A.U. D-types dominate the outer parts of the belt, and the Trojan region. (From Gradie and Tedesco in *Science* (1982) **216** 1405.)

compositional types with increasing distance from the Sun, gives us a clue to the origin of the asteroid belt.

According to current theories of the origin of the solar system, the planets should have lower and lower densities as we go farther from the Sun. This is indeed observed, at least in a general way. The inner planets contain metals, and dense, refractory silicates. The outer planets, and their satellites, contain ices and volatile compounds such as methane and ammonia gases, and low-density carbonaceous material. This is simply because the heat of the young Sun prevented volatile materials from condensing into the inner planets, while the colder temperatures in the outer solar system allowed the retention of those volatiles. (We will not discuss here the complicated subject of planetary atmospheres!) The asteroid belt is a transitional region. The innermost asteroids are silicaceous, while the outer ones are carbonaceous, and even contain substantial amounts of water of hydration. (This is the kind of water found in clay-like materials. We do not expect to find free water or ice exposed on the surfaces of the asteroids.)

The asteroids therefore fit in very well with our ideas of the formation of the solar system. The separation of asteroid types is also an argument against the theory that the asteroids are remnants of an exploded planet. Such an explosion would probably have caused a complete mixing of compositional types, rather than the gradual transition from S- to C-types. It is not known why the asteroids failed to coalesce into a single planet, but this is probably the result of the disturbing influence of Jupiter.

As we continue our journey among the asteroids we will encounter many interesting, unusual, and puzzling objects. We will find that Vesta is covered with igneous rock, perhaps even lava flows. This asteroid is only 576 km in diameter. How did it ever get hot enough to be melted, while other asteroids of similar size, or larger, remained cool enough to retain water of hydration?

We will encounter some asteroids, such as **Psyche**, that are composed of almost pure nickel–iron alloy. Are these the bare, exposed cores of much larger, differentiated, proto-asteroids?

Everywhere, we will find asteroids that resemble the satellites of Mars — dusty, pocked with craters, and grooved with fracture lines. Evidence of catastrophic collisions abounds, and we may occasionally see what appears to be two asteroids stuck together. We may see asteroids with satellites, or even clouds of satellites surrounding them.

We will end our imaginary journey, with the hope that some spacecraft will undertake a real journey of this type in the near future. Such trips are now on the drawing boards at NASA. The asteroids contain clues to the origin of the solar system which cannot be obtained by visiting the larger planets.

ORIGIN

We will now go back in time, to see how the asteroids originated. We must be aware from the start, however, that the following scenario is dependent partly on speculation, and that our speculations are constantly changing as new observational facts refine our knowledge of the asteroids.

Astronomers are agreed that the Sun and the entire solar system condensed from one large nebula. We can see stars being born within such nebulae in parts of our galaxy. Astronomers differ, however, on the exact mechanism by which the planets, comets, and asteroids condensed.

The collapse of the solar nebula may have been triggered, or at least been accompanied, by a nearby supernova explosion. When stars reach the end of their lifetimes, some of them explode and become **supernovae**. Supernovae create many heavy elements, such as metals, and spew those elements into interstellar space. They also create shock waves which can disturb nearby nebulae. The major planets may have condensed at the same time as the Sun, or they may have been gradually built-up by the agglomeration of smaller **planetesimals**. In either case, meter- to kilometer-size planetesimals did exist throughout the solar system. Most of them were removed by collision with the planets, or by ejection out of the solar system. The planetesimals remaining between the orbits of Mars and Jupiter, although in relatively stable orbits, were prevented from accumulating into a single planet, perhaps because of perturbations from Jupiter. Planetesimals in the outer parts of the solar system either collided with the large outer planets, or were thrown out to the fringes of the solar system, to form the present 'Oort Cloud' of comets. These planetesimals among the outer planets will be discussed further in Chapter 5.

The young asteroid belt must have contained much more mass than the present belt. The abundant craters on all of the inner planets, and on almost all satellites, are vivid evidence of the depletion of the original asteroids and planetesimals.

In the primordial asteroid belt, the asteroids were presumably moving at small velocities relative to each other. Low-velocity collisions allowed small objects to accumulate into larger ones, but this process was stopped, either by the aforementioned perturbations by Jupiter, or by some other mechanism. At present, these low-velocity collisions do not occur. The asteroids now encounter each other with average velocities of about 5 km/sec. These higher-velocity collisions can only result in fragmentation, not agglomeration. It is for this reason that some astronomers speculate that a fairly massive object passed through the primordial asteroid belt, depleting the population of the belt, preventing the formation of a single planet, and creating the present relatively high velocities.

At any rate, it seems certain that collision and fragmentation have been the dominant processes in the evolution of the present asteroid belt. Almost every asteroid we see today must be a fragment of a larger original body. Only the very largest asteroids could have survived major collisions. The collisions are still continuing today, although the time-scale is so long that we will probably never actually see two asteroids collide.

The collisions produce craters on the larger bodies, and the ejecta from these craters cover the asteroids with fragmented rocks and dust. This fragmented layer is the regolith. Eventually, the deeper regoliths were compacted into solid layers called **breccias**. (This is the same process by which sedimentary rocks were created on the Earth. For example, compacted sand eventually becomes sand*stone*.) Many of the meteorites we see today are composed of brecciated rock; that is, they have a fragmented internal structure, often containing inclusions of different types of materials. These meteorites must once have been part of the compacted regoliths of asteroids.

It seems clear that some of the asteroids were heated enough to melt, early in the history of the solar system. The mechanism for this heating is not well understood, nor do we understand why other, similar asteroids, were *not* heated.

All of this sounds like a very violent history for the asteroids, and indeed, some meteorites show evidence of violent shock, melting, and metamorphosis. Even the carbonaceous chondrite meteorites, which were once thought to be unaltered, 'pristine' material, show evidence of chemical and physical alteration.

THERMAL HISTORY OF THE ASTEROIDS

We have previously mentioned the interesting nature of Vesta, which seems to have a basaltic surface. This asteroid must have undergone considerable heating and melting early in its history. Yet Vesta, with a diameter of 576 km, is too small to have been melted by any of the 'classical' processes — i.e. radioactive decay of uranium, potassium, and thorium; and gravitational collapse. On the other hand, asteroids which are similar in size to Vesta, or even larger, have *not* undergone drastic heating.

This paradox is still unexplained, but we are at least beginning to solve some of the mysteries.

The most important theoretical development has been the 'discovery' of two possible methods of heating small bodies. These are the radioactive decay of aluminum-26, and heating by electrical induction.

RADIOACTIVE HEATING

If radioactive decay is to be an important factor in the heating of small bodies, it requires the presence of sufficient quantities of an energetically radioactive isotope, with a half-life of 10^5 to 10^8 years. The best candidate for such an isotope is ^{26}Al, which decays to ^{26}Mg, with a half-life of 7×10^5 years. ^{26}Al has, in fact, been found in some meterorites, but it is not certain that enough of this material existed to have melted some of the asteroids. The fact that some asteroids were melted, some were merely heated, and others remained relatively cool, requires that the concentration of the isotope was not uniform throughout the primordial solar nebula. Other elemental abundance anomalies also suggest that the solar nebula was quite heterogeneous.

ELECTRICAL HEATING

When the Sun was just formed, it probably was rapidly rotating, and was emitting an intense 'solar wind' of electrically-charged plasma. Such conditions are known to exist in T-Tauri stars, which are stars in the process of formation. The electrical and magnetic fields produced under such conditions could generate electrical currents through the objects in the solar system. These currents could have heated the bodies to their melting point, if the electrical conductivity of the planetesimals was high enough.

Both of these heating mechanisms are capable of melting objects much smaller than Vesta — objects only a few kilometers in diameter could have been melted in this way. This possibility has given rise to new scenarios for the formation of the solar system.

The traditional view of asteroid evolution has been that a few asteroids were formed, which were large enough to become **differentiated**. That is, their interiors were heated by uranium–potassium–thorium radioactivity, and the heavier minerals separated from the lighter ones. Iron and nickel settled to the core of the asteroid. This core was covered by a silicate **mantle,** and a basaltic **crust**. The iron, stony, and basaltic meteorites that we find were thought to have come from separate layers of differentiated asteroids which had fragmented after violent collisions.

There are some difficulties with this traditional view. Among them is the fact of the great diversity among the iron meteorites. There are at least a dozen different types of such 'irons'. It is known that the irons cooled very slowly, and must have been buried many kilometers deep within their parent bodies. Under the older views of asteroid heating mechanisms, this would mean that there had to be dozens of very large asteroids in the primordial belt, and that each of those asteroids suffered catastrophic collisions. This seems implausible, even allowing for a much more massive asteroid belt than the present one.

THE 'RAISIN-BREAD' THEORY

The possibility that much smaller asteroids could have been melted allows a new interpretation of asteroid evolution.

After the asteroidal planetesimals formed, melted, and differentiated, they collided with each other, relatively gently, to form larger asteroids. The cores of the original planetesimals were then distribured 'like raisins in bread' throughout the larger asteroids. This allows for several different types of materials to be included within a few large asteroids. This scenario can account for the diversity of compositions and for the cooling rates observed in the iron meteorites. It still does *not* account for the fact that the majority of asteroids, both large and small, are *undifferentiated* carbonaceous chondrites. If one assumes that only the asteroids in the inner parts of the belt contained enough radioactive material at the time of their formation, and that partial mixing of asteroid types occurred later, it would help to explain the different evolutionary tracks of the asteroids. In my opinion, however, the great diversity of asteroid and meteorite compositions has not yet been adequately explained.

SUMMARY OF ASTEROIDAL EVOLUTION

(1) Small planetesimals condensed from the solar nebula 4.5 billion years ago. Materials with high melting points condensed first, volatile materials last. In the innermost parts of the belt, the temperatures were too high for volatiles to condense at all. Instead, the volatile elements chemically reacted with the high-temperature planetesimals. Low-melting-point carbonaceous material condensed in all but the inner parts of the belt.

(2) The asteroids grew from agglomeration of the small planetesimals. At this time, the relative velocities of the asteroids must have been small. That is, the asteroids were moving slowly with respect to each other, so that collisions produced coalescence, rather than destruction of the colliding bodies. Such gentle collisions generally cannot occur in the present asteroid belt.

(3) The decay of radioactive materials heated some of the early asteroids. Some asteroids melted, some were only metamorphosed, some were not affected at all, depending on their initial composition. The isotope of aluminium, which may have produced this heating, may have come from a nearby supernova explosion, which is thought to have triggered the initial collapse and fragmentation of the solar nebula. Alternatively, electrical induction could have been the primary cause of heating. The asteroids continued to accumulate into larger bodies. Some of these bodies had molten cores and were thoroughly differentiated, others acquired pockets (raisins) of differentiated material, others remained relatively uniform and undifferentiated.

(4) Something happened to 'stir up' the asteroids. This may have been caused directly by gravitational perturbations from Jupiter, or by the passage of a large body, or bodies, through the asteroid belt. These disturbing bodies could have been large planetesimals which later coalesced with the proto-Jupiter. At any rate, the orbital eccentricities and inclinations of the asteroids increased, causing the average relative velocity of the asteroids to increase to its present value of 5 km/sec. Agglomeration could no longer occur, the asteroids were not able to accumulate into a single planet, and collisions produced only fragmentation of the existing asteroids.

(5) Destructive collision became the dominant evolutionary process. This resulted

in fragmentation of most of the asteroids during the past 4.5 billion years, and the production of craters and regoliths on the surviving pieces.

The above scenario probably represents the consensus of most astronomers today. No one regards this picture as anything more than the 'most plausible' theory, however. There are enormous gaps in our understanding of the asteroids, and new observations could change our ideas at any time. It is worthwhile, therefore, to mention an alternative to the 'consensus view'.

THE EXPLODED PLANET THEORY

The hypothesis that the asteroids are the fragments of an exploded planet was first formulated as soon as it became clear that there were *many* asteroids, instead of the single planet that was expected from Bode's Law. Olbers proposed this idea shortly after he discovered the second asteroid, Pallas, a year after the discovery of Ceres. This theory accorded with the views of planetary formation held at that time. The theory of Laplace held that the planets formed directly from the solar nebula, without the prior accumulation of smaller planetesimals. Nowadays, the planetesimal theory is in greater favor, so we think of the asteroids as left-over planetesimals. There are still a few astronomers, however, who think that a single, large planet between Mars and Jupiter exploded. The chief proponents of this theory are M. Ovenden and T. Van Flandern. Their main arguments are theoretical considerations concerning the stability of the solar system, and a possible grouping of long-period comet orbits. In their theory of the origin of the solar system, a planet between Mars and Jupiter is *required.* That planet exploded, possibly because of a collision with yet another planet. The debris of the explosion then formed the asteroid belt.

The main arguments against this theory are the difficulty of destroying a large planet, and the smoothness of the distribution of asteroid types from S-asteroids in the inner belt, to C-types in the outer belt. A catastrophic explosion would probably have scattered the various types randomly throughout the belt. The orbits of the asteroids would not be as circular as they are now. Furthermore, most scientists feel that the asteroids can be explained as the natural outcome of solar system formation as we currently understand it. If the existence of the asteroids can be explained as the failure of planetesimals to coalesce, why introduce the extra complication of trying to *explode* an entire planet?

WHAT THE ASTEROIDS LOOK LIKE

The 'conventional' scenario, described previously, was arrived at largely through the study of meteorites, theoretical studies of the orbital dynamics of the solar system, and ground-based observations of the asteroids. If this picture of the composition and evolution of the asteroids is correct, it should correctly predict what we will find when we finally reach the asteroids with spacecraft.

We should find that most of the asteroids are covered with regoliths — several kilometers deep on the largest asteroids, and thin, or absent, on the smallest ones.

The asteroids will probably have major internal fractures, and may show the linear striations visible on Mar's satellite, Phobos.

If we can investigate the interiors of the asteroids, we should find that the C-type objects are fairly homogeneous and undifferentiated, while other types will have a layered, differentiated interior, or a spotty, raisin-bread structure. Some asteroids should prove to be the stripped cores, or the outer fragments, of differentiated asteroids which were destroyed.

Most of all, we hope and expect to find that the asteroids have the mineralogical compositions that we have inferred for them. That is, we expect the C-asteroids to be carbonaceous, the S-asteroids to be silicaceous, and the M-asteroids to be metallic. It is far from certain that our inferences are correct!

SOME REPRESENTATIVE ASTEROIDS AND FAMILIES

To further illustrate the state of our knowledge of the asteroids, it may be helpful to describe a few individual asteroids and Hirayama families in greater detail. We will choose the most prominent asteroids and families, not because they are necessarily the most interesting, but simply because they are the most thoroughly studied.

Ceres

Ceres is by far the largest asteroid, having a diameter of about 940 km, and a mass of 1.17×10^{24} gm. This one object contains about one-third of the combined mass of the entire asteroid belt. Although Ceres is so large, and was the first asteroid to be discovered, it is not the brightest asteroid. Vesta appears brighter because it has a higher albedo, and can come somewhat closer to the Earth than Ceres. The mean visual magnitude of Ceres at opposition is 6.9, which means that it is not normally visible with the naked eye, but can easily be seen with binoculars.

The diameter of Ceres was recently derived from the occultation of a star. Prior to that occultation, it was thought that Ceres was over 1000 km in diameter.

The reflection spectrum of Ceres is relatively flat at visual wavelengths, which implies a carbonaceous chondritic composition. (See Fig. 13.) Other possible compositions can also produce such a spectrum, however. A major advance in the study of Ceres occurred in 1977, when L. Lebofsky detected an absorption band in the spectrum of Ceres at a wavelength of 3 microns (which is in the infrared). This absorption band is produced by water of hydration, which is water imbedded in the mineral structure of the asteroid. The discovery of water of hydration on Ceres strengthens the identification of Ceres with the carbonaceous chondrite meteorites. The spectrum of Ceres has now been observed from the ultraviolet to the far infrared, and even at radio wavelengths, and it is clear that Ceres closely resembles the C-2 type carbonaceous chondrites. This is not quite the same as saying that Ceres *is* a C-2 carbonaceous chondrite, however! It is quite possible that none of our meteorites have exactly the same composition as the surface of Ceres.

The density of Ceres is estimated to be 2.6 gm/cm^3. This, again, is consistent with a carbonaceous chondritic composition containing 10–15 per cent water (by weight).

The mass of Ceres was estimated from its gravitational effect on the asteroid Pallas. The result is that Ceres has a mass of 5.9×10^{-10} solar masses, or 1.17×10^{24} gm. This, combined with the diameter derived from radiometric and polarimetric observations, yields the density mentioned above.

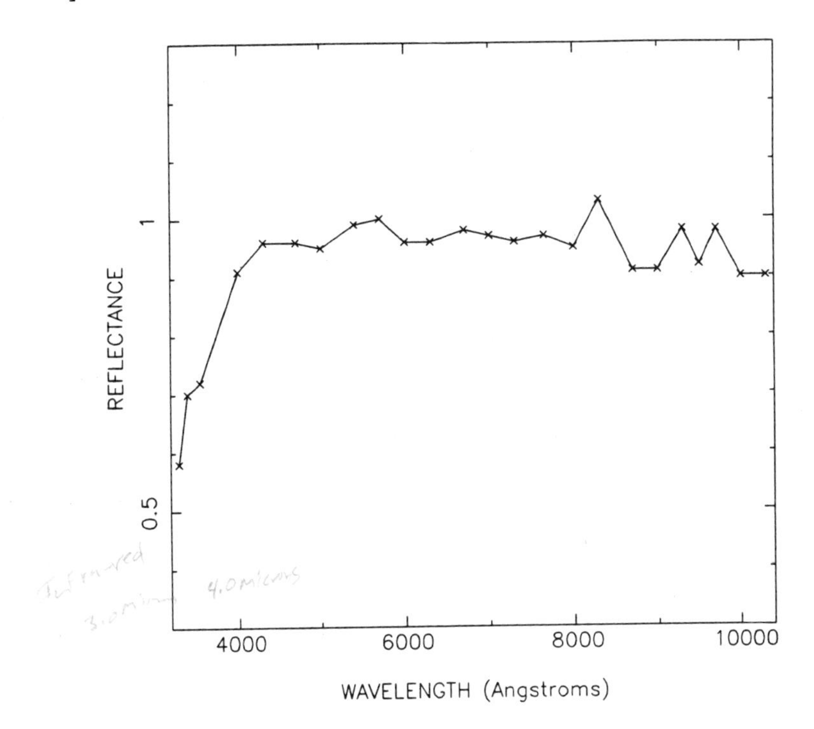

Fig. 13 — The reflectance spectrum of Ceres.
Except for wavelengths shorter than 4000 ångstroms, the spectrum of Ceres is rather flat. This is typical of C-type asteroids. Compare it with the spectrum of Vesta in Fig. 10.

Ceres rotates with a period of slightly more than 9 hours. The amplitude of the light curve is only 0.04 mag., and its color remains constant, which implies that Ceres is practically spherical, and is uniformly gray over its entire surface. Its albedo is about 9 per cent.

Pallas

Pallas is another carbonaceous-type asteroid with a spectrum very similar to that of Ceres. Its diameter is quite well determined because of a well-observed stellar occultation on May 29, 1978. These observations are so accurate that we can say that Pallas is not quite spherical. A tri-axial ellipsoid fitted to the observations has diameters of 559 km, 525 km, and 532 km; for a mean diameter of 538±12 km. This value for the diameter is about 10 per cent smaller than the values derived from radiometry and polarimetry. Pallas has a mass of 2.26×10^{23} gm. Combining this with the diameter given above, we derive a density of 2.6 gm/cm^3.

Perhaps the most unusual characteristic of Pallas is its orbit. The orbit of Pallas is inclined 35 degrees to the ecliptic. This is an extraordinarily high inclination for a main-belt object. The orbital eccentricity of Pallas is also large (0.23).

An object as large as Pallas could not have accreted at such a high inclination, nor is it likely that it could have been pushed into its present position by an enormous collision. This leaves only two possibilities for the origin of Pallas's orbit. Either

Pallas was perturbed into its present orbit by a massive object passing through the asteroid belt, or Pallas originated elsewhere, and was perturbed into its present orbit by encounters with Jupiter, Saturn, etc. Astronomers presently favor the former idea, since the passage of a massive object is also invoked to 'stir up' the other asteroids into their present high velocities.

Vesta

Vesta has been mentioned several times in this book because of the puzzling composition of this object. It seems certain that Vesta's surface is covered with igneous rock — that is, rock which was once molten. Two methods have been proposed for heating small bodies like Vesta — heating by electrical induction, and heating by relatively short-lived radioactivity. We still do not understand why other asteroids were *not* heated by these same processes.

Vesta is the brightest asteroid, having a visual magnitude of 6.0 at average oppositions. This means that Vesta can be seen with the naked eye, under optimum conditions. Vesta's brightness is due to its high albedo. This asteroid reflects about one-fourth of the sunlight which strikes it. Compare this with Ceres's reflectivity of only 9 per cent. The spectrum of Vesta shows the characteristic absorption of the mineral pyroxene. It is different enough from other asteroid types to be classified as type U (unclassifiable).

Vesta has a diameter of 576 km, a mass of 2.75×10^{23} gm, and a density of 3.1 gm/cm^3. This asteroid has a very well-determined mass, because of its frequent close approaches to another asteroid (197) Arete. If we can accurately measure the diameter of Vesta, by means of stellar occultations, then Vesta will also have the best-determined density of any asteroid.

Vesta rotates fairly rapidly, with a period of only 5.34 hours. As Vesta rotates, its color changes slightly. This means that Vesta's surface is not completely homogeneous in composition. Of all the asteroids, Vesta probably has the most interesting surface features. Astronomers long for the day when we can obtain closeup photographs of this object!

Vesta is quite possibly the source of the **eucrite** meteorites. Of all the known asteroids, Vesta seems to be the only one having the proper composition. (See, however, the discussion of the **Budrosa family**, below.) Yet Vesta is not in a location which would allow meteoroids to be ejected into Earth-crossing orbits. Apparently the only way that fragments of Vesta could reach the Earth is by an improbable double-collision scenario. That is, a collision could knock fragments off the surface of Vesta, then the fragments could suffer a second collision and be pushed into an Earth-crossing orbit.

It is possible that the eucritic meteorites come from a smaller asteroid which has not yet been discovered, or at least not yet studied. Our knowledge of asteroids in unusual orbits is still spotty, and it is precisely those objects which are most capable of delivering meteoroids to the Earth.

Some asteroid families

Ever since the discovery of asteroid families by Hirayama, in 1918, it was speculated that these families are the fragments of individual 'parent' asteroids which broke up as a result of catastrophic collisions, but remained in similar orbits. Now the observational evidence shows that many families are indeed composed of homogeneous, or mineralogically-related members, thus supporting the fragmentation hypothesis.

A glance at Fig. 4 shows the obvious existence of the Eos, Koronis, and Themis families. Rigorous statistical investigations show that almost half of all the asteroids occur in families. More than 100 separate families have been identified. If asteroid families are really fragments of disrupted parent bodies, then the family members provide us with the means of examining the interiors of the parent asteroids. Asteroid families are, in effect, 'dissected' asteroids.

The **Koronis** family is composed of S-type asteroids. The members of this group are very similar in color, spectrum, and albedo. There seems little doubt that all of these objects came from a single, homogeneous parent body. The minimum size of the parent body was about 90 km. The largest surviving member of this family, (208) Lacrimosa, is about 45 km in diameter.

The **Themis** family is composed of C-type asteroids, while the **Eos** family members are intermediate between C and S. Both of these families show slight differences among their members. These differences could be the result of slight inhomogeneities in the parent bodies, shock-induced metamorphism caused by the disrupting collision, or differences in 'weathering' of the fragments after the breakup.

The **Budrosa** family is located at a heliocentric distance of 2.9 A.U., and has an inclination of 6 degrees. It contains only six objects. Asteroid (338) Budrosa itself is an M-type (metallic) object. The remaining members are of unusual types. The Budrosa family is sometimes mentioned as an alternative to Vesta as the parent body of the eucrite meteorites. This family is near the 2:5 resonance with Jupiter, which makes it relatively easy for meteoroids to be perturbed into Earth-crossing orbits. The eucrites are basaltic rocks, and the parent body of the eucrites probably consisted of a basaltic crust, a silicate mantle, and possibly a metallic core. Apparently, the Budrosa family comes from a highly differentiated parent body. Nevertheless, no actual basaltic objects have yet been detected within the Budrosa family. Conversely, meteorites resembling another member of this family (349) Dembowska, are either extremely rare, or completely absent on the Earth. It is highly improbable that only the basaltic components of the Budrosa family were transported to the Earth. So, the origin of the eucrite meteorites remains in doubt.

The **Phocaea** group is not a true family. The members of this group are merely separated from the main belt by resonance gaps. It is not surprising, therefore, that the asteroids in this group do not resemble each other. The Phocaea asteroids are scattered over a fairly large area around a=2.4, i=23.

Establishing the existence of small families is not a clear-cut process. Different statistical methods produce somewhat different results. On the whole, however, the agreement between various authorities is quite good.

CAPTURED ASTEROIDS

On either side of the asteroid belt lie the planets Mars and Jupiter. Mars has two satellites — Phobos and Deimos — which resemble in every way typical C-type asteroids. Jupiter has eight outer satellites which travel in peculiar, highly-eccentric

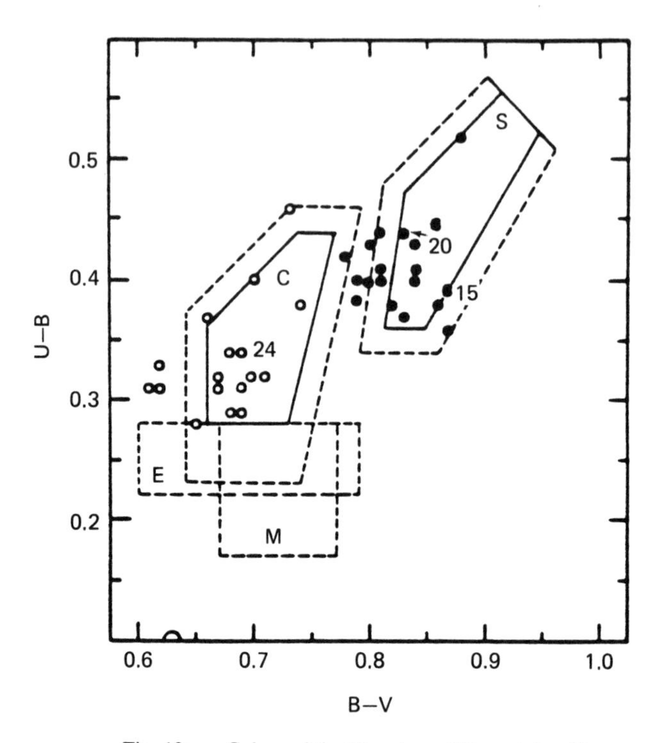

Fig. 13a — Colors of the Themis and Koronis families.
Note that the Themis family members are loosely clustered in the region of C-type asteroids, while the Koronis family consists of S-types. In spite of some scatter in the colors of individual family members, it is thought that both families originated from the fragmentation of larger parent bodies. (From Gradie *et al.*, in *Asteroids,* Tom Gehrels, ed. (1979) p. 368.)

orbits. Is it possible that these satellites of Mars and Jupiter were once asteroids, snatched out of the main belt by the gravitational pull of the planets?

According to current theories of the formation of the solar system, C-type objects should not have formed in the immediate vicinity of Mars. They only begin to appear in the middle regions of the asteroid belt. For this reason, it has been suggested that Phobos and Deimos were once independent asteroids which were perturbed into Mars-crossing orbits. Mars then captured these asteroids into their current satellite orbits.

Similarly, some of the outer satellites of Jupiter seem to have compositions similar to the Trojan asteroids. (See Chapter 5.) These outer satellites fall into two distinct groups. The outermost four satellites all travel in retrograde orbits, that is, they move in the opposite direction from most of the satellites in the solar system. They all have similar inclinations, and are at roughly the same distance from Jupiter — between 21 and 24 million kilometers from that planet.

The other group of small satellites lies at about 11 million kilometers from Jupiter. These satellites travel in direct orbits (not retrograde), and they all have similar inclinations of about 27 degrees.

It is tempting to speculate that these two groups of satellites were formed by the breakup of two bodies which were captured by Jupiter at some time in the distant past. The fact is, however, that such captures are simply impossible under ordinary circumstances! There is no way for a heliocentric orbit to be changed into a planetocentric orbit, without something to change the velocity of the heliocentric object. The planet's gravity alone cannot do the job.

When we send space probes to orbit around the other planets, it is always necessary to fire on-board rockets on the spacecraft, to put the probe into orbit around the planet. There is no possible combination of approach angles and velocities which would enable the spacecraft to go into a satellite orbit without that final rocket burn to adjust its velocity and direction. How then could Phobos, Deimos, and the Jovian satellites have been captured into their present orbits?

The answer is that the circumstances in the early solar system were different than they are now. The still-forming planets were surrounded by extensive atmospheres or 'envelopes'. These envelopes could have provided enough drag to slow down a passing asteroid, and capture it into a satellite orbit. This possibility does not prove that these satellites are captured asteroids, it merely provides a mechanism for such captures.

Not much is known about the outer satellites of Jupiter. They are small and faint, and they were not examined by the Voyager spacecraft. The satellites of Mars, on the other hand, were thoroughly studied by the Viking orbiters. While it is not certain that Phobos and Deimos are captured asteroids, and their present environment near Mars is different from the environment in the asteroid belt, these two satellites probably are good examples of what 'true' asteroids really look like. Both of these satellites are quite irregular in shape. They are often called 'potato-shaped'. Phobos is 27 km across, in its longest dimension, while Deimos has a maximum diameter of 15 km. Both satellites appear to be covered with a thick layer of dust, or regolith. On Phobos, this regolith is estimated to be 100 — 200 meters deep, and the depth on Deimos is probably similar. Deimos is liberally sprinkled with craters, but Phobos is truly spectacular in this regard. In spite of the small size of Phobos, it has one crater 10 km in diameter, and two others 5 km wide. The largest crater is named Stickney. The impact which caused this crater must have come very close to shattering the entire satellite into millions of pieces. Indeed, Phobos is covered with many long, linear grooves. It is thought that these grooves are fracture lines, caused by the cracking of Phobos when it was hit by the object which caused the crater. The impact must have created a considerable amount of heat within Phobos. This heat may have released water vapor from the interior of the satellite, which then issued from the cracks, and caused the beaded appearance of the grooves.

The cratered and fractured surface of Phobos certainly indicates that this object has had a violent history. We fully expect that the asteroids in the Main Belt have had similar histories. Voyager photographs of Saturn's satellite, **Mimas**, and of **Miranda**, a satellite of Uranus, show that these objects have suffered impacts even more devastating than the one which created Stickney. We can only wonder at what we will find when we finally have close-up photographs of the asteroids. But for now, Phobos and Deimos provide our best indications of what the asteroids really look like.

5

Asteroids beyond the Main Belt

We have mentioned previously that the vast majority of asteroids lie within the belt between 2.0 and 3.5 A.U. from the Sun. There are some very interesting asteroids outside of this belt, however.

As we travel outward from the main belt, we encounter the **Hilda** asteroids, at about 4.0 A.U. Still further out, at 4.26 A.U., we meet the appropriately-named **Thule.** (The phrase *Ultima Thule* was used by the Romans to denote a remote, unknown territory.) Thule may be regarded as the last outpost of the main belt.

The Hilda asteroids and Thule occur at the 3:2 and 4:3 resonances with Jupiter, respectively. In previous chapters, we stated that *gaps* occur in the asteroid belt at such resonances. At these two outer resonances, however, we have *concentrations* of asteroids instead. (I call Thule a concentration of asteroids, although it is the *only* asteroid at that location!) it is not known exactly why some resonances produce depletions, while other resonances produce enhancements of the asteroid population. It is important to note that, while the *orbits* of Thule and the Hildas are relatively close to the *orbit* of Jupiter, the individual asteroids themselves never get close to the giant planet. The Hildas have fairly eccentric orbits, and their distance from the Sun ranges between 3.4 and 4.6 A.U. during each orbit. When the Hildas are closest to Jupiter's orbit,. Jupiter is *always* somewhere else. As Jupiter approaches, the Hilda asteroids move toward their perihelion points. Thus, the eccentricity and orientation of the orbits of the Hildas keep them safely away from Jupiter's grasp. The orbit of Thule is not very eccentric, but it always manages to stay at least 1.1 A.U. away from Jupiter.

Still farther from the Sun, we reach the asteroids which actually lie in the same orbit as Jupiter, but 60 degrees ahead of, or behind, that planet. These are the **Trojan** asteroids.

As mentioned in a previous chapter, J. L. Lagrange found that two objects could be within the same orbit if those objects formed an equilateral triangle with the Sun. This means that the angle, Jupiter–Sun–asteroid, must equal 60 degrees for the Trojan asteroids. Lagrange did this theoretical work long before any asteroids had been discovered! The points 60 degrees preceding or following Jupiter, or any other plant, are called the **Lagrangian points,** 'L4' and 'L5', respectively. So far, no 'Trojan' asteroids have been found in the orbits of any of the other planets.

The Trojan asteroids do not actually remain at the precise Lagrangian points. Instead, they oscillate to and fro, or **librate,** about these points. Typically, these asteroids move from about 45 degrees from Jupiter, out to 80 degrees, then back again. For this reason, the Lagrangian points are often called **libration points.** One such libration cycle takes about 150–200 years. A few Trojans have been found as much as 100 degrees away from Jupiter.

As we mentioned in Chapter 4, the asteroid belt is a transition zone between the dense inner planets and the gaseous outer planets. The inner parts of the belt contain many rocky, silicaceous asteroids, while low-density carbonaceous asteroids predominate in the outer parts of the belt. As we move even farther out, to the Hildas and Trojans, we find still other compositional types, so far not matched with any meteorite types. These compositional classes are called P- and D-types. Roughly 40 per cent of the Hildas and 70 per cent of the Trojans belong to these types.

In 1980, J. Gradie and J. Veverka proposed an intriguing theory for the composition of the D-type asteroids, (which were called 'RD-type' at that time). They suggested that these asteroids may contain an organic compound similar to kerogen — a derivative of coal tar! These astronomers believe that kerogen could easily be produced by non-biological processes in the colder regions of the original solar nebula. Indeed, some carbonaceous meteorites contain organic substances similar to kerogen.

Remember that, in chemistry, the word 'organic' simply means 'carbon-bearing'. It does not imply any connection with living organisms, even though living things are composed of organic compounds. During the past decade, it has become increasingly clear that nature produces organic substances throughout the universe. Interstellar nebulae contain formaldehyde and ethyl alcohol; amino acids have been found in meteorites; and it has even been suggested that comets carry viruses! The Trojan asteroids may yet provide further evidence of the ease with which organic compounds can be produced.

One particular Trojan asteroid deserves special mention. The asteroid **Hektor** is by far the largest of the Trojans. Its light curve shows very great variations during the asteroid's rotation. This implies a very elongated shape. Taken at face value, the light curve suggests that Hektor is a cylinder, 150 km wide, by 300 km long. It is difficult to understand how such a large, aspherical object could have formed. Astronomers A. F. Cook, W. K. Hartmann, and others, suggest another shape for Hektor. They believe that Hektor might actually be *two* asteroids in contact. In the Trojan clouds, collisions between asteroids are generally more gentle than collisions in the Main Belt, because the Trojans move more slowly with respect to one another. If two large Trojans collide with each other, they might stick together, instead of shattering into fragments. The area around the zone of contact might be brightened, in the same way that lunar craters are often bright. In this way, we can explain the light curve of Hektor. What an intriguing sight it must be!

HIDALGO

The Trojans are the most distant groups of asteroids, but there are two individual objects which travel well beyond the orbit of Jupiter, and even beyond Saturn. These objects are so unusual that they may not be related to the other asteroids at all.

The first of these objects is **Hidalgo**, discovered in 1920 by Walter Baade. Hidalgo travels from the inner edge of the asteroid belt, at 2.0 A.U., to just beyond the orbit of Saturn, at 9.7 A.U. The orbit is obviously eccentric, (e=0.66), and highly inclined (i=42 degrees). This object seems to be a D-type, but it shows color variations during its ten-hour rotation period. The diameter of Hidalgo is estimated to be about 40–60 kilometers.

Because of its peculiar orbit, it is often theorized that Hidalgo is an extinct cometary nucleus. The main objection to this hypothesis was that Hidalgo seemed to be much larger and darker than we thought comets could be. However, when Comet Halley was visited by spacecraft in 1986, we found that comets are not 'white snowballs'. Indeed, the surface of Halley is as dull and black as anything known in the solar system! Although Hidalgo is considerably larger than Comet Halley, it is entirely possible that this 'asteroid' actually started its life as a comet. No trace of cometary activity has ever been seen on Hidalgo, however, so the question is still open. An alternate theory is that Hidalgo originated among the Trojans, or in the outer parts of the asteroid belt. However, if it is possible to transport ordinary asteroids into orbits like Hidalgo's, there should be other such objects. So far, Hidalgo remains unique. No other asteroid makes excursions from the asteroid belt to the regions beyond Jupiter.

CHIRON

For 57 years, Hidalgo was the only asteroid known to travel beyond the orbit of Jupiter. In 1977, however, the writer of this book discovered an object which never comes nearer than 8.5 A.U. from the Sun. This object, called **Chiron** (kai'ron), goes out as far as the orbit of Uranus, at 18.9 A.U. Chiron is called an asteroid only for lack of a better term. Unlike Hidalgo, Chiron *never* goes near the asteroid belt, and it probably originated in the outer reaches of the solar system.

I will describe the discovery of Chiron in some detail, as an example of how our knowledge of the solar system increases, step by step.

For several years, I conducted a 'Solar System Survey', using the 48-inch Schmidt telescope at Palomar Observatory, in California. The purpose of this survey was simply to search for any unusual objects in the solar system. I expected to find Apollo asteroids, comets, and maybe even some surprises. Each month, I photographed an area of the sky near the 'opposition point' of the ecliptic. This point is exactly 180 degrees away from the Sun. At opposition, the planets and asteroids move at their maximum retrograde velocities, that is — westward. Their apparent speed is a function of their distance from the Earth. Very near-by asteroids can move several *degrees* per day, while distant objects, like Chiron, move only a few *minutes* of arc per day. The fast-moving objects can easily be spotted by the long trails that they make on a photograph during the long exposure time. For slow-moving objects, it is necessary to take two photographs of each field, one day apart. The two photographs are then compared under a special microscope, to find the moving objects.

A typical photograph contains 200–300 ordinary asteroids. These asteroids are so numerous that it is impossible to keep track of all of them. So, I only looked for the objects which were moving much faster, or much slower, than the main-belt asteroids.

In this way, I found a very slow-moving object in October, 1977. From its motion, I was able to deduce that this object must be near the orbit of Uranus. Yet Uranus and its satellites were on the other side of the sky. I realized that nothing else big enough to be seen was known at that distance! After more photographs were taken, it became possible to compute an orbit. This confirmed that the object was nearly 18 A.U. from the Sun at the time of discovery. Its perihelion distance was only 8.5 A.U. Clearly, the object was no ordinary asteroid. Its brightness implied a diameter of a few hundred kilometers. The orbit of Chiron is more like cometary orbits than anything else, but it is at least 10 times larger than typical comets. On the other hand, it is 10 times smaller than the smallest planet.

When this discovery was reported to the news media, they immediately hailed it as a 'possible tenth planet'. I did not help matters any, because I didn't know *what* to call it! It is curious how people need to label things, and place everything in pigeonholes. Chiron simply does not fit into any pigeonhole. it remains unique, and giving it a label will not bring us any closer to an understanding of the true nature of this remarkable object.

In searching for a name for the new object, I found that the names of the mythological Centaurs had not yet been used for any asteroids. Chiron was the most famous of the Centaurs. He was the son of Saturn, and the grandson of Uranus, which ties in well with Chiron's orbit. This seemed like an excellent name for my new object! Now, for people who like to label things, we can say that Chiron is a 'Centaurian asteroid'!

After a preliminary orbit had been determined, astronomers started searching through old plate archives to try to find pre-discovery images of Chiron. In 1977, Chiron was near aphelion, and therefore was quite faint. It becomes relatively bright at perihelion, every fifty years. The last perihelion passage before discovery occurred in 1945. Sure enough, several photographs were found from the years 1941–1948 which contained images of Chiron. Then, astronomers W. Liller and L. Chaisson looked at plates taken during the previous perihelion passage in 1895. Incredibly, they found extremely faint images of Chiron on those 'ancient' plates. So, a few weeks after it was discovered, we had observations of Chiron spanning more than 80 years!

The positions of Chiron which were obtained from those photographic plates made it possible to compute a very accurate orbit for Chiron. These calculations showed that the orbit is unstable. Chiron can make some extremely close approaches to Saturn, and its orbit is drastically changed at such encounters. It is most probable that Chiron's orbit will slowly evolve inward; that is, closer to the Sun. If this happens, Chiron will eventually collide with one of the planets. There is also a possibility that a close approach to Saturn will eject Chiron out of the solar system entirely. In either case, Chiron's life expectancy as a member of the solar system is only about a million years.

In the years after Chiron was discovered, astronomers attempted to learn more about this object. Detailed studies were hampered by Chiron's faintness, however. Infrared observations have shown that Chiron has a moderately dark, rocky or dusty surface. Its diameter is about 200–300 kilometers, and its shape is nearly spherical. The characteristics of Chiron, and those of Hidalgo, are shown in Table 5.1.

Table 5.1 — Characteristics of Hidalgo and Chiron

	Hidalgo	Chiron
Orbital		
semi-major axis	5.86 A.U.	13.70 A.U.
perihelion	2.0 A.U.	8.5 A.U.
aphelion	9.7 A.U.	18.9 A.U.
eccentricity	0.66	0.38
inclination	42.4 degrees	6.9 degrees
period of revolution	14.2 years	50.7 years
Physical		
surface type	D?	C?
diameter	40–60 km	200–300 km
rotation period	10.06 hours	5.9 hours

As Chiron slowly approaches perihelion, in 1996, it will become bright enough to be studied in greater detail. Perhaps then we will learn whether Chiron has any similarity to comets, or to any other known objects in the solar system.

It is often speculated that Chiron is simply a large comet. It does not *look* like a comet, only because it is so far from the Sun that its gases remain frozen. The main argument against this 'cometary' theory is that Chiron is unique. If comets can be as big and bright as Chiron, why don't we see any more of them? After discovering Chiron, I continued my survey of the solar system for eight more years. I found no other objects as distant as Chiron.

Chiron could, nevertheless, be a freak comet. Or it could be an escaped satellite of Saturn or Uranus.

When Chiron makes its close approach to Saturn, it actually comes closer to that planet than Saturn's outermost satellite, Phoebe. Furthermore, Chiron and Phoebe are approximately the same size, and have similar spectra. These similarities with Saturn's outer satellite can certainly lead one to suspect a common origin for the two objects. Perhaps Chiron was once a satellite of Saturn. On the other hand, Chiron also can come close to Uranus (see Fig. 14). Was Chiron once a satellite of Uranus? It has even been theorized that both Chiron and the planet Pluto were once satellites of *Neptune*. A massive object may have passed close to the Neptunian system, and ripped Chiron and Pluto from their orbits.

It is also possible that Chiron is a planetesimal left over from the formation of the solar system. Slowly, its orbit may have evolved inward from the outer reaches of the

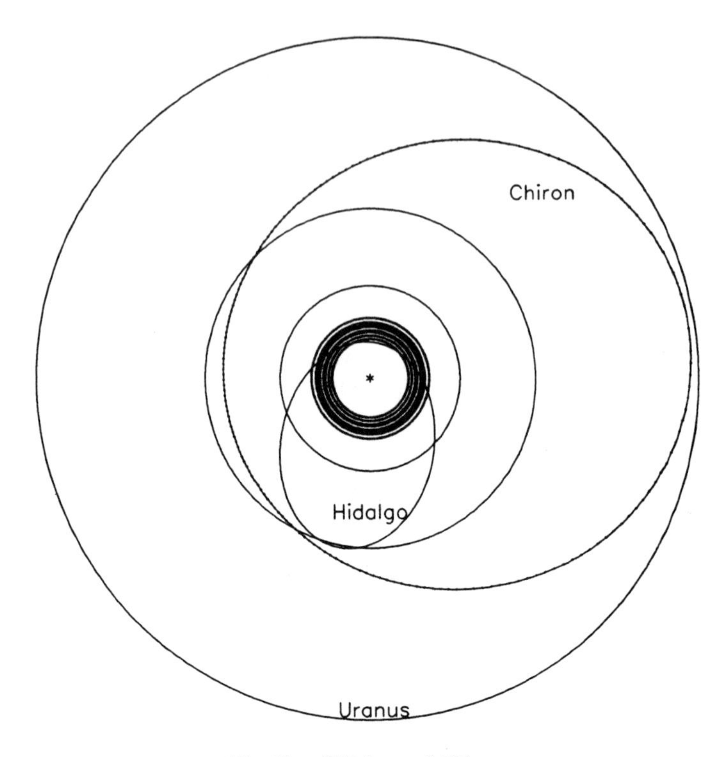

Fig. 14 — Hidalgo and Chiron.
The inner band here represents the asteroid belt. Outside of the belt are the planets Jupiter, Saturn, and Uranus. Hidalgo travels from the asteroid belt to the orbit of Saturn. Chiron ranges from Saturn to Uranus.

solar system. If this is true, it would be difficult to explain why there are no other planetesimals like Chiron.

In 1988, Chiron was observed to flare up in brightness by a factor of two. This temporary brightening was not accompanied by any changes in the spectrum. Perhaps we are seeing the start of some comet-like activity in Chiron? In popular writings about astronomy, it is often implied that we can explain everything, or that we are on the verge of explaining everything. This is not the case. Chiron is only one example of the many unexplained phenomena of nature.

The region beyond the Main Belt of asteroids seems full of surprises and mysteries. Who knows what else may be discovered there?

6

Asteroids near the Earth

In the last chapter, we talked about asteroids which lie *beyond* the main belt. There are also some asteroids which are *nearer* than the main belt asteroids. Most of these asteroids are relatively small — of the order of one kilometer in diameter — but their importance is out of proportion to their size. Some of these asteroids come close to the Earth. Sometimes they even collide with us, and become meteorites. Until Man landed on the Moon, and brought back samples, the fallen fragments of these asteroids gave us our only direct evidence of the physical structure of bodies in the solar system. These nearby asteroids may someday become of practical importance to us. Their utilization will be discussed in Chapter 7.

The innermost asteroids are, somewhat arbitrarily, divided into three groups:

(1) The **Amor** asteroids cross the orbit of Mars, but they do not quite reach the Earth's orbit. Their perihelion distances are less than 1.3 A.U. Asteroids with perihelia just outside this limit are simply called 'Mars-crossers'.
(2) The **Apollo** asteroids actually cross the orbit of the Earth. One or two of them even get closer to the Sun than Mercury. Their *average* distance from the Sun is always more than 1 A.U., however.
(3) The **Aten** asteroids cross the Earth's orbit, *and* their average distance from the Sun is *less* than 1 A.U.

These three groups of asteroids are not completely separate from one another. Perturbations can cause an Amor asteroid to become an Apollo, and *vice versa.*

As a group, these objects are collectively referred to as 'Apollo–Amor–Aten objects', 'near-Earth asteroids', or simply, 'Apollos'. I will use the last of these designations.

Almost 100 of these Apollo objects are now known, and it is estimated that there are about 1000 of them larger than 1 kilometer in diameter.

COLLISIONS AND CRATERING

Since all of the above types of asteroids cross the orbits of the inner planets, they can occasionally collide with those planets. The Earth is one target of this bombardment, although erosion has obliterated most traces of ancient impacts. It is estimated that the Earth receives a direct hit from a kilometer-size asteroid every 300 000 years or so. When such a body hits the Earth at a speed of 30 kilometers per second, the collision has as much energy as a 100 000 megaton bomb, and can create a crater 20 kilometers in diameter. Even much smaller asteroids can cause considerable local devastation. Meteor Crater in Arizona was probably caused by a meteorite only 100 meters in diameter. The crater is 1300 meters across, and 200 meters deep. Such 'small' collisions must occur every few thousand years. If the Earth's weather did not quickly erode most craters, the Earth's surface would be just as heavily cratered as the Moon's.

The estimates of numbers of asteroids, and of cratering rates, come from two lines of evidence: astronomical observations, and geological surveys.

Astronomers find Apollo-type asteroids sometimes deliberately, sometimes by accident. From the results of the deliberate searches, we can obtain some idea of the total numbers of these asteroids. Of course, we will never find every asteroid that comes near the Earth. Most of these objects are only visible for a few weeks, every few years. The asteroid Hermes was only seen for 4 days in 1932, and never seen again. However, from the number of asteroids we do find, we can estimate the total number of asteroids that must exist. Such statistical studies lead to the estimate that there are about 1000 Earth-crossing asteroids larger than one kilometer in diameter, and that one of them collides with us every 200 000 or 300 000 years. Smaller objects must be even more numerous, but they are almost impossible to discover until they hit us!

The geological clues to the frequency of asteroid collisions come from searches for the remains of ancient craters. Small craters, like the one in Arizona, do not last more than half a million years. This is a very short time, geologically speaking. Larger craters, of course, are not so easily obliterated. Aerial surveys show many faint circular features scattered all over the world. Geological investigation of these features often shows that they are the remains of ancient impacts. By counting the craters, and estimating their ages, we find that the Earth gets hit by a kilometer-size object two or three times each million years. This agrees fairly well with the astronomical evidence.

ASTEROIDS AND THE DINOSAURS

About 65 million years ago, the dinosaurs vanished from the face of the Earth. These reptiles had dominated the land, the sea, and the air. Then, in a moment of geological time, they vanished. We have all heard the facile explanations of their demise: 'The dinosaurs were over-specialized'; 'Sneaky little mammals ate the dinosaurs' eggs'; and so forth. None of these explanations hold water. Far from being over-specialized, the dinosaurs had adapted to a wide range of environments. Furthermore, although the dinosaurs receive most of our attention, many other species became extinct at about the same time. Both plants and animals disappeared *en masse*. This period of time marks the transition from the **Cretaceous** to the **Tertiary** geological epochs. It is often called the Cretaceous–Tertiary 'boundary'. Paleontologists refer to this period as a time of 'mass extinctions'. Besides occurring at the Cretaceous–Tertiary boundary, mass extinctions also occurred at the Cenomanian–Turonian boundary, 90 million years ago, and at the Eocene–Oligocene boundary, 36 million years ago.

In 1979, several scientists in the United States, and in the Netherlands, proposed that the mass extinctions were caused by the impact of a large asteroid. This asteroid created a cloud of dust and steam that blocked out the Sun for several years. Plants could not carry out photosynthesis, and animals starved to death. The evidence that these scientists cited was the existence of a layer of clay in the geological record at the Cretaceous-Tertiary (K–T) boundary. This clay is rich in the element **iridium**, which is rare on Earth, but relatively abundant in meteorites. This theory touched off a debate which seems to increase in volume each year. Early in the debate, paleontologists pointed out that the dinosaurs and other creatures did not disappear in a year or two. Instead, they gradually vanished during a period of a million years or more. This may be a short time by geological standards, but it hardly seems like a sudden event to most of us! Proponents of a catastrophic collision have modified their theories somewhat, to spread the destruction out in time. They now speak of 'showers' of comets, with large comets striking the Earth every few thousand years, and the whole duration of this shower being over a million years. Some theories of this type postulate that the comet showers are produced by passing stars, or by an unknown planet at the edge of the solar system.

Many geologists explain the abundance of iridium at the K–T boundary by postulating a period ot terrestrial volcanism. 'The *volcanoes* wiped out the dinosaurs, not asteroids', say these geologists. Others agree that a large asteroid probably did hit the Earth at that time, but they see no reason to believe that this a steroid caused the demise of the dinosaurs.

Whatever the outcome of this debate, some things are certain: large asteroids *do* strike the Earth, and the results of these collisions can be globally devastating. We do not know if asteroid or comet collisions have ever caused the extinction of whole species of animals, but it is a distinct possibility. It is also becoming apparent that such collisions would have a deleterious effect on the Earth's climate, at least for several years. It is interesting, and sobering, to note that these studies of the effects of asteroid collisions have a direct bearing on studies of the 'nuclear winter' that would be caused by a global thermonuclear war. One large asteroid traveling at 30 kilometers per second has far more power than all the world's nuclear arsenals. Unfortunately, man goes to war far more often than such asteroids strike the Earth.

ORIGIN OF THE EARTH-CROSSING ASTEROIDS

Because the Apollos, Atens, and Amors collide with the Earth, and with the other planets, these asteroids have limited life expectancies. An average asteroid of this type can only be expected to survive for a few tens of millions of years before striking a planet, or being permanently ejected from the solar system. This is very much less than the age of the solar sytem. Why are there any Apollos left? There are only two possible answers to this question: either the original population of Apollos was very large, or 'new' Apollos are constantly being injected into the Earth's vicinity.

If the original population of Earth-crossing asteroids was very large, and we now see only the remnants of this population, then the rate of meteorite bombardment would have been much greater in the past. As the population became depleted, fewer asteroids would be left to produce impacts. Studies of the cratering rates on the Earth and Moon, however, show that the rate of cratering has been roughly constant for the last 3 billion years. It must be, therefore, that new Apollos are constantly being created, to keep the population of these asteroids fairly constant. Only about 15 new asteroids are needed every million years, to keep the population of Earth-crossers at a steady level. There are only two possible sources of such asteroids — the main belt, and the comets. It is possible that both of these sources contributed to the present population of Earth-crossing asteroids.

The asteroids in the main belt are in stable orbits. Those orbits will not change appreciably unless the asteroids are disturbed by some internal or external influence. If two asteroids collide, the resulting fragments go into different orbits, but they do *not* have enough velocity to escape from the main belt and reach the orbit of the Earth. That would require an impossibly energetic collision. How then, can we move an asteroid out of the main belt, and into a planet-crossing orbit?

We have already seen that there are gaps in the asteroid belt, which are caused by perturbations from Jupiter. It is possible for an asteroid collision to knock large fragments into these gaps, or into various other resonance regions. From there, Jupiter can perturb the fragments into Mars-crossing orbits. Mars, in turn, can perturb these objects into Earth-crossing paths. In addition to the Kirkwood gaps, and possibly even more important for the formation of Apollos, is the inner edge of the belt, which is largely dominated by Mars. Collisions near the inner edge of the belt could nudge asteroids close enough to Mars to be removed from the belt. Remember, however, that Mars is a small planet, and its gravitational pull is small. It is probable that a large fraction of the Earth-crossing asteroids originated from relatively gentle collisions within the asteroid belt, followed by perturbations by Jupiter or Mars.

Comets, on the other hand, are *not* in stable orbits. They often pass near Jupiter, and their orbits are drastically changed during such encounters. Furthermore, the orbits of the comets are changed by non-gravitational forces. The comets' escaping gases give a slight push to the comets, which slowly changes their orbits.

It is also known that comets gradually decay. Eventually, all of their frozen gases are vaporized, leaving a core of solid particles. We know of several nearly-extinct comets which show little evidence of cometary activity, such as tails. These comets look almost like asteroids. It is easy to see, then, that comets could provide a source of Apollo asteroids. The only stumbling block to this theory is that we really know very little about comets. We do not know how many comets contain non-volatile cores of sufficient size. Although we know of a few comets that look almost like asteroids, we also know of several comets that have disappeared without trace!

The diversity of meteorite types found on the Earth is evidence for a diversity of sources for the Earth-crossing asteroids. If we assume that all comets were formed in the same way, and in the same general region of the solar system, then cometary nuclei should all resemble each other. The asteroids, on the other hand, are known to have a variety of compositional types within the main belt. Nevertheless, if comets do have solid cores capable of surviving passage through the Earth's atmosphere, then at least some meteorites must come from comets.

COMPOSITION OF THE APOLLOS

How can we determine the composition of an Apollo asteroid? The answer is quite simple: go to your local museum and look at the meteorites! You may even be able to touch one of these former asteroids! It does not require a detailed chemical analysis to show that some of the meteorites are metallic, and some are stony. The metallic ones are mostly made of iron and nickel; some of the stony ones contain quantities of carbon. We have already described the composition of meteorites, in Chapter 3, and we will not dwell on it again here. The question now is: 'Do the compositions of the Apollo asteroids that we see in the sky match the compositions of the meteorites that we find on the ground?' The answer, by and large, is 'yes'. However, there are some meteorite types for which no counterpart is seen among the Apollos, and there are some Apollos that do not resemble any known meteorites.

If you glance back at Table 3.1, you will see that there are many different types, and subtypes, of meteorites. There are more known meteorite types than there are well-studied Apollo asteroids. It is not surprising, therefore, that some meteorite types are not represented among the known Apollos.

Over 30 near-Earth asteroids have been studied in some detail with ground-based techniques, and with the IRAS satellite. S-, C-, and M-types have been see, as well as a number of 'unclassified' objects. There is a preponderance of S-types, but this is probably the result of observational selection. C-type asteroids have darker surfaces than S-type, and so the C-types are more difficult to discover.

Some of the Apollo asteroids are as small as 0.3 km in diameter, while the largest is about 40 km in diameter. These asteroids tend to have relatively rapid rotation rates, irregular shapes, and bare surfaces, with little or no regolith. This may be taken as additional evidence for a collisional origin for the Apollos. If these asteroids are 'shrapnel', resulting from collisions among the main-belt asteroids, we would expect them to have irregular shapes and, on the average, rapid rotation. Furthermore, since the Apollos have relatively short lifetimes in their planet-crossing orbits, they do not have time to accumulate regoliths.

The similarity in compositional types between the Main Belt asteroids, the Apollos, and the meteorites, strongly suggests that most of the Apollos and the meteorites originally came from the main belt, but it does not rule out the possibility that some Apollos are extinct comets.

SOME UNUSUAL APOLLO ASTEROIDS

There is one Apollo asteroid which does not resemble any known type of meteorite. Furthermore, it does not resemble any other asteroid, or any comet! This intriguing object is (2201) Oljato.

(2201) Oljato

Oljato is an Apollo-type asteroid with a very large eccentricity of 0.71. Its spectrum is different from the spectrum of any other object in the solar system. As shown in Fig. 15, Oljato's reflectance spectrum shows a peak at about 6000 ångstroms, together with strong ultraviolet reflectance. The spectrum resembles no meteorite or asteroid

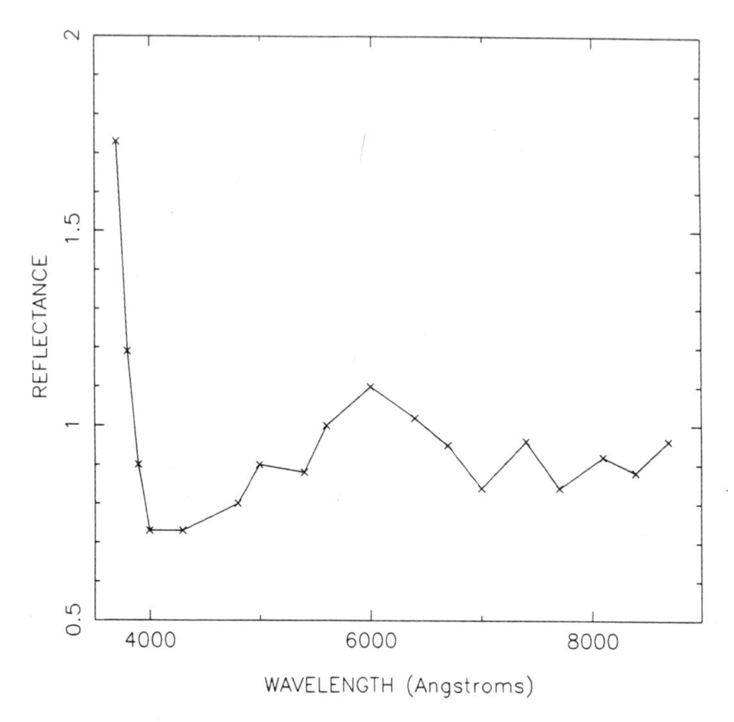

Fig. 15 — The reflectance spectrum of Oljato.
Note the high reflectivity in the ultraviolet, and the peak at 6000 ångstroms. This spectrum is unlike that of any other asteroid, comet, or meteorite. (The data are from: McFadden, Gaffey, and McCord (1984) *Icarus* **59** 25.)

types, nor does this asteroid show any trace of cometary activity. There is a possible connection between Oljato and a meteor stream. There are hints that the shape of Oljato is very irregular, and that its spectrum may be variable.

The highly eccentric orbit, and unique spectrum of Oljato, point toward a cometary origin. But, if comets can turn into objects like this, why are there no other asteroids or comets that resemble Oljato? Perhaps further observations will allow us to fit Oljato somewhere into the evolutionary scheme of things, but so far Oljato remains one of the more enticing mysteries of asteroid science.

1983TB

Another asteroid which shows possible links to the comets is 1983TB. This asteroid was discovered in an unusual way. It was not found during any ground-based searches, but was detected by the InfraRed Astronomical Satellite (IRAS), In October, 1983. It was immediately noticed because of its rapid motion across the sky. The IRAS satellite had already found several comets by this time. IRAS scientists notified other astronomers of their discovery, and photographs were taken with ground-based telescopes. These photographs showed that the object looked completely asteroidal. There was no trace of any cometary 'fuzz'. Spectra also showed no trace of cometary activity. All in all, 1983TB seemed to be a typical Apollo-type asteroid. Soon, however, Dr Fred Whipple noticed that the orbit of 1983TB was identical with the orbits of the **Geminid** meteors. There is no other well-established case of a meteor shower traveling in the orbit of an asteroid, although a few such cases are suspected, (including Oljato, above). On the other hand, there are many cases of *comets* within such orbits, and it is practically certain that the comets *produce* the meteors. Is 1983TB the remnant of a comet, or can asteroids produce meteor showers? At present, the evidence is ambiguous. If 1983TB is an ex-comet, its composition is hardly different from many ordinary asteroids. This object appears to be an S-type asteroid, with an albedo of more than 10 per cent. From the spacecraft observations of Halley's Comet, we now know that cometary nuclei are much darker than this.

So here we have two asteroids suspected of being extinct comets. Other asteroids with possible links to the comets are Chiron and Hidalgo, discussed in the previous chapter. Yet these asteroids are all different from each other, and, in most respects, different from any observed comets! These objects remind us of how little we really know about the solar system.

RADAR STUDIES OF 1986DA

Another interesting Earth-approaching asteroid is 1986DA, discovered in February 1986, by M. Kizawa, of Shizuoka, Japan. 1986DA is one of the growing number of asteroids which have been studied with radar.

1986DA is an Amor asteroid, having a semi-major axis of 2.82, an eccentricity of 0.586, and an inclination of 4.3 degrees. Its period of revolution is 4.73 years.

Radiometry and photometry indicated that 1986DA is an M-type asteroid, possibly having a bare metallic surface. Two months after this asteroid was discovered, Dr S. J. Ostro and his colleagues sent radar beams to 1986DA from the giant radio-radar antenna at Arecibo, Puerto Rico. They found that this object is an excellent reflector of radar — twice as reflective as any other known asteroid. This high reflectivity confirms that 1986DA is very rich in metal. It may have the same composition as some of the iron meteorites. Dr Ostro's group points out that this fact may be of more than purely scientific interest. It may have economic consequences as well! Since asteroids like 1986DA are relatively easy to reach with space probes, there is a possibility that such objects could be economically exploited. An object which is rich in iron, nickel, and other important metals would be an obvious candidate for such exploitation. We will discuss this topic more thoroughly in the next chapter.

7

Utilization of the asteroids

It seems unlikely that mankind will ever solve its problems of war, pollution, and over-population. Yet even if we do solve these problems, one insurmountable barrier will remain to the continued existence of our technologically advanced, industrial civilization. We live on a planet whose resources are limited. Earth is an island in space. The resources of this island are vast, yet we are depleting them rapidly. No amount of conservation and recycling will stop this depletion in the long run. Eventually, all of the oil, coal, gas, and precious metals will be gone. If we are cautious, we may not reach this barrier for centuries, but ultimately we *will* reach it. Furthermore, we are *not* cautious. We never have been cautious, and we probably never will be.

Many people see no way out of our long-range decline. At least since the time of Thomas Malthus, two hundred years ago, people have pointed out the ultimate barriers to our continued growth. In his book, *An Essay on the Principle of Population,* Malthus stated that population tends to grow faster than our ability to feed that population. In prosperous times, population increases. If the population increases too fast, prosperity disappears, and the growth in population is checked by war, famine, and disease. If we look at parts of Asia and Africa today, we can see the effects of over-population. People are, quite literally, starving by the millions.

And yet, in a larger sense, Malthus was wrong. The Earth's population is very much larger than it was in Malthus' time, but the global level of prosperity is higher than Malthus could have dreamed. There seems to be enough food to feed the whole world, if only we could distribute it properly. Problems of under-development, and politics, prevent the distribution of food to the starving, while people in other parts of the world have problems disposing of their surpluses. Nevertheless, in the Western world at least, we seem to have overcome the predicted limits to our growth. How have we been able to provide so much of the Earth's huge population with food, energy, automobiles and airplanes? The answer is: technology. Thomas Malthus could not have predicted the development of *petroleum* energy, much less *nuclear* energy! Equally vital has been the 'Green Revolution' in agriculture. Fertilizers and insecticides are essential to modern agriculture. These chemicals require energy, and often petroleum products, for their production.

Though technology sometimes seems to cause as many problems as it solves, we literally could not live without it. It is the growth of technology that has forestalled a Malthusian tragedy throughout the world.

Now we are reaching a new crisis. The raw materials which made the technological revolution possible are rapidly disappearing. The fuel shortages of the 1970s may have been largely artificial, but there is no doubt that the world's oil will eventually be depleted, probably in the next century. Our coal reserves may last longer, but they will not last forever. Iron is plentiful, but not chromium, titanium, or platinum. What will happen when all of these things are gone?

There seems to be only one solution for 'Island Earth'. The answer is — 'Get off the island!'

I have called the Earth an 'island'. It has also been described as a 'pebble in the sky', and a 'grain of sand in the universe'. The point is, the Earth has limitations, but the universe does not. Our Sun will continue shining for billions of years, and its vast energy will be dissipated uselessly into space, while our oil wells dry up. There is carbon in the asteroids, and methane gas on Jupiter. Can we make use of this limitless energy? Can we draw upon the resources of the stars and planets? Yes, we can, if we have the will and the vision.

MINING THE ASTEROIDS

The mining and processing of minerals on the Earth is costly, both in terms of economics and of pollution. Just think of the processes which are necessary to produce simple pig iron. Not only do the mines themselves create huge scars on the Earth, but the smelting and processing of the ores produces soot, fumes, and acid rain. As we exhaust the high-grade ores, it becomes necessary to expend even greater efforts to extract metals from low-grade ores. In the case of some other metals, we are often dependent on supplies from politically unstable parts of the world. What we really need is an abundant source of minerals that could be processed without causing pollution, and that could be available to any country having the technological ability to obtain it. The Moon and the asteroids provide such a source.

Since the Moon is so close to us, and since it has already been partially explored, it is probable that our exploitation of space will begin there. The Moon can be reached in a few days. It is rather rich in aluminium and titanium, but it cannot really be said to contain high-grade ores of any other metals. There is an enormous amount of oxygen locked into the Moon's minerals, and this could be extracted using solar power. But the Moon is deficient in carbon and nitrogen, and, as far as we can tell, it contains no water at all. The asteroids have all of these materials in abundance.

It is estimated that a kilometer-size, nickel–iron asteroid contains 7 billion tons of iron, 1 billion tons of nickel, and enough cobalt to supply the entire world for 3000 years. The total value at today's prices would be about 5 *trillion* dollars! ($\$5 \times 10^{12}$). Clearly, a substantial investment in asteroid exploration could be economically justified. In addition to these metals, the asteroids also contain that most precious commodity — water. For these reasons, a mature program of space industrialization will inevitably utilize the asteroids.

It would be possible to set up a mining camp on an asteroid, extract the metals from the ores, and ship the metals back to Earth. However, it would probably be more economical to bring the entire asteroid into orbit around the Earth. An asteroid in Earth-orbit would always be accessible, and the crews working on the asteroid could be replaced and resupplied frequently. A mining station on a distant asteroid would have to be self-sufficient for a year or more, and would be much more difficult to reach.

There are several ways in which an asteroid could be transported into an orbit around the Earth. The one thing that makes it possible is the availability of practically unlimited energy from the Sun. Solar electric cells, or solar boilers, could be used to power a variety of propulsion systems. Perhaps the most promising system would be to use an electro-magnetic 'mass driver' to expel pieces of the asteroid at high velocities. In other words, the asteroid would provide the fuel to propel itself. A mass driver is a device which accelerates specially-designed 'buckets' along a track at extremely high velocities. The mass driver is powered by electricity, and it works best in a vacuum. Mass drivers have already been built which achieved payload accelerations of 30*g*, (30 times the force of gravity on the Earth's surface). It has been proposed that mass drivers could be built on the Moon's surface, and could fire buckets of minerals into space, for later collection and delivery to Earth. Mass drivers could also be used as rocket engines. A rocket works by expelling gases out of a nozzle at high velocity. The expulsion of the gases in one direction pushes the rocket away in the opposite direction. Placed on an asteroid, a mass driver would fire bits of the asteroid away, thus pushing the asteroid itself into a different orbit. The power for the mass driver would come from a thin solar panel which would have to have a diameter approximately equal to the diameter of the asteroid itself. About one quarter of the asteroid would be consumed as fuel.

Another way to move an asteroid is with sails. Yes, *sails!* These sails would not be powered by the wind. Indeed, they would require the total absence of air in order to function. The sails get their 'push' from sunlight.

Light is a form of energy. When light strikes an object, it imparts a small push against that object. The force that light exerts is tiny, but measurable. The amount of this force depends on the surface area of the object which receives the light. The greater the area, the greater the push. It is possible to construct a 'sail' which is so large, and so light, that it could be significantly moved by sunlight. This sail could also pull an asteroid along with it! There have already been extensive studies of possible designs for solar sails, and a working model of a sail should be placed into orbit within the next few years.

Another technique which would help guide the asteroid to Earth is called 'gravity assist'. This is the technique which propelled the Voyager space probes from Jupiter, to Saturn, to Uranus, to Neptune, and beyond. In gravity assist, one uses the gravity of a planet to change the orbit of a passing body. It is sometimes called the 'slingshot effect'.

Calculations have been made for several possible asteroid retrieval missions. Typically, they involve close flybys of the Moon, the Earth, and sometimes Venus. The entire mission, from Earth launch to 'parking' the asteroid in Earth orbit, would take three or four years. The same solar panels that provide propulsive power for the asteroid retrieval would furnish the energy for mining the asteroid after arrival.

ENERGY FROM SPACE

We have described how the asteroids could provide the Earth with minerals at relatively low cost (though requiring a very high initial investment). The power for moving the asteroids into Earth orbit, and for extracting its ores, would come from solar energy. Solar energy is inexhaustible, and it is practically free. Why not use this source of energy to supply power to the Earth? During the oil crisis of the 1970s, there were many attempts to harness solar power, wind power, ocean wave power, etc. In the southern United States, many homes now have solar heaters to provide their hot water. Aside from such small-scale successes, experimentation with 'free' energy almost disappeared when petroleum power became cheap again. Nevertheless, it is only a question of time before oil truly becomes a scarce commodity. When that happens, we will again look to the Sun. The trouble with solar power, of course, is that it is unavailable at night and during cloudy weather. These limitations do not exist in space. Why not collect solar energy in space, and then beam it down to Earth? It is entirely feasible.

In 1973, Dr Peter Glaser, of the Arthur D. Little Company, in Cambridge, Massachusetts, spoke to the United States Congress. Dr Glaser talked about the possibility of collecting solar energy in space, and sending it down to Earth in the form of microwaves. Huge antennas on the Earth would collect the microwaves beamed down from space, and convert that energy into electricity. Dr Glaser's proposal drew little attention at that time. A few years later, Dr Gerard K. O'Neill incorporated this concept into a far grander scheme.

SPACE HABITATS

Dr O'Neill suggested that the surface of a planet is *not* the best place for human beings to live. Artificial 'colonies' in space could provide people with living conditions far better, in many respects, than conditions on the Earth. For example, consider the case of agriculture. On Earth, farming practically ceases during the Winter months. What a waste of time and land! In addition, droughts, storms, and insects devastate many crops. In space, the Sun shines constantly. The climate within a space habitat would be artificially controlled, of course. This means that the seasons could be different in different parts of the habitat. Fruits would always be 'in season', and would be grown under optimum conditions. With care, insect pests could be kept out of the space colony.

The inhabitants of these colonies would economically justify their existence by providing electrical power to the Earth, via microwaves. They could also manufacture certain pharmaceutical chemicals, crystals, the other items that can only be produced in zero gravity.

These space habitats would be several *kilometers* in diameter. they would mostly consist of huge solar panels to collect the sunlight, and convert it into microwave energy. The microwaves would be focused onto antennas on the Earth's surface, where they would be converted into electricity. Since the power satellites would be in sunlight almost constantly, they could transmit the microwaves 24 hours a day. Beams of the right wavelengths would not be stopped by clouds.

Construction of these enormous satellites will require millions of tons of materials. It would not make economic sense to transport all of these materials from the surface of the Earth. It would be far better to obtain them from space. In spite of all the appeal the asteroids have from an economic standpoint, our exploitation of space will probably start with the Moon, for the reasons mentioned previously. Eventually, however, we will turn to the asteroids for their metals, their water, and for their carbon and nitrogen. A small asteroid, 'parked' in an orbit near the space colony, could provide almost all of the materials needed to support the colonists, and to build still more colonies.

At first glance, these projects sound like wild fantasies. Yet, many studies have been undertaken which show that these ideas are not so wild at all. Indeed, the things we have discussed so far are quite tame compared to the ideas of some future-minded individuals. These ideas are exciting. Almost equally exiciting is the fact that they are entirely feasible! At least in principle, these projects could be accomplished with existing technologies. They would be enormously expensive, however, and they will not become real possibilities until we become economically and materially self-sufficient in space.

COLONIZING THE SOLAR SYSTEM

In his book, *The High Frontier*, Dr O'Neill paints a glowing picture of life in the space colonies.

Imagine a self-sufficient colony in space. Everything needed for the colonists' survival, with minor exceptions, is produced within the huge space habitat. The inhabitants live in a park-like setting, with trees, gardens, and even flowing rivers. Areas are set aside for agriculture, for manufacturing, and for recreation. Gravity is provided by rotating the whole structure. This means that there is no gravity in the center of the structure, but Earth-normal gravity along the rim. The colony may house anywhere from 10 000 to one million people. Raw materials and energy are abundant. Indeed, the colony 'exports' energy down to the Earth's surface.

Many of the colony's inhabitants were born in space. They have never set foot on the Earth. From time to time, as the population grows, new colonies are constructed. Individual families could even construct their own free-floating 'mini-habitats'. These little habitats could move into different orbits at will, using either mass drivers, solar sails, or chemical propulsion. They could even leave Earth-orbit entirely. With no economic or emotional ties to Earth, the colonists may well decide to move elsewhere. Perhaps an orbit around Saturn would provide a pleasant setting. The entire solar system will be available for exploration and colonization. All of this will be made possible by energy from the Sun, and metals, water, and carbon from the asteroids.

Meanwhile, the people who live on Earth will be free of many of today's problems. Clean, abundant energy will be provided by the power satellites. There will be no fuel shortages, and no need for smoke-belching power plants. With abundant minerals and energy, many of the causes of political tensions will be removed.

Does this all sound too good to be true? Perhaps it is. And yet — it is *possible*!

CANDIDATES FOR ASTEROID MISSIONS

In choosing an asteroid for a retrieval mission, we need to select an asteroid which is easy to reach. As an indicator of the amount of energy needed to reach an object, space engineers use the term ΔV/,('delta-V'). ΔV is the velocity which must be imparted to a spacecraft in order to match its orbit to the orbit of the object. The higher the ΔV, the greater the amount of energy needed to reach the object. Of all the asteroids, the Apollos, Amors, and Atens have the lowest ΔVs. this is simply because their orbits are close to the Earth's orbit, so they are relatively easy to reach. This is what makes the Apollo asteroids such desirable targets for retrieval missions. some of the Apollos, in fact, are almost as easy to reach as the Moon, in terms of the amount of energy required. Nevertheless, a mission to an asteroid will take months or years, instead of the few days required to reach the Moon. A manned mission to an asteroid would have to be self-sufficient for at least two or three years. This means that the spacecraft would have to have a 'closed life support system'. In other words, all wastes will have to be recycled, and food will have to be grown on-board.

One way of reducing the amount of time needed to reach the asteroids is to propel the spacecraft continuously. This can be accomplished with a low-thrust ion engine, or with a solar sail. In both cases, the propulsive energy would come from sunlight.

It is impossible to predict the manner in which we will utilize the resources of space. However, it is clear that we cannot fully develop this frontier by lifting all of the needed materials from the surface of the Earth. We must learn to obtain those materials from space. It is for this reason that the asteroids may play a major role in the future of mankind. They are treasures, waiting to be seized.

8

The future of asteroid research

In this book, I have mentioned how remarkable it is that we have been able to learn so much about the asteroids even though they are so small and distant from us. Even in our largest telescopes, the asteroids are only little points of light. Yet, we have been able to learn much about their composition, structure, and origin. In the next few years, however, our knowledge of the asteroids will take a quantum leap forward. We will soon have close-up photographs and measurements of these bodies. Eventually, we will have actual samples of the asteroids to study in our laboratories, here on Earth. If all goes well, the first voyages to the asteroids will take place in the 1990s. Even before then, telescopes in space will have given us a clearer view of these little objects.

THE HUBBLE SPACE TELESCOPE

The Hubble Space Telescope, to be launched in 1989 or 1990, will be a moderately large telescope (2.4 meters in diameter), in orbit around the Earth. Free of the blurring effects of Earth's atmosphere, the HST will view celestial objects with unprecedented clarity. Even this telescope will not be able to show much surface detail on any of the asteroids. It will, however, be able to see if any of the asteroids have satellites. Most important of all, it will be able to view the asteroids at all wavelengths, from the ultraviolet to the near-infrared. The importance of this capability cannot be over-emphasized. The ultraviolet portion of the spectrum, invisible from the Earth's surface, may prove to be the key to unambiguously identifying the surface composition of individual asteroids. These studies from Earth orbit will help pave the way for space probes to the asteroids themselves.

THE SHUTTLE INFRARED TELESCOPE FACILITY (SIRTF)

During the IRAS mission of 1983, many thousands of asteroids were observed in the infrared. IRAS (the InfraRed Astronomical Satellite) was a small survey telescope, which scanned almost the entire sky. The analysis of these observations is still going on, but in the near future we should have a catalog listing the albedos and diameters of thousands of asteroids as small as 10 kilometers in diameter.

Surveys, such as this, need to be followed up by detailed investigations. This will be the job of SIRTF. SIRTF will be a large telescope mounted on the Space Shuttle. it will have the capability of obtaining infrared spectra of very small asteroids. While the Hubble Space Telescope will observe from the ultraviolet to the near-infrared portions of the spectrum, SIRTF will be able to view the *far* infrared — the portion of the spectrum which comes from the thermal emission of objects, and thus is an indication of their temperatures. With the Space Telescope and SIRTF, we will finally be able to obtain complete spectral coverage of all known asteroids. Among other things, we will be able to examine, in detail, entire asteroid families, from their largest members to their smallest.

BEING THERE

Although much can be learned from the surface of the Earth, and even more from Earth orbit, there is no substitute for 'being there'. Only by actually landing on an asteroid can we do a chemical analysis of its composition, sift through its surface material, and probe its interior. Recognizing the importance of asteroid research to our understanding of the solar system as a whole, NASA has begun work on one asteroid mission, and is planning others. Similar plans are also underway at the European Space Agency.

GALILEO

Our first close-up view of an asteroid will probably come from a spacecraft bound for Jupiter. The much-delayed **Galileo** mission is scheduled to be launched in 1989. After an incredibly circuitous voyage, the space probe will arrive at Jupiter in 1995. Along the way, Galileo will fly by two asteroids — **(951)Gaspra** and **(243)Ida.**

Galileo's fantastic voyage will begin with a launch in October or November, 1989. It will not be sent directly to Jupiter, but instead will head for Venus first. After swinging around Venus, in February 1990, the spacecraft will be propelled back to the Earth. It will pass Earth in December 1990, circle around the Sun, and pass the Earth again two years later. This second pass near the Earth will finally send the spacecraft toward Jupiter, where it will arrive in December 1995.

The purpose of this convoluted flight plan is to use the gravity of Venus and the Earth to boost the spacecraft into a Jupiter-crossing orbit with a minimal expenditure of fuel. The same principle was used to send the Voyager spacecraft past Jupiter, Saturn, Uranus, and Neptune.

After the first flyby of the Earth, Galileo will swing out as far as the inner asteroid belt. There, it will pass within 900 kilometers of asteroid Gaspra. This flyby will occur in October, 1991. At this time, we should obtain our first close view of the surface of an asteroid. In addition to its cameras, Galileo carries a battery of instruments which will study the asteroid's composition and magnetic field. Gaspra is an S-type asteroid, about 16 kilometers in diameter.

On the final leg of its flight to Jupiter, Galileo will pass by the 30-kilometer-wide asteroid Ida. This will happen in August 1993. Ida is another S-type asteroid.

Finally, as Galileo approaches Jupiter, an atmospheric probe will detach itself from the spacecraft, and plunge into Jupiter's dense atmosphere. The main body of the spacecraft will go into orbit around Jupiter for a long-term study of Jupiter and its satellites.

CRAF

The Comet-Rendezvous Asteroid-Flyby mission (CRAF) will be the first of a new generation of 'low-cost' spacecraft. To be launched in the early 1990s, CRAF is among NASA's highest priorities for solar system exploration. During this mission, the spacecraft will fly by one or two asteroids, and then rendezvous with a comet. (A **rendezvous** involves matching the orbit of the spacecraft with the orbit of the target object. In this way, the spacecraft can stay near its objective for a long time — days or weeks. During a **flyby,** on the other hand, the spacecraft merely flies past its target at a high speed. Obviously, one can gather more information during a rendezvous than during a quick flyby.) As it flies past an asteroid, CRAF will take close-up photographs of its surface, and make measurements of its magnetic field, mass, and infrared reflectivity.

EXPLORATION OF THE MAIN BELT

The CRAF mission is primarily designed to study a comet. Any asteroid it encounters will merely be a secondary target. A few years after CRAF is launched, there will be a mission dedicated solely to studying several Main Belt asteroids in detail. The 'Mainbelt Asteroid Multiple Orbiter/Flyby' mission will not only fly by several asteroids, it will also rendezvous with at least one, and possibly two asteroids. In the words of the Solar System Exploration Committee of the NASA Advisory Council. The objectives of the MAMOF mission are to:

> characterize asteroids of various types, including determinations of size, shape, rotation, albedo, mass, density, surface morphology, surface composition, magnetic field and solar wind interaction;
>
> provide a more detailed study of one or two selected Mainbelt asteroids, emphasizing elemental and mineralogical composition and detailed morphology.

MAMOF will consist of two separate spacecraft. Each one will visit four or five asteroids. During the estimated six- to nine-year lifetime of this dual mission, we will probably learn more about the asteroids than we have learned in the past two hundred years!

NEAR-EARTH ASTEROIDS

Following the exploration of the main belt, the next asteroid mission will probably study an Apollo–Amor asteroid. The Near-Earth Asteroid Rendezvous mission (NEAR) will study the relationships between the Apollos and the other asteroids and comets. Are the Apollos fragments of main belt asteroids, or are they extinct comets, or both? Can we use these objects to provide us with a cheap source of raw materials for construction in space? These are the questions that will be addressed by the NEAR mission. The answers to these questions will determine the future course of asteroid exploration. It is likely that several missions to near-Earth asteroids will be needed, in order to obtain a thorough understanding of this probably-diverse group of objects. At present, however, only one such mission is planned.

REMOTE SENSING

It may be a long time before we will be able to actually retrieve a portion of an asteroid, to be studied in our laboratories. Until then, we will have to be content with **remote sensing** from the flyby and rendezvous missions.

When we think of past space probes, such as the Voyager spacecraft to the outer planets, what we remember most are the spectacular photographs of those planets and satellites. The incredible canyons of Mars, the volcanoes of Io, and the tortured terrain of Miranda were all discovered on images taken by the spacecraft cameras. But remote sensing involves far more than the mere taking of pictures. We are, of course, terribly interested in what these object look like. But we also want to know what they are made of. What is their chemical composition? Do they have magnetic fields? What are their masses and densities? These are questions that can be answered by other instruments on the spacecraft.

Some of the instruments on these spacecraft are similar to the instruments used on ground-based telescopes — spectrographs and polarimeters, for example. On the Earth, these instruments are hampered by the limitations imposed by the Earth's atmosphere, and by the great distances to the objects we are studying. These limitations disappear when the instruments are on board a spacecraft, in close proximity to the planet, satellite, or asteroid.

Some of the instruments used on spacecraft are unlike anything used by ground-based astronomers, though versions of them are used in laboratories here. Among these more exotic devices are X-ray fluorescence spectrometers, gamma-ray spectrometers, and magnetometers.

As you might expect, magnetometers measure magnetic fields. It is entirely possible that the iron-rich asteroids are somewhat magnetized. There will be no way to find this out until we can put a magnetometer near such an asteroid. Measuring the magnetic fields would give us much information about the internal structure and composition of the asteroids.

Gamma-ray spectrometers measure the gamma rays emitted by radioactive isotopes, and by 'induced' radioactivity. An analysis of the gamma rays emitted by an object can give an accurate measure of the abundances of the elements within the upper two meters of the surface of that object. The gamma ray emission from most solar system objects is very weak, so the spectrometer has to stay in the object's vicinity for an hour or more, in order to gather enough data. For this reason, flyby missions are not suitable, and rendezvous are needed.

When X-rays from the Sun strike the surface of an asteroid, they cause the minerals within the asteroid to emit X-rays at different wavelengths. Analysis of these emissions is another way of determining the chemical composition of an asteroid's surface. Like gamma-ray spectroscopy, X-ray fluorescence spectroscopy requires a fairly long time for data-gathering, so rendezvous missions will be needed.

The combination of high-energy spectroscopy, magnetometry, and imaging can give us a remarkably exact description of the composition of the asteroids.

After the flyby and rendezvous missions have been accomplished, we will probably want to investigate the asteroids in even more detail. The swiftness with which this is done will depend on what we learn from the earlier missions. If we find that the asteroids are indeed a promising source of raw materials, it is probable that asteroid exploration will receive a high priority. In any case, it seems likely that we will finally bring a sample of an Apollo asteroid to the Earth, early in the twenty-first century. After that, the dreams described in the previous chapter may finally become realities.

GROUND-BASED RESEARCH

When NASA's Solar System Exploration Committee published its recommendations for space exploration to the year 2000, it wisely included recommendations for ground-based research. There are so very many asteroids, and our spacecraft can only visit a few of them. How can we pick the best targets? The answer, of course, is to study as many asteroids as possible from the ground. Not only do we need to study the known asteroids as well as we possibly can, we need to find even more asteroids. This is particularly true for the Apollo asteroids.

THE SPACEWATCH CAMERA

Searches for asteroids will probably use the 'old-fashioned' technique of photography for many years yet. Photography is effective, and relatively inexpensive. On the other hand, it is inefficient and subjective. How many asteroids were never discovered, because some astronomer was daydreaming at the wrong time?

An ideal system would use a *machine* to find the asteroids. Such machines are already in use.

Dr Tom Gehrels, of the University of Arizona, has played an important role in asteroid research. He has also stimulated many other astronomers to enter this field, and is now exploring the frontiers of automated asteroid searches.

After several years of planning and preparation, Dr Gehrels made the first observations with his 'Spacewatch Camera' in May, 1983. His system uses a 91-cm (36-inch) telescope with a CCD camera attached to it. A CCD (Charge-Coupled Device) is a solid state imaging device, many times more efficient than photographic materials. The output of a CCD is a stream of digital data which can be recorded on a magnetic disk or tape.

The procedure used is as follows. The camera scans across a particular strip of the sky for 20 minutes. Then, the telescope is moved back to the starting point, and scans that strip again. The computer which is attached to the system 'remembers' the position of every object seen during the first scan, and compares this with the positions found during the second scan. Anything which has changed its position during the 20-minute interval is probably an asteroid. At present, the comparison of the two scans is done visually, by subtracting the two images and displaying the result on a screen. The ultimate goal is to have the computer do all the comparisons, and to locate moving objects automatically. It would then ring a bell to alert the astronomer that an asteroid has been found!

The biggest problem with CCDs is that they are quite small; typically, 1 cm square. The obvious solution to this problem is to use bigger CCDs. Indeed, 5 cm CCDs are now available. At this point, however, one runs into another problem — getting a computer big enough to store and process all that data. Computers and computer memories are getting more and more powerful, and less and less expensive, so it is only a question of time before these automated search techniques will compete successfully with the labor-intensive manual techniques.

There is much yet to be learned about the asteroids, and the asteroids have much to tell us about the rest of the solar system. Each year brings new advances in asteroid research, and it looks like the pace of new advances will continue to accelerate. The first close-up photographs of the asteroids are just a few years away. The next decade will be an exciting time for astronomers, not to mention the next century!

Appendix A
Asteroids and the amateur astronomer

There are many things that amateur astronomers can do for sheer enjoyment, but relatively little can be done of real scientific importance. The asteroids provide unique opportunities for *both* types of endeavor.

Amateurs can measure the positions of asteroids, compute their orbits and ephemerides, measure their colors and rotation rates, and make genuine contributions to science by observing occultations.

CELESTIAL MECHANICS

The most noticeable thing about asteroids is the fact that they *move*. Measuring their positions and predicting where they will be in the future is therefore of fundamental importance. In principle, these activities are within the capability of many amateurs, but they provide a formidable challenge. To be of much value, measurements of asteroid positions must be done with considerable precision. It is usually necessary to measure asteroid positions to an accuracy of about one arcsecond. This means that the scale of photographs must be large enough, and that the measuring apparatus must be suitably precise. Generally, the amateur will have to build his own equipment. In order to convert the measured positions on a photograph to celestial coordinates, a certain amount of computing must be done. These computations are fairly simple, however, and can easily be done with a small computer, a calculator, or by hand.

An excellent pair of articles about the measurement of asteroid positions is contained in *Sky and Telescope,* September 1982, pp. 279–284.

After an asteroid's positions have been measured on at least three separate days, it becomes possible to compute its orbit. This job is generally left to professionals, but it can certainly be done by the mathematically-inclined amateur. The only real difficulty comes in learning how to do it. Understandable books about celestial mechanics are quite rare! See the bibliography section for books which may be of help.

The computation of ephemerides from known orbital elements is a somewhat simpler problem, which often lends itself to a 'cookbook' approach. That is, you can obtain the formulae, or complete computer programs, and simply 'plug in' the appropriate numbers. Again, check the bibliography.

PHOTOMETRY

Amateur astronomers can now purchase commercially available photometers, which are perfectly suitable for measuring the brightness and color of many asteroids. With such equipment, the amateur can determine rotation rates, pole orientations, and compositional classes. Most of the asteroids accessible to amateurs' telescopes have already been observed, so the amateur is not likely to make scientifically valuable contributions. Nevertheless, the enjoyment that can be derived from making such observations can be great.

There is one area in which photometry by amateurs *can* be of scientific importance. This is the observation and timing of occultations. In this field, the amateur can often do *more* than most professionals.

The occultation of stars by the asteroids provides our most accurate means of measuring the diameters of the asteroids. Occultations can only be observed along narrow paths across the Earth. The width of a path is equal to the diameter of the occulting asteroid. Usually, the location of the path is not known very accurately until a few days before the occultation occurs. Therefore, portable equipment is required, and the observer must be prepared to move his equipment from one place to another with little advance notice. In addition, teams of observers should be stationed at intervals across the path of the occultation, so that the diameter of the asteroid can be measured along several 'chords'. Professional astronomers rarely have the flexibility, or the manpower, to observe these occultations without assistance from amateurs.

The important thing in observing occultations is not really the photometry, but the *timing*. It is often feasible to observe an occultation visually, without any electronic equipment, as long as the observer has a means of accurately noting the times of the beginning and end of the occultation.

Anyone interested in making observations of asteroids, should contact the following organizations:

Minor Planets Section
Association of Lunar and Planetary Observers
c/o Frederick Pilcher
Illinois College
Jacksonville, Illinois 62650
USA

or:

Minor Planets Section

British Astronomical Association
c/o A. J. Hollis
85 Forest Rd.
Cuddington, Northwich
Cheshire CW8 2ED
England

and:

International Occultation Timing Association
6 N 106 White Oak Lane
St. Charles, Illinois 60174
USA

A year's membership in 'IOTA' costs $12 in North America, and $17 elsewhere. Membership in IOTA includes a subscription to the *Occultation Newsletter* which contains detailed information about asteroid and lunar occultations.

A publication called *Tonight's Asteroids* is indispensable for anyone who is seriously interested in asteroids. To obtain this publication, send at least four stamped, self-addressed envelopes, or a sufficient International Money Order, to:

Joseph F. Flowers
Route 4, Box 446
Wilson, North Carolina 27893
USA

Appendix B
The numbered asteroids

The following table contains data for 3445 asteroids. Listed, in order, are the asteroid's number, name, compositional type, semi-major axis of the orbit, eccentricity, inclination, perihelion distance, period, absolute magnitude (B(1,0)), radius, and rotation rate. A rotation rate of zero means that the rate is unknown.

APPENDIX B

The numbered asteroids

No.	Name	Type	A (A.U.)	E	I (deg.)	q (A.U.)	Period (years)	Mag.	Rad. (km)	Rot. (hours)
1	CERES	C	2.7671905	0.07835778	10.60449	2.5503595	4.6031866	4.35	470.0	9.07500
2	PALLAS	CU	2.7724760	0.23383491	34.80188	2.1241744	4.6163816	5.08	294.0	7.81100
3	JUNO	S	2.6677494	0.25799987	13.00439	1.9794705	4.3573012	6.47	124.0	7.21000
4	VESTA	U	2.3624704	0.08970533	7.14061	2.1505442	3.6311948	4.24	288.0	5.34200
5	ASTRAEA	S	2.5746117	0.19071350	5.36111	2.0835986	4.1311183	8.26	60.0	16.81200
6	HEBE	S	2.4252641	0.20191789	14.78574	1.9355600	3.7769270	6.85	102.0	7.27500
7	IRIS	S	2.3861809	0.22930604	5.51142	1.8390152	3.6859975	6.61	104.0	7.13500
8	FLORA	S	2.2014072	0.15632550	5.88982	1.8572710	3.2662585	7.61	81.0	12.34800
9	METIS	S	2.3868725	0.12176260	5.58298	2.0962408	3.6876001	7.49	79.0	5.07900
10	HYGIEA	C	3.1341634	0.12016398	3.83965	2.7575498	5.5485873	6.32	215.0	17.49500
11	PARTHENOPE	S	2.4515281	0.10029997	4.62541	2.2056398	3.8384447	7.79	78.0	7.83000
12	VICTORIA	S	2.3336399	0.22035946	8.37924	1.8194002	3.5649278	8.37	68.0	8.65400
13	EGERIA	C	2.5776548	0.08697759	16.50068	2.3534567	4.1384449	7.56	122.0	7.04500
14	IRENE	S	2.5871568	0.16506664	9.11086	2.1601033	4.1613488	7.54	75.0	9.35000
15	EUNOMIA	S	2.6438670	0.18496472	11.76158	2.1548450	4.2989206	6.40	130.0	6.08100
16	PSYCHE	M	2.9234552	0.13385805	3.09458	2.5321271	4.9985571	6.91	124.0	4.19600
17	THETIS	S	2.4681058	0.13770881	5.59067	2.1282258	3.8774450	8.93	49.0	12.27500
18	MELPOMENE	S	2.2960522	0.21794643	10.13956	1.7956359	3.4791460	7.45	81.0	11.57200
19	FORTUNA	C	2.4421968	0.15863468	1.57364	2.0547798	3.8165505	8.16	99.0	7.44500
20	MASSALIA	S	2.4078932	0.14537703	0.70463	2.0578408	3.7364213	7.59	67.0	8.09800
21	LUTETIA	M	2.4355896	0.16135731	3.06930	2.0425892	3.8010724	8.56	54.0	8.17300
22	KALLIOPE	M	2.9099596	0.09796403	13.69665	2.6248882	4.9639845	7.52	87.0	4.14800
23	THALIA	S	2.6286302	0.23096313	10.15297	2.0215135	4.2618113	7.87	58.0	12.30800
24	THEMIS	C	3.1300230	0.13421845	0.75856	2.7099161	5.5375962	7.90	114.0	8.38000
25	PHOCAEA	S	2.4015017	0.25418034	21.58601	1.7910872	3.7215540	9.01	36.0	9.94500
26	PROSERPINA	S	2.6556256	0.08930968	3.56345	2.4184525	4.3276315	8.78	44.0	12.00000
27	EUTERPE	S	2.3481417	0.17123073	1.58541	1.9460676	3.5982094	8.11	58.0	8.50000
28	BELLONA	S	2.7816868	0.14802636	9.40919	2.3699238	4.6394057	8.27	62.0	15.69500
29	AMPHITRITE	S	2.5552933	0.07322169	6.10667	2.3681905	4.0847092	7.02	100.0	5.39000
30	URANIA	S	2.3653100	0.12678652	2.09197	2.0654204	3.6377435	8.66	47.0	13.68600
31	EUPHROSYNE	C	3.1461239	0.22764701	26.34468	2.4299183	5.5803795	7.73	135.0	5.53100
32	POMONA	S	2.5856571	0.08441851	5.52301	2.3673799	4.1577311	8.63	46.0	9.44300
33	POLYHYMNIA	S	2.8623216	0.34050575	1.88855	1.8876846	4.8425889	9.51	31.0	18.60100
34	CIRCE	C	2.6868303	0.10687097	5.48574	2.3996861	4.4041324	9.46	56.0	15.00000
35	LEUKOTHEA		3.0028589	0.21988705	8.03537	2.3425691	5.2035818	9.68	33.7	0.00000
36	ATALANTE	C	2.7445779	0.30501673	18.48659	1.9074357	4.5468783	9.46	60.0	9.93000
37	FIDES	S	2.6420364	0.17714019	3.07121	2.1740255	4.2944565	8.26	48.0	7.33000
38	LEDA	CU	2.7423198	0.15158775	6.95243	2.3266177	4.5412683	9.57	59.0	13.00000
39	LAETITIA	S	2.7676494	0.11373660	10.37950	2.4528663	4.6043315	7.45	78.0	5.13800
40	HARMONIA	SX	2.2669325	0.04673239	4.25852	2.1609933	3.4131696	8.17	58.0	9.13600
41	DAPHNE	C	2.7711277	0.26851508	15.78591	2.0270381	4.6130142	8.16	102.0	5.98800
42	ISIS	S	2.4399810	0.22577658	8.54226	1.8890903	3.8113570	8.58	47.0	13.59000
43	ARIADNE	SX	2.2025924	0.16857211	3.46922	1.8312967	3.2688966	9.02	42.0	5.75300
44	NYSA	E	2.4225149	0.15131171	3.71008	2.0559602	3.7705066	7.96	34.0	6.42200
45	EUGENIA	C	2.7207904	0.08358135	6.60141	2.4933832	4.4878945	8.11	122.0	5.69900
46	HESTIA	F	2.5251079	0.17241392	2.33416	2.0897441	4.0125446	9.30	82.0	21.04000
47	AGLAJA	C	2.8801374	0.13298693	4.98251	2.4971168	4.8878717	8.88	79.0	0.00000
48	DORIS	CX	3.1131577	0.06822712	6.53906	2.9007559	5.4928999	7.97	123.0	11.90000
49	PALES	CX	3.0816524	0.23587982	3.17964	2.3547528	5.4097285	8.72	88.0	10.40000
50	VIRGINIA	C	2.6475759	0.28877592	2.84088	1.8830197	4.3079696	10.16	44.0	0.00000
51	NEMAUSA	CU	2.3653538	0.06624741	9.96842	2.2086554	3.6378450	8.45	76.0	7.78500
52	EUROPA	C	3.1075892	0.10305788	7.45844	2.7873278	5.4781690	7.29	146.0	5.63100
53	KALYPSO	C	2.6187272	0.20280042	5.15656	2.0876482	4.2377505	9.70	55.0	26.55000
54	ALEXANDRA	C	2.7121646	0.19650435	11.78303	2.1792126	4.4665689	8.68	88.0	7.04000
55	PANDORA	CMEU	2.7589211	0.14541167	7.19739	2.3577418	4.5825682	8.55	56.4	4.80400
56	MELETE	P	2.5993972	0.23221974	8.08680	1.9957658	4.1909161	9.29	72.0	16.00000
57	MNEMOSYNE	S	3.1480839	0.11771372	15.22652	2.7775111	5.5855951	8.15	58.0	12.29000
58	CONCORDIA	C	2.7027831	0.04329615	5.06668	2.5857632	4.4434142	9.81	52.0	0.00000
59	ELPIS	C	2.7126052	0.11697348	8.64418	2.3953023	4.4676580	8.76	82.0	13.69000
60	ECHO	S	2.3952658	0.18281546	3.59861	1.9573741	3.7070680	9.68	26.0	25.20800
61	DANAE	S	2.9864850	0.16275103	18.20824	2.5004315	5.1610794	8.73	44.0	11.50000
62	ERATO	C	3.1125698	0.18576132	2.22726	2.5343747	5.4913440	9.67	32.0	0.00000
63	AUSONIA	S	2.3951671	0.12600312	5.77862	2.0933685	3.7068391	8.67	46.0	9.29700
64	ANGELINA	E	2.6852169	0.12490597	1.30824	2.3498173	4.4001656	8.61	30.0	8.75200
65	CYBELE	CPF	3.4337718	0.10508107	3.54978	3.0729475	6.3629303	7.75	154.0	6.07000
66	MAJA	C	2.6456652	0.17393441	3.04867	2.1854930	4.3033066	10.31	46.0	0.00000
67	ASIA	S	2.4208515	0.18734875	6.01385	1.9673080	3.7666235	9.38	30.0	15.89000
68	LETO	S	2.7829115	0.18516640	7.96384	2.2676098	4.6424699	7.99	64.0	14.84800
69	HESPERIA	M	2.9759197	0.17004269	8.55721	2.4698863	5.1337156	7.97	54.0	5.65500
70	PANOPAEA	C	2.6135967	0.18447790	11.58902	2.1314459	4.2253032	9.03	76.0	14.00000
71	NIOBE	S	2.7534571	0.17508546	23.29441	2.2713668	4.5689611	8.30	53.0	28.80000
72	FERONIA	U	2.2660818	0.12036769	5.41782	1.9933188	3.4112487	9.99	48.0	8.09800

No.	Name	Type	A (A.U.)	E	I (deg.)	q (A.U.)	Period (years)	Mag.	Rad. (km)	Rot. (hours)
73	KLYTIA		2.6649060	0.04196042	2.37888	2.5530853	4.3503366	10.19	28.0	0.00000
74	GALATEA	C	2.7780325	0.23922776	4.06843	2.1134501	4.6302667	9.62	56.0	9.00000
75	EURYDIKE	M	2.6734552	0.30449915	4.99261	1.8593903	4.3712873	9.91	24.0	8.92000
76	FREIA	P	3.4040089	0.17108653	2.11587	2.8216288	6.2803817	8.81	98.0	9.79000
77	FRIGGA	M	2.6690910	0.13270622	2.42740	2.3148861	4.3605886	9.57	33.0	9.01200
78	DIANA	C	2.6223910	0.20488612	8.66095	2.0850995	4.2466469	9.11	72.0	7.22000
79	EURYNOME	S	2.4441001	0.19269559	4.62869	1.9731328	3.8210127	9.07	40.0	5.97900
80	SAPPHO	SU	2.2953789	0.20032939	8.66064	1.8355470	3.4776156	9.11	42.0	14.05000
81	TERPSICHORE	C	2.8553362	0.20886590	7.80739	2.2589538	4.8248725	9.47	61.0	0.00000
82	ALKMENE	S	2.7590153	0.22341850	2.83815	2.1426003	4.5828028	9.24	33.0	12.99900
83	BEATRIX	C	2.4314442	0.08416295	4.97622	2.2268066	3.7913723	9.51	59.0	10.16000
84	KLIO	C	2.3622074	0.23616666	9.32651	1.8043327	3.6305883	10.39	44.0	0.00000
85	IO	C	2.6541002	0.19397318	11.96316	2.1392758	4.3239031	8.60	74.0	6.83400
86	SEMELE	C	3.1081262	0.21280754	4.79475	2.4466934	5.4795885	9.60	56.0	16.63400
87	SYLVIA	P	3.4842789	0.08329299	10.87419	3.1940627	6.5038323	7.88	141.0	5.18300
88	THISBE	C	2.7689497	0.16345111	5.21681	2.3163619	4.6075768	8.01	105.0	6.04200
89	JULIA	S	2.5517888	0.18063158	16.11995	2.0908551	4.0763087	7.70	84.0	11.38000
90	ANTIOPE	C	3.1504493	0.16233833	2.23581	2.6390107	5.5918913	9.18	64.0	0.00000
91	AEGINA	C	2.5894365	0.10596689	2.11720	2.3150420	4.1668501	9.86	52.0	6.02500
92	UNDINA	U	3.1962066	0.08654488	9.87870	2.9195912	5.7141581	7.51	92.0	15.94000
93	MINERVA	C	2.7569706	0.14079385	8.56383	2.3688061	4.5777092	8.55	84.0	5.97900
94	AURORA	C	3.1605694	0.08132598	8.00993	2.9035330	5.6188574	8.47	95.0	7.22000
95	ARETHUSA	C	3.0715537	0.14350434	12.93104	2.6307726	5.3831587	8.80	114.0	8.68800
96	AEGLE	U	3.0492260	0.14060307	16.00830	2.6204956	5.3245683	8.65	56.0	0.00000
97	KLOTHO	MP	2.6681166	0.25867859	11.76186	1.9779319	4.3582006	8.44	54.0	35.00000
98	IANTHE	C	2.6873202	0.18782806	15.56376	2.1825659	4.4053369	9.80	53.0	0.00000
99	DIKE	C	2.6644919	0.19529341	13.87725	2.1441343	4.3493223	10.40	40.0	30.00000
100	HEKATE	SU	3.1004195	0.15543094	6.39222	2.6185184	5.4592209	8.79	42.0	10.00000
101	HELENA	S	2.5832782	0.14007466	10.16834	2.2214262	4.1519942	9.17	36.0	16.00000
102	MIRIAM	C	2.6603513	0.25316143	5.14807	1.9868529	4.3391881	10.16	45.0	0.00000
103	HERA	S	2.7028306	0.07904059	5.41583	2.4891973	4.4435310	8.75	48.0	23.74000
104	KLYMENE	C	3.1556456	0.14815402	2.82073	2.6881242	5.6057324	9.23	67.0	9.00000
105	ARTEMIS	C	2.3718631	0.17815299	21.48851	1.9493086	3.6528718	9.33	63.0	30.00000
106	DIONE	U	3.1602468	0.18219934	4.62014	2.5844519	5.6179967	8.50	70.0	0.00000
107	CAMILLA	C	3.4788816	0.08322890	9.93210	3.1893382	6.4887266	7.88	106.0	4.85000
108	HECUBA	SX	3.2362649	0.06334246	4.32336	3.0312719	5.8219185	9.24	35.0	8.00000
109	FELICITAS	C	2.6956618	0.29870781	7.87926	1.8904465	4.4258642	9.93	38.0	26.30000
110	LYDIA	M	2.7323987	0.07925171	5.96795	2.5158515	4.5166469	8.78	38.0	10.92700
111	ATE	C	2.5939474	0.10128826	4.91194	2.3312111	4.1777434	8.92	78.0	22.20000
112	IPHIGENIA	C	2.4345722	0.12800092	2.60417	2.1229446	3.7986913	10.76	34.0	0.00000
113	AMALTHEA	SX	2.3772893	0.08665541	5.03928	2.1712844	3.6654143	9.64	24.0	9.93500
114	KASSANDRA	C	2.6753588	0.13879590	4.94511	2.3040299	4.3759570	9.36	66.0	20.00000
115	THYRA	S	2.3800864	0.19198848	11.57972	1.9231373	3.6718853	8.56	46.0	7.24400
116	SIRONA	S	2.7742763	0.13773456	3.56435	2.3921626	4.6208782	8.80	40.0	12.03000
117	LOMIA	C	2.9896538	0.02239825	14.92833	2.9226909	5.1692958	8.91	78.0	0.00000
118	PEITHO		2.4383466	0.16121136	7.74690	2.0452573	3.8075285	9.87	32.9	7.78000
119	ALTHAEA	S	2.5807042	0.08090604	5.76291	2.3719096	4.1457901	9.59	30.0	0.00000
120	LACHESIS	C	3.1138444	0.06452914	6.96219	2.9129107	5.4947176	8.68	86.0	30.00000
121	HERMIONE	C	3.4430120	0.14313433	7.55403	2.9501987	6.3886309	8.34	100.0	8.87000
122	GERDA	S	3.2146690	0.05228177	1.63974	3.0466003	5.7637401	8.84	42.0	0.00000
123	BRUNHILD	S	2.6941676	0.12123912	6.41071	2.3675289	4.4221849	10.00	24.0	10.04000
124	ALKESTE	S	2.6294420	0.07811652	2.95730	2.4240391	4.2637858	9.21	37.0	9.92100
125	LIBERATRIX	CMEU	2.7437093	0.08010462	4.66322	2.5239255	4.5447202	9.79	31.4	3.96900
126	VELLEDA	S	2.4392004	0.10649252	2.91839	2.1794438	3.8095286	10.39	21.0	0.00000
127	JOHANNA		2.7596314	0.06397280	8.24244	2.5830898	4.5843377	9.46	37.1	12.00000
128	NEMESIS	CEU	2.7498429	0.12642765	6.24566	2.4021866	4.5599685	8.48	58.2	39.00000
129	ANTIGONE	UX	2.8714263	0.20790824	12.22736	2.2744331	4.8657136	8.02	56.0	4.95700
130	ELEKTRA	C	3.1139982	0.21967603	22.89754	2.4299273	5.4951243	8.09	87.0	5.25000
131	VALA	S	2.4319308	0.06810766	4.95526	2.2662976	3.7925105	10.94	17.0	0.00000
132	AETHRA	SU	2.6126838	0.38469076	25.07521	1.6076084	4.2230892	10.21	19.0	0.00000
133	CYRENE	SX	3.0674195	0.13674100	7.22585	2.6479776	5.3722939	9.23	36.0	12.67000
134	SOPHROSYNE	C	2.5646181	0.11606679	11.58026	2.2669511	4.1070886	9.56	58.0	17.14000
135	HERTHA	M	2.4286063	0.20422934	2.29580	1.9326136	3.7847364	9.07	40.0	8.40000
136	AUSTRIA	M	2.2870545	0.08550401	9.57399	2.0915022	3.4587152	10.78	23.0	11.50000
137	MELIBOEA	C	3.1105809	0.22398382	13.44215	2.4138610	5.4860818	8.98	75.0	20.00000
138	TOLOSA	S	2.4477067	0.16470441	3.20736	2.0445588	3.8294737	9.89	26.0	10.10300
139	JUEWA	C	2.7791915	0.17608726	10.93328	2.2898114	4.6331644	8.80	86.0	30.00000
140	SIWA	C	2.7329092	0.21446396	3.18875	2.1467986	4.5179124	9.33	52.0	22.00000
141	LUMEN	CX	2.6657870	0.21569481	11.91510	2.0907905	4.3524938	9.55	58.0	0.00000
142	POLANA	U	2.4185653	0.13614197	2.23811	2.0892971	3.7612891	11.11	26.0	0.00000
143	ADRIA	C	2.7605312	0.07144794	11.47547	2.5632968	4.5865803	10.14	45.0	0.00000
144	VIBILIA	C	2.6566458	0.23265947	4.81440	2.0385518	4.3301253	8.91	66.0	13.81000
145	ADEONA	C	2.6724358	0.14537337	12.61947	2.2839348	4.3687873	9.21	68.0	8.10000
146	LUCINA	C	2.7187893	0.06654107	13.09423	2.5378783	4.4829445	8.96	70.0	18.54000
147	PROTOGENEIA	CMEU	3.1344264	0.03513270	1.92308	3.0243053	5.5492859	9.62	34.0	0.00000
148	GALLIA	S	2.7724092	0.18636665	25.33112	2.2557247	4.6162148	8.64	46.0	20.66400
149	MEDUSA	U	2.1748059	0.06516278	0.93989	2.0330894	3.2072346	11.85	13.0	26.00000
150	NUWA	CEU	2.9843166	0.12449787	2.19174	2.6127756	5.1554589	9.21	41.8	8.14000
151	ABUNDANTIA	S	2.5909624	0.03635473	6.45719	2.4967687	4.1705337	10.39	21.0	0.00000
152	ATALA	S	3.1459887	0.07227970	12.16022	2.9185977	5.5800195	9.49	31.0	5.28200
153	HILDA	P	3.9732363	0.14193875	7.83996	3.4092803	7.9198437	8.40	111.0	0.00000
154	BERTHA		3.1795099	0.09586639	21.12833	2.8753376	5.6694412	8.37	64.1	12.00000
155	SCYLLA	CMEU	2.7578304	0.27547833	11.38869	1.9981078	4.5798507	12.29	10.0	0.00000

No.	Name	Type	A (A.U.)	E	I (deg.)	q (A.U.)	Period (years)	Mag.	Rad. (km)	Rot. (hours)
156	XANTHIPPE	C	2.7283523	0.22579697	9.72649	2.1122985	4.5066171	9.69	54.0	22.50000
157	DEJANIRA		2.5808754	0.19718046	12.13428	2.0719771	4.1462030	12.34	10.3	0.00000
158	KORONIS	S	2.8707948	0.05335279	0.99957	2.7176299	4.8641081	10.57	18.0	0.00000
159	AEMILIA	C	3.1038117	0.10730023	6.12827	2.7707722	5.4681835	9.17	70.0	30.00000
160	UNA	C	2.7283790	0.06723409	3.83013	2.5449390	4.5066838	10.01	48.0	5.58000
161	ATHOR	M	2.3793795	0.13787901	9.04847	2.0513129	3.6702490	9.89	24.0	7.28800
162	LAURENTIA	C	3.0170035	0.18194087	6.09706	2.4680872	5.2403917	9.77	54.0	11.87000
163	ERIGONE	C	2.3659329	0.19299768	4.80570	1.9093134	3.6391809	10.63	36.0	16.14300
164	EVA	C	2.6339722	0.34458810	24.48255	1.7263367	4.2748098	9.64	56.0	13.66000
165	LORELEY	C	3.1310995	0.06952454	11.20057	2.9134111	5.5404534	8.14	114.0	7.60000
166	RHODOPE	U	2.6853127	0.21235833	12.02666	2.1150641	4.4004016	10.86	20.0	9.50000
167	URDA	S	2.8526413	0.03293223	2.21059	2.7586975	4.8180442	10.24	22.0	16.00000
168	SIBYLLA	C	3.3794060	0.04879129	4.62514	3.2145205	6.2124162	8.96	75.0	0.00000
169	ZELIA	S	2.3583970	0.13046041	5.50530	2.0507195	3.6218073	10.55	19.0	0.00000
170	MARIA	SU	2.5521545	0.06403087	14.42548	2.3887379	4.0771856	10.62	20.0	0.00000
171	OPHELIA	CFPD	3.1421349	0.11820383	2.54412	2.7707224	5.5697694	9.26	56.0	13.40000
172	BAUCIS	S	2.3796375	0.11368852	10.01963	2.1090999	3.6708460	9.96	32.0	0.00000
173	INO	C	2.7422085	0.20946968	14.23424	2.1677990	4.5409918	8.76	84.0	5.93000
174	PHAEDRA	S	2.8605764	0.14436758	12.13311	2.4476020	4.8381615	9.50	31.0	0.00000
175	ANDROMACHE	U	3.1935079	0.22637576	3.21994	2.4705751	5.7069230	9.29	40.0	0.00000
176	IDUNA	U	3.1888850	0.16481049	22.56645	2.6633234	5.6945353	9.42	65.0	0.00000
177	IRMA	C	2.7685857	0.23760951	1.39242	2.1107433	4.6066685	10.49	38.0	0.00000
178	BELISANA	S	2.4595449	0.04563144	1.90188	2.3473125	3.8572888	10.75	18.0	12.33000
179	KLYTAEMNESTRA	S	2.9739437	0.10937161	7.81271	2.6486785	5.1286030	9.29	34.0	11.17300
180	GARUMNA	S	2.7225802	0.16702732	0.86896	2.2678349	4.4923239	11.41	13.0	0.00000
181	EUCHARIS	S	3.1380885	0.20629807	18.73556	2.4907067	5.5590138	8.86	41.0	16.00000
182	ELSA	S	2.4163983	0.18600912	2.00400	1.9669262	3.7562356	10.20	19.0	80.00000
183	ISTRIA	SU	2.7934239	0.34895161	26.41780	1.8186541	4.6687999	10.86	16.0	11.77000
184	DEJOPEJA	MP	3.1773894	0.08578485	1.14410	2.9048173	5.6637702	9.29	40.1	6.70000
185	EUNIKE	C	2.7393727	0.12662470	23.24120	2.3925004	4.5339494	8.49	94.0	10.83000
186	CELUTA	U	2.3623459	0.14998451	13.18418	2.0080307	3.6309080	10.02	25.0	19.60000
187	LAMBERTA	C	2.7284267	0.24018057	10.62660	2.0731115	4.5068016	9.13	72.0	10.65000
188	MENIPPE	SU	2.7606599	0.17884205	11.73155	2.2669380	4.5869012	9.94	27.1	0.00000
189	PHTHIA	S	2.4497645	0.03637644	5.17964	2.3606508	3.8343036	10.49	21.0	0.00000
190	ISMENE	C	3.9683847	0.16563134	6.16856	3.3110957	7.9053416	8.64	62.0	0.00000
191	KOLGA		2.8946793	0.09118217	11.51730	2.6307361	4.9249368	10.01	28.5	0.00000
192	NAUSIKAA	S	2.4026802	0.24705462	6.81648	1.8090869	3.7242939	8.40	50.0	13.62200
193	AMBROSIA		2.6031146	0.29321539	12.04007	1.8398412	4.1999092	10.77	21.2	0.00000
194	PROKNE	C	2.6165137	0.23856135	18.53216	1.9923148	4.2323790	8.60	96.0	15.67000
195	EURYKLEIA	C	2.8781016	0.04396859	6.96785	2.7515554	4.8826904	9.97	48.0	0.00000
196	PHILOMELA	S	3.1134765	0.02732452	7.25727	3.0284021	5.4937434	7.65	80.0	8.33300
197	ARETE	S	2.7402172	0.16143627	8.80756	2.2978466	4.5360460	10.36	21.0	0.00000
198	AMPELLA	S	2.4589400	0.23025073	9.28228	1.8927673	3.8558660	9.40	33.0	0.00000
199	BYBLIS		3.1872833	0.16041967	15.27207	2.6759801	5.6902452	9.86	32.3	0.00000
200	DYNAMENE	C	2.7377956	0.13385139	6.89420	2.3713379	4.5300350	9.23	68.0	19.00000
201	PENELOPE	MP	2.6783736	0.18017629	5.75188	2.1957941	4.3833561	9.09	43.8	3.74700
202	CHRYSEIS	S	3.0717478	0.10138112	8.84601	2.7603304	5.3836684	8.57	48.0	0.00000
203	POMPEJA	C	2.7368579	0.06010264	3.18016	2.5723655	4.5277076	9.76	53.0	46.60000
204	KALLISTO	S	2.6702869	0.17572410	8.28789	2.2010531	4.3635192	9.96	25.0	0.00000
205	MARTHA	C	2.7762604	0.03437020	10.68833	2.6808398	4.6258368	10.19	43.0	0.00000
206	HERSILIA	C	2.7406113	0.04000120	3.77589	2.6309836	4.5370250	9.65	56.0	0.00000
207	HEDDA	C	2.2846222	0.02901406	3.80370	2.2183361	3.4531991	10.95	31.0	0.00000
208	LACRIMOSA	U	2.8914497	0.01516274	1.75696	2.8476974	4.9166970	9.82	24.0	0.00000
209	DIDO	C	3.1415195	0.06738997	7.18873	2.9298127	5.5681334	9.20	68.0	8.00000
210	ISABELLA	C	2.7212081	0.12417694	5.26755	2.3832970	4.4889283	10.16	43.0	0.00000
211	ISOLDA	C	3.0457432	0.15498073	3.87434	2.5737116	5.3154488	8.87	83.0	18.37500
212	MEDEA	C	3.1067722	0.11856090	4.27458	2.7384305	5.4760084	9.27	66.0	0.00000
213	LILAEA	FX	2.7530730	0.14365627	6.80244	2.3575766	4.5680051	9.75	52.0	0.00000
214	ASCHERA	E	2.6108501	0.03094039	3.43281	2.5300696	4.2186446	10.31	12.0	6.83500
215	OENONE	S	2.7661068	0.03382372	1.69304	2.6725469	4.6004829	10.70	18.0	0.00000
216	KLEOPATRA	M	2.7955377	0.24951619	13.10931	2.0980058	4.6740999	8.04	47.0	5.39000
217	EUDORA	CMEU	2.8753247	0.30430996	10.46891	2.0003347	4.8756256	10.94	19.6	0.00000
218	BIANCA	S	2.6665945	0.11571566	15.21576	2.3580277	4.3544717	9.61	29.0	6.63600
219	THUSNELDA	SM	2.3542869	0.22365534	10.84973	1.8277380	3.6123440	10.57	20.0	30.00000
220	STEPHANIA	CEU	2.3487113	0.25730082	7.60009	1.7443860	3.5995185	12.18	11.1	0.00000
221	EOS	S	3.0139985	0.09818666	10.87410	2.7180641	5.2325640	8.74	56.0	10.43600
222	LUCIA	C	3.1284506	0.14776047	2.16038	2.6661892	5.5334239	10.25	42.0	7.00000
223	ROSA		3.0850296	0.13051476	1.94426	2.6823878	5.4186239	11.06	30.5	0.00000
224	OCEANA	M	2.6446781	0.04284703	5.85041	2.5313613	4.3008986	9.59	36.0	18.93300
225	HENRIETTA	C	3.3704231	0.27761140	20.91436	2.4347551	6.1876626	9.52	60.0	0.00000
226	WERINGIA		2.7105310	0.20492405	15.94650	2.1550779	4.4625344	10.91	19.9	15.00000
227	PHILOSOPHIA		3.1396489	0.21015306	9.14138	2.4798419	5.5631609	10.02	30.0	0.00000
228	AGATHE	S	2.2011900	0.24191456	2.53970	1.6686901	3.2657752	13.85	4.2	0.00000
229	ADELINDA	C	3.4040234	0.15985911	2.10948	2.8598592	6.2804217	10.44	39.0	0.00000
230	ATHAMANTIS	S	2.3819399	0.06075284	9.44661	2.2372303	3.6761751	8.62	62.0	23.99000
231	VINDOBONA		2.9217076	0.15104368	5.11878	2.4804020	4.9940758	10.49	24.2	0.00000
232	RUSSIA	C	2.5498855	0.17813858	6.09900	2.0956526	4.0717492	11.22	27.0	0.00000
233	ASTEROPE	S	2.6608210	0.09927253	7.67663	2.3966744	4.3403373	9.27	33.0	19.70000
234	BARBARA	S	2.3850074	0.24465489	15.36997	1.8015037	3.6832790	10.23	22.0	26.50000
235	CAROLINA	S	2.8798089	0.06340933	9.03985	2.6972022	4.8870354	9.96	26.0	17.56000
236	HONORIA	S	2.8014667	0.18718643	7.67695	2.2770703	4.6889782	9.31	34.0	24.00000
237	COELESTINA		2.7648928	0.07317110	9.74987	2.5625825	4.5974550	10.30	26.2	0.00000
238	HYPATIA	C	2.9080348	0.08814463	12.39625	2.6517072	4.9590602	8.99	77.0	8.90000

No.	Name	Type	A (A.U.)	E	I (deg.)	q (A.U.)	Period (years)	Mag.	Rad. (km)	Rot. (hours)
239	ADRASTEA		2.9655411	0.23538403	6.17498	2.2675002	5.1068830	11.72	13.7	0.00000
240	VANADIS	C	2.6649046	0.20707569	2.10594	2.1130674	4.3503327	9.94	49.0	0.00000
241	GERMANIA	C	3.0466964	0.10352983	5.50780	2.7312725	5.3179445	8.52	100.0	0.00000
242	KRIEMHILD		2.8625443	0.11983279	11.32187	2.5195177	4.8431549	10.27	26.7	0.00000
243	IDA	S	2.8608384	0.04206682	1.13600	2.7404921	4.8388257	11.05	13.0	0.00000
244	SITA		2.1740732	0.13741261	2.84283	1.8753281	3.2056141	13.40	6.3	0.00000
245	VERA	S	3.0856490	0.20823297	5.18719	2.4431152	5.4202561	8.82	42.0	14.38000
246	ASPORINA	UX	2.6948030	0.10883446	15.62037	2.4015155	4.4237494	9.84	35.0	16.22200
247	EUKRATE	C	2.7389104	0.24465835	25.02958	2.0688131	4.5328021	9.04	72.0	12.10000
248	LAMEIA	U	2.4712808	0.06183436	4.03490	2.3184707	3.8849297	11.18	17.6	0.00000
249	ILSE		2.3780718	0.21669859	9.64701	1.8627471	3.6672244	12.16	11.2	42.62000
250	BETTINA	EMP	3.1543908	0.12247711	12.88045	2.7680502	5.6023889	8.29	64.1	5.10500
251	SOPHIA		3.0935378	0.10474240	10.51897	2.7695131	5.4410553	11.20	17.4	0.00000
252	CLEMENTINA		3.1570907	0.08485026	10.07146	2.8892107	5.6095829	10.62	22.8	0.00000
253	MATHILDE		2.6492851	0.26269102	6.70022	1.9533417	4.3121419	11.39	16.0	0.00000
254	AUGUSTA	S	2.1950049	0.12131913	4.51550	1.9287088	3.2520201	13.16	5.7	6.00000
255	OPPAVIA	CMEU	2.7450528	0.08020593	9.48836	2.5248835	4.5480590	11.32	15.6	0.00000
256	WALPURGA		3.0011427	0.06255227	13.32475	2.8134146	5.1991220	10.97	19.4	0.00000
257	SILESIA	C	3.1084554	0.12757073	3.65010	2.7119074	5.4804597	10.49	39.0	0.00000
258	TYCHE	S	2.6166167	0.20244849	14.26039	2.0868866	4.2326288	9.32	34.0	10.04000
259	ALETHEIA	CMEU	3.1435397	0.11273573	10.73871	2.7891505	5.5735054	8.72	51.4	0.00000
260	HUBERTA	C	3.4362860	0.12647425	6.39996	3.0016842	6.3699198	9.96	49.0	0.00000
261	PRYMNO	F	2.3313329	0.08967887	3.63608	2.1222618	3.5596433	10.28	42.0	8.00000
262	VALDA	U	2.5526969	0.21406549	7.72792	2.0062525	4.0784850	12.78	8.6	0.00000
263	DRESDA	U	2.8878484	0.07574962	1.31053	2.6690948	4.9075141	11.58	11.6	0.00000
264	LIBUSSA	S	2.7994940	0.13422835	10.42644	2.4237225	4.6840262	9.47	32.0	0.00000
265	ANNA		2.4234159	0.26474169	25.63542	1.7818366	3.7726099	12.69	8.8	0.00000
266	ALINE	C	2.8024864	0.15867274	13.39366	2.3578081	4.6915383	9.37	64.0	0.00000
267	TIRZA		2.7727516	0.10094497	6.00436	2.4928563	4.6170697	11.65	14.2	5.90000
268	ADOREA	F	3.1054265	0.12514506	2.43988	2.7167978	5.4724512	9.16	61.0	6.10000
269	JUSTITIA		2.6136878	0.21706629	5.47031	2.0463443	4.2255239	11.22	17.3	0.00000
270	ANAHITA	S	2.1984792	0.15008916	2.36526	1.8685113	3.2597444	9.88	26.0	15.06000
271	PENTHESILEA	C	3.0084088	0.10221099	3.54711	2.7009163	5.2180147	10.78	33.0	0.00000
272	ANTONIA		2.7769682	0.03031511	4.45038	2.6927841	4.6276059	11.70	13.8	0.00000
273	ATROPOS	C	2.3945773	0.16234516	20.41393	2.0058293	3.7054698	11.35	26.2	25.00000
274	PHILAGORIA		3.0424058	0.12270638	3.68150	2.6690831	5.3067145	11.11	18.2	0.00000
275	SAPIENTIA	C	2.7709706	0.16387112	4.78146	2.3168886	4.6126223	9.76	54.0	0.00000
276	ADELHEID	CMEU	3.1162508	0.07069053	21.62916	2.8959613	5.5010881	9.44	37.2	6.32000
277	ELVIRA	S	2.8854730	0.08946371	1.15754	2.6273279	4.9014606	11.09	14.2	0.00000
278	PAULINA		2.7531383	0.13477209	7.81680	2.3820920	4.5681677	10.39	25.3	0.00000
279	THULE	D	4.2709308	0.01079156	2.33849	4.2248406	8.8264036	9.58	65.0	0.00000
280	PHILIA		2.9430523	0.10981760	7.45039	2.6198535	5.0489025	11.86	12.9	0.00000
281	LUCRETIA	U	2.1878366	0.13170826	5.30548	1.8996805	3.2361031	13.35	5.5	4.34800
282	CLORINDE	U	2.3390367	0.08154589	9.03505	2.1482978	3.5773013	11.86	12.0	6.42000
283	EMMA	C	3.0450871	0.15007661	7.98072	2.5880909	5.3137312	9.67	56.0	6.82000
284	AMALIA	C	2.3594918	0.22145563	8.05319	1.8369690	3.6243298	10.97	30.0	8.50000
285	REGINA		3.0876217	0.20393480	17.59014	2.4579482	5.4254546	11.85	12.9	0.00000
286	ICLEA	CX	3.1944540	0.04148909	17.92879	3.0619190	5.7094588	10.02	46.0	0.00000
287	NEPHTHYS	S	2.3525565	0.02257356	10.03378	2.2994509	3.6083620	9.28	34.0	7.60300
288	GLAUKE	S	2.7569325	0.20960902	4.32955	2.1790545	4.5776143	11.04	15.0	0.00000
289	NENETTA	UX	2.8728232	0.20412947	6.68711	2.2863953	4.8692641	10.72	23.0	0.00000
290	BRUNA		2.3371861	0.25864121	22.31740	1.7326934	3.5730569	13.15	7.1	0.00000
291	ALICE		2.2217648	0.09338671	1.85569	2.0142815	3.3116705	12.70	6.9	4.32000
292	LUDOVICA		2.5295391	0.03181368	14.94608	2.4490652	4.0231118	10.94	19.6	0.00000
293	BRASILIA	U	2.8615770	0.10689252	15.58340	2.5556960	4.8407001	10.92	19.0	0.00000
294	FELICIA		3.1304221	0.25093186	6.31094	2.3448994	5.5386553	11.18	17.6	0.00000
295	THERESIA	S	2.7963424	0.17085405	2.70543	2.3185759	4.6761184	11.33	13.0	0.00000
296	PHAETUSA	S	2.2283998	0.16025136	1.74956	1.8712957	3.3265164	13.70	4.6	0.00000
297	CAECILIA		3.1645679	0.14479619	7.55716	2.7063506	5.6295238	10.35	25.8	0.00000
298	BAPTISTINA		2.2634919	0.09556648	6.28662	2.0471778	3.4054022	12.28	8.4	0.00000
299	THORA		2.4342928	0.06093534	1.60300	2.2859583	3.7980375	12.75	8.5	0.00000
300	GERALDINA		3.2073476	0.03958039	0.74794	3.0803995	5.7440610	10.95	19.5	0.00000
301	BAVARIA		2.7253671	0.06623389	4.89066	2.5448554	4.4992228	11.10	18.2	0.00000
302	CLARISSA	F	2.4047151	0.11223122	3.41330	2.1348310	3.7290263	11.76	21.0	12.00000
303	JOSEPHINA		3.1185575	0.06975838	6.88863	2.9010119	5.5071974	9.90	31.7	0.00000
304	OLGA	C	2.4036236	0.22043142	15.80306	1.8737895	3.7264876	10.75	34.0	18.30000
305	GORDONIA	S	3.0957530	0.19353345	4.43046	2.4966211	5.4469004	9.96	25.0	0.00000
306	UNITAS	S	2.3574841	0.15117621	7.27217	2.0010884	3.6197045	9.87	26.0	8.75000
307	NIKE	C	2.9074962	0.14276500	6.12344	2.4924073	4.9576826	11.04	29.0	0.00000
308	POLYXO	DU	2.7500138	0.03826585	4.35569	2.6447823	4.5603938	9.13	69.0	12.03200
309	FRATERNITAS		2.6652448	0.11451972	3.71340	2.3600218	4.3511662	11.39	16.0	0.00000
310	MARGARITA		2.7622120	0.11500827	3.16472	2.4445348	4.5907698	11.55	14.8	0.00000
311	CLAUDIA	S	2.9006512	0.00321004	3.22570	2.8913400	4.9401851	11.08	14.0	0.00000
312	PIERRETTA	S	2.7808359	0.16267823	9.02236	2.3284545	4.6372771	10.00	24.0	0.00000
313	CHALDAEA	C	2.3767881	0.18038628	11.62225	1.9480482	3.6642551	9.87	61.0	10.08000
314	ROSALIA		3.1446531	0.18854780	12.57660	2.5517356	5.5764666	10.85	20.5	0.00000
315	CONSTANTIA		2.2411952	0.16796120	2.42764	1.8647614	3.3552086	13 77	4.2	0.00000
316	GOBERTA		3.1604650	0.15353256	2.34332	2.6752307	5.6185789	10.87	33.3	0.00000
317	ROXANE	MU	2.2868030	0.08488338	1.76570	2.0926912	3.4581442	10.42	19.0	8.16000
318	MAGDALENA	C	3.2004569	0.07648679	10.62074	2.9556642	5.7255597	10.24	42.0	0.00000
319	LEONA		3.3660672	0.24157378	10.62216	2.5529137	6.1756711	11.30	27.3	0.00000
320	KATHARINA		3.0120940	0.11234676	9.35336	2.6736951	5.2276053	11.69	11.0	0.00000
321	FLORENTINA	U	2.8862319	0.04493119	2.59304	2.7565501	4.9033947	11.15	14.0	2.87000

No.	Name	Type	A (A.U.)	E	I (deg.)	q (A.U.)	Period (years)	Mag.	Rad. (km)	Rot. (hours)
322	PHAEO		2.7846389	0.24435519	8.02393	2.1041980	4.6467934	10.10	28.9	0.00000
323	BRUCIA	S	2.3811924	0.30111423	24.25057	1.6641816	3.6744452	10.92	16.0	10.00000
324	BAMBERGA	C	2.6801097	0.34130710	11.14189	1.7653692	4.3876185	7.76	126.0	29.43000
325	HEIDLEBERGA	MEU	3.1936438	0.17931464	8.55732	2.6209767	5.7072868	9.76	32.3	6.70000
326	TAMARA	C	2.3173094	0.18972057	23.74269	1.8776681	3.5275731	10.10	40.0	0.00000
327	COLUMBIA		2.7782493	0.06354467	7.14713	2.6017063	4.6308084	11.30	16.6	0.00000
328	GUDRUN		3.1013727	0.11996910	16.11951	2.7293038	5.4617391	9.91	32.9	15.00000
329	SVEA	C	2.4760385	0.02237133	15.91337	2.4206462	3.8961535	10.62	36.0	0.00000
330	ADALBERTA		2.4691260	0.25140992	6.76039	1.8483630	3.8798492	14.00	6.3	0.00000
331	ETHERIDGEA	C	3.0275538	0.09488816	6.05091	2.7402747	5.2679033	10.60	36.0	0.00000
332	SIRI		2.7743924	0.09120238	2.85154	2.5213611	4.6211686	10.42	24.9	7.00000
333	BADENIA	C	3.1210513	0.17648080	3.79426	2.5702455	5.5138044	10.49	39.0	0.00000
334	CHICAGO	C	3.8686552	0.04133402	4.66270	3.7087481	7.6092186	8.38	100.0	0.00000
335	ROBERTA	F	2.4720824	0.17790985	5.09058	2.0322745	3.8868198	9.85	48.0	8.00000
336	LACADIERA	C	2.2516150	0.09450317	5.64653	2.0388303	3.3786345	10.74	35.0	13.70000
337	DEVOSA	CX	2.3826740	0.13849750	7.85995	2.0526795	3.6778748	9.77	54.0	4.61000
338	BUDROSA	M	2.9143612	0.02251717	6.03959	2.8487382	4.9752522	9.71	40.0	6.00000
339	DOROTHEA	U	3.0133014	0.10332746	9.94661	2.7019446	5.2307487	10.41	22.0	0.00000
340	EDUARDA	S	2.7455091	0.11734237	4.68637	2.4233446	4.5491929	11.04	15.0	7.70000
341	CALIFORNIA	S	2.1991913	0.19415680	5.66577	1.7722033	3.2613282	12.50	6.8	0.00000
342	ENDYMION	C	2.5686398	0.12962221	7.32049	2.2356870	4.1167531	11.16	26.0	0.00000
343	OSTARA	CU	2.4113009	0.23078261	3.27697	1.8548146	3.7443559	12.50	16.0	0.00000
344	DESIDERATA	C	2.5918283	0.31666425	18.47599	1.7710890	4.1726246	9.08	74.0	10.53000
345	TERCIDINA	C	2.3253641	0.06173772	9.74294	2.1818016	3.5459816	9.74	54.0	12.37100
346	HERMENTARIA	S	2.7953660	0.10018537	8.75766	2.5153112	4.6736698	8.39	51.0	26.00000
347	PARIANA	EMP	2.6139369	0.16243219	11.71095	2.1893494	4.2261281	9.97	30.7	4.06000
348	MAY		2.9696393	0.07179634	9.76374	2.7564299	5.1174726	10.61	22.9	0.00000
349	DEMBOWSKA	R	2.9245498	0.09170438	8.25518	2.6563556	5.0013647	7.15	82.0	4.70100
350	ORNAMENTA	C	3.1179321	0.14941481	24.82652	2.6520669	5.5055408	9.31	66.0	0.00000
351	YRSA	S	2.7656357	0.15420105	9.19798	2.3391719	4.5993080	10.07	24.0	0.00000
352	GISELA	SX	2.1943109	0.14965288	3.37971	1.8659259	3.2504780	11.07	15.0	6.70000
353	RUPERTO-CAROLA		2.7352738	0.32748058	5.69700	1.8395247	4.5237775	12.19	11.0	0.00000
354	ELEONORA	S	2.7970023	0.11581514	18.42991	2.4730673	4.6777744	7.55	77.0	4.27700
355	GABRIELLA		2.5392301	0.10433004	4.27276	2.2743123	4.0462537	11.56	14.8	0.00000
356	LIGURIA	C	2.7554743	0.23989087	8.23699	2.0944612	4.5739832	9.24	78.0	31.82000
357	NININA	C	3.1468782	0.07997123	15.08960	2.8952186	5.5823865	9.71	55.0	20.00000
358	APOLLONIA		2.8779757	0.15135756	3.54957	2.4423723	4.8823700	10.29	26.5	0.00000
359	GEORGIA	C	2.7280209	0.15585154	6.77006	2.3028545	4.5057960	10.05	46.0	7.30000
360	CARLOVA	C	3.0035210	0.17675696	11.71123	2.4726279	5.2053027	9.31	69.0	6.21000
361	BONONIA	D	3.9451175	0.21563749	12.66539	3.0944023	7.8359184	8.89	90.0	0.00000
362	HAVNIA	C	2.5798669	0.04437170	8.05761	2.4653938	4.1437726	9.96	49.0	18.00000
363	PADUA	C	2.7490392	0.07327180	5.94753	2.5476120	4.5579691	10.01	48.0	15.00000
364	ISARA	SMRU	2.2203674	0.14948080	6.00587	1.8884650	3.3085465	10.99	16.0	9.15500
365	CORDUBA	C	2.8040898	0.15462044	12.78926	2.3705201	4.6955652	10.17	44.0	0.00000
366	VINCENTINA		3.1410718	0.06543359	10.58183	2.9355402	5.5669436	9.54	37.4	0.00000
367	AMICITIA	S	2.2191408	0.09608764	2.94420	2.0059087	3.3058052	12.01	9.9	0.00000
368	HAIDEA	EMP	3.0612407	0.21103139	7.78418	2.4152226	5.3560696	11.03	18.8	0.00000
369	AERIA	EMP	2.6492474	0.09638796	12.73295	2.3938918	4.3120499	9.51	36.6	0.00000
370	MODESTIA	C	2.3240988	0.09095720	7.86289	2.1127052	3.5430877	11.61	22.0	0.00000
371	BOHEMIA	U	2.7271149	0.06298378	7.39083	2.5553508	4.5035515	9.84	33.0	0.00000
372	PALMA	CEU	3.1449690	0.26290542	23.84526	2.3181396	5.5773067	8.43	62.4	8.67000
373	MELUSINA	CU	3.1111698	0.14928633	15.44837	2.6467147	5.4876399	10.08	45.0	0.00000
374	BURGUNDIA	S	2.7800975	0.08120107	8.99670	2.5543506	4.6354303	10.07	24.0	0.00000
375	URSULA	C	3.1267421	0.10011628	15.95125	2.8137045	5.5288920	8.36	100.0	16.83000
376	GEOMETRIA	S	2.2886717	0.17097031	5.42907	1.8973769	3.4623842	10.62	19.0	7.74000
377	CAMPANIA	CMEU	2.6904380	0.07589760	6.67398	2.4862401	4.4130058	9.88	31.3	15.00000
378	HOLMIA	S	2.7762461	0.12775563	6.99441	2.4215651	4.6258011	10.96	16.0	0.00000
379	HUENNA	MP	3.1278222	0.19523846	1.67007	2.5171511	5.5317569	9.78	31.0	6.60000
380	FIDUCIA	C	2.6789494	0.11318443	6.16887	2.3757341	4.3847694	10.44	39.0	0.00000
381	MYRRHA	C	3.2114575	0.10993045	12.54785	2.8584204	5.7551050	9.31	75.0	0.00000
382	DODONA	EMP	3.1135707	0.17772475	7.40636	2.5602121	5.4939928	9.75	32.3	0.00000
383	JANINA	C	3.1343637	0.17947191	2.65790	2.5718334	5.5491195	10.90	20.0	6.40000
384	BURDIGALA	S	2.6510172	0.14782223	5.60855	2.2591379	4.3163714	10.65	18.0	0.00000
385	ILMATAR	S	2.8470333	0.12576383	13.56753	2.4889796	4.8038430	8.58	48.0	62.35000
386	SIEGENA	C	2.8955772	0.16918072	20.27720	2.4057014	4.9272289	8.31	102.0	9.76300
387	AQUITANIA	S	2.7403295	0.23615064	18.07891	2.0931990	4.5363250	8.46	56.0	16.00000
388	CHARYBDIS	C	3.0031362	0.06340603	6.45907	2.8127191	5.2043023	9.51	60.0	9.49000
389	INDUSTRIA	S	2.6086817	0.06505633	8.14456	2.4389703	4.2133899	9.26	35.0	11.00000
390	ALMA	U	2.6507344	0.13191678	12.14879	2.3010581	4.3156810	11.35	16.0	0.00000
391	INGEBORG	S	2.3203406	0.30657482	23.16069	1.6089826	3.5344968	12.24	8.8	0.00000
392	WILHELMINA		2.8821833	0.14152084	14.31815	2.4742942	4.8930812	10.85	20.5	17.00000
393	LAMPETIA	C	2.7773860	0.33142146	14.87908	1.8569006	4.6286502	9.40	58.0	38.70000
394	ARDUINA		2.7603166	0.23030242	6.21746	2.1246090	4.5860453	10.97	19.4	0.00000
395	DELIA	C	2.7867281	0.08405897	3.34872	2.5524786	4.6520233	11.43	25.0	0.00000
396	AEOLIA		2.7408133	0.16033851	2.55067	2.3013554	4.5375266	10.83	20.7	15.00000
397	VIENNA	S	2.6363099	0.24552161	12.84949	1.9890389	4.2805018	10.40	25.0	15.48000
398	ADMETE		2.7375767	0.22577316	9.54256	2.1195054	4.5294914	11.86	12.9	0.00000
399	PERSEPHONE		3.0512004	0.07659792	13.12432	2.8174849	5.3297410	10.22	27.4	0.00000
400	DUCROSA		3.1444242	0.10215627	10.52200	2.8232014	5.5758576	11.31	16.6	0.00000
401	OTTILIA	X	3.3426046	0.05120915	5.95375	3.1714325	6.1112142	10.43	40.8	0.00000
402	CHLOE	S	2.5564034	0.11581290	11.84420	2.2603388	4.0873713	10.08	23.0	0.00000
403	CYANE	S	2.8150420	0.09523218	9.13776	2.5469594	4.7231021	10.49	20.0	0.00000
404	ARSINOE	C	2.5912485	0.20328309	14.11602	2.0644915	4.1712246	9.89	50.0	8.93000

No.	Name	Type	A (A.U.)	E	I (deg.)	q (A.U.)	Period (years)	Mag.	Rad. (km)	Rot. (hours)
405	THIA	C	2.5785286	0.25056827	11.90296	1.9324312	4.1405492	9.39	63.0	10.08000
406	ERNA	EMP	2.9148126	0.18085995	4.20989	2.3876395	4.9764075	11.40	15.9	0.00000
407	ARACHNE	C	2.6252172	0.06905684	7.53596	2.4439280	4.2535138	9.82	52.0	0.00000
408	FAMA		3.1715183	0.13467900	9.09186	2.7443812	5.6480799	10.69	22.0	0.00000
409	ASPASIA	C	2.5760262	0.07028655	11.23736	2.3949661	4.1345229	8.47	97.0	9.03000
410	CHLORIS	C	2.7249980	0.24076626	10.94943	2.0689104	4.4983087	9.18	68.0	32.50000
411	XANTHE		2.9344690	0.11523529	15.33828	2.5963144	5.0268307	10.14	28.4	0.00000
412	ELISABETHA	MEU	2.7645628	0.04192025	13.76142	2.6486716	4.5966315	10.26	26.9	0.00000
413	EDBURGA	MEU	2.5810101	0.34637353	18.76842	1.6870164	4.1465273	11.12	17.1	0.00000
414	LIRIOPE	EMP	3.4937789	0.08369524	9.58635	3.2013662	6.5304503	10.59	37.9	0.00000
415	PALATIA	C	2.7882624	0.30476531	8.13765	1.9384967	4.6558657	10.20	46.0	0.00000
416	VATICANA	S	2.7878895	0.22062899	12.91499	2.1728003	4.6549315	8.94	41.0	0.00000
417	SUEVIA		2.8028040	0.13294588	6.60061	2.4301827	4.6923356	10.24	27.1	0.00000
418	ALEMANNIA	M	2.5935347	0.11806018	6.83353	2.2873414	4.1767459	10.82	16.0	0.00000
419	AURELIA	F	2.5912960	0.25751272	3.94953	1.9240043	4.1713395	9.31	63.0	0.00000
420	BERTHOLDA	P	3.4111736	0.04683183	6.70055	3.2514222	6.3002205	9.45	75.0	0.00000
421	ZAHRINGIA		2.5412898	0.28167230	7.76227	1.8254789	4.0511775	12.93	7.9	0.00000
422	BEROLINA	CMEU	2.2284617	0.21431383	4.99396	1.7508715	3.3266549	11.74	10.3	20.00000
423	DIOTIMA	C	3.0686934	0.03082432	11.22196	2.9741030	5.3756409	8.25	104.0	4.62000
424	GRATIA		2.7729523	0.11032435	8.20760	2.4670281	4.6175714	10.71	21.8	0.00000
425	CORNELIA		2.8865173	0.06061542	4.05599	2.7115500	4.9041219	9.36	40.6	0.00000
426	HIPPO	C	2.8887303	0.10390503	19.53290	2.5885766	4.9097624	9.40	63.0	0.00000
427	GALENE		2.9716108	0.11841372	5.12881	2.6197312	5.1225696	10.48	24.3	0.00000
428	MONACHIA		2.3084269	0.17797139	6.19956	1.8975929	3.5073104	13.03	7.5	0.00000
429	LOTIS		2.6069007	0.12184051	9.50527	2.2892747	4.2090759	10.72	21.7	13.57800
430	1984 AQ 1		2.8415990	0.25743881	14.62951	2.1100612	4.7900958	13.50	1.0	0.00000
431	NEPHELE	CFPD	3.1230750	0.18561450	1.83067	2.5433869	5.5191679	9.98	39.0	0.00000
432	PYTHIA	S	2.3700950	0.14447223	12.11041	2.0276821	3.6487877	10.10	24.0	8.28700
433	EROS	S	1.4582421	0.22294241	10.82795	1.1331381	1.7609396	11.88	11.0	5.27000
434	HUNGARIA	E	1.9442884	0.07365116	22.51276	1.8010893	2.7110717	11.91	5.7	26.51000
435	ELLA	U	2.4491639	0.15502042	1.81480	2.0694933	3.8328934	11.24	16.0	0.00000
436	PATRICIA		3.2054293	0.05641318	18.50657	3.0246010	5.7389088	11.02	18.9	0.00000
437	RHODIA		2.3847730	0.24950023	7.36700	1.7897716	3.6827357	11.61	14.4	12.00000
438	ZEUXO	U	2.5533092	0.06899929	7.40141	2.3771327	4.0799527	10.67	20.0	0.00000
439	OHIO	C	3.1325703	0.07251295	19.13894	2.9054184	5.5443573	10.70	35.0	0.00000
440	THEODORA		2.2098596	0.10815861	1.59674	1.9708443	3.2850883	12.81	6.6	0.00000
441	BATHILDE	M	2.8054125	0.08218958	8.12954	2.5748370	4.6988883	9.35	32.0	10.35000
442	EICHSFELDIA		2.3451133	0.06995354	6.07009	2.1810641	3.5912507	10.87	20.3	0.00000
443	PHOTOGRAPHIA	S	2.2162831	0.03972710	4.23341	2.1282365	3.2994218	11.31	16.0	16.20000
444	GYPTIS	CS	2.7723894	0.17300585	10.26469	2.2927499	4.6161656	8.76	83.0	6.21400
445	EDNA	C	3.1984177	0.19111846	21.44762	2.5871408	5.7200885	10.20	43.0	0.00000
446	AETERNITAS	A	2.7880850	0.12384663	10.62694	2.4427903	4.6554217	10.05	26.0	0.00000
447	VALENTINE	U	2.9863746	0.04606593	4.79910	2.8488045	5.1607928	10.23	27.0	0.00000
448	NATALIE	CU	3.1500380	0.17084169	12.74461	2.6118803	5.5907965	11.19	27.0	0.00000
449	HAMBURGA	C	2.5544126	0.16861390	3.09385	2.1237032	4.0825977	10.43	38.0	0.00000
450	BRIGITTA	U	3.0152037	0.10146089	10.16758	2.7092783	5.2357025	11.34	12.8	0.00000
451	PATIENTIA	C	3.0632637	0.07017221	15.23287	2.8483076	5.3613796	7.56	140.0	9.72700
452	HAMILTONIA		2.8456903	0.01328512	3.22732	2.8078849	4.8004446	13.35	5.1	0.00000
453	TEA	S	2.1826906	0.10939097	5.55486	1.9439240	3.2246921	11.88	10.0	0.00000
454	MATHESIS	CU	2.6264746	0.11219880	6.31228	2.3317873	4.2565703	10.13	44.0	7.66000
455	BRUCHSALIA	C	2.6579525	0.29196385	12.02947	1.8819265	4.3333211	9.86	52.0	0.00000
456	ABNOBA		2.7876794	0.17892782	14.40570	2.2888861	4.6544061	10.97	19.4	0.00000
457	ALLEGHENIA		3.0841663	0.18395276	12.95087	2.5168254	5.4163499	12.94	7.8	0.00000
458	HERCYNIA	S	2.9903829	0.24499583	12.63700	2.2577517	5.1711864	10.78	17.0	0.00000
459	SIGNE		2.6211803	0.20842890	10.28188	2.0748506	4.2437067	11.55	14.8	6.38000
460	SCANIA		2.7169323	0.10407174	4.62715	2.4341764	4.4783521	11.85	12.9	0.00000
461	SASKIA	U	3.1061759	0.15836672	1.44381	2.6142612	5.4744325	11.38	24.1	0.00000
462	ERIPHYLA	S	2.8734672	0.08487206	3.19101	2.6295903	4.8709021	10.37	15.0	0.00000
463	LOLA	C	2.3978119	0.22041488	13.56026	1.8692985	3.7129805	12.80	13.0	0.00000
464	MEGAIRA		2.8047357	0.20233709	10.16177	2.2372334	4.6971874	10.60	21.3	0.00000
465	ALEKTO		3.0868971	0.21041425	4.65557	2.4373701	5.4235449	10.80	20.9	0.00000
466	TISIPHONE	C	3.3644061	0.07131110	19.12873	3.1244867	6.1711006	9.11	70.0	0.00000
467	LAURA		2.9440527	0.10639007	6.45239	2.6308348	5.0514770	11.94	12.4	0.00000
468	LINA	C	3.1421614	0.18885663	0.44123	2.5487432	5.5698400	10.61	35.0	8.30000
469	ARGENTINA		3.1546690	0.17911789	11.71423	2.5896113	5.6031299	9.90	31.7	0.00000
470	KILIA		2.4047122	0.09499955	7.24392	2.1762657	3.7290199	11.30	16.6	0.00000
471	PAPAGENA	S	2.8895807	0.22964451	14.93898	2.2260044	4.9119310	7.61	72.0	7.11300
472	ROMA	S	2.5429397	0.09423607	15.82123	2.3033030	4.0551233	10.05	24.0	0.00000
473	NOLLI		2.9790897	0.25564000	27.78000	2.2175152	5.1419210	11.10	18.2	0.00000
474	PRUDENTIA		2.4528713	0.21232264	8.81831	1.9320712	3.8415999	11.82	13.1	8.57100
475	OCLLO		2.5932522	0.38129279	18.79470	1.6044638	4.1760635	12.32	10.4	0.00000
476	HEDWIG	MP	2.6499417	0.07329208	10.92365	2.4557219	4.3137450	9.60	56.0	0.00000
477	ITALIA	S	2.4158072	0.18917134	5.29670	1.9588057	3.7548571	11.27	14.0	0.00000
478	TERGESTE	S	3.0154862	0.08363741	13.16433	2.7632787	5.2364388	9.12	37.0	0.00000
479	CAPRERA		2.7225075	0.21834429	8.68037	2.1280637	4.4921436	10.77	21.2	0.00000
480	HANSA	S	2.6440783	0.04559656	21.32884	2.5235174	4.2994356	9.45	32.0	0.00000
481	EMITA	C	2.7410512	0.15740101	9.84189	2.3096070	4.5381174	9.72	54.0	0.00000
482	PETRINA	S	2.9987097	0.09443481	14.49294	2.7155271	5.1928005	9.96	25.0	0.00000
483	SEPPINA	S	3.4230244	0.04385097	18.69728	3.2729213	6.3330803	9.57	37.0	0.00000
484	PITTSBURGHIA		2.6695495	0.05768254	12.50396	2.5155630	4.3617120	11.37	16.1	0.00000
485	GENUA		2.7538445	0.18950982	13.85818	2.2319639	4.5699253	9.66	35.4	17.59000
486	CREMONA		2.3528893	0.16319196	11.08281	1.9689168	3.6091278	12.06	11.7	0.00000
487	VENETIA	S	2.6694703	0.08650225	10.24692	2.4385552	4.3615179	9.28	34.0	0.00000

No.	Name	Type	A (A.U.)	E	I (deg.)	q (A.U.)	Period (years)	Mag.	Rad. (km)	Rot. (hours)
488	KREUSA	C	3.1455100	0.17880657	11.50950	2.5830719	5.5787458	8.76	84.0	40.00000
489	COMACINA	C	3.1641605	0.03241884	12.94874	3.0615821	5.6284361	9.31	65.0	0.00000
490	VERITAS	C	3.1672964	0.09937355	9.27887	2.8525510	5.6368060	9.42	64.0	0.00000
491	CARINA		3.1870968	0.08790752	18.90688	2.9069271	5.6897459	9.91	31.6	0.00000
492	GISMONDA	CFPD	3.1106100	0.17962280	1.63027	2.5518737	5.4861588	10.98	24.0	0.00000
493	GRISELDIS		3.1255100	0.16311526	15.21617	2.6156914	5.5256238	11.75	13.5	0.00000
494	VIRTUS	C	2.9852507	0.06358376	7.08797	2.7954371	5.1578798	9.98	49.0	0.00000
495	EULALIA		2.4883285	0.12980478	2.28583	2.1653316	3.9251981	11.45	15.5	0.00000
496	GRYPHIA	S	2.1985948	0.07968430	3.78946	2.0234013	3.2600014	12.92	6.5	0.00000
497	IVA	M	2.8558135	0.29763886	4.82613	2.0058124	4.8260827	10.94	15.0	4.62000
498	TOKIO	S	2.6524050	0.22156627	9.52396	2.0647216	4.3197613	9.89	36.0	0.00000
499	VENUSIA	C	3.9994695	0.21347681	2.08590	3.1456754	7.9984083	10.43	46.0	0.00000
500	SELINUR		2.6131558	0.14649577	9.77454	2.2303395	4.2242341	10.47	24.4	0.00000
501	URHIXIDUR		3.1516602	0.15094514	20.94282	2.6759324	5.5951161	10.10	28.9	0.00000
502	SIGUNE	S	2.3855262	0.17674173	24.99946	1.9639041	3.6844807	11.87	10.0	10.50000
503	EVELYN	C	2.7247567	0.17555799	5.02142	2.2464039	4.4977117	10.10	46.0	0.00000
504	CORA		2.7228122	0.21537670	12.90929	2.1363819	4.4928975	11.16	17.7	0.00000
505	CAVA	U	2.6869864	0.24417213	9.81233	2.0308993	4.4045162	10.04	30.0	8.17900
506	MARION	C	3.0449965	0.14023504	16.98026	2.6179812	5.3134937	9.81	52.0	0.00000
507	LAODICA		3.1574821	0.09010932	9.53100	2.8729634	5.6106262	10.56	23.4	0.00000
508	PRINCETONIA	C	3.1616654	0.02333321	13.33961	3.0878937	5.6217804	9.21	70.0	52.78000
509	IOLANDA	S	3.0631850	0.09307390	15.37971	2.7780824	5.3611727	9.51	30.0	0.00000
510	MABELLA	C	2.6087971	0.19318062	9.52638	2.1048279	4.2136693	10.74	34.0	19.40000
511	DAVIDA	C	3.1757350	0.17736657	15.93329	2.6124656	5.6593475	7.19	162.0	5.16700
512	TAURINENSIS	S	2.1898766	0.25390026	8.75492	1.6338663	3.2406299	12.16	11.0	5.93000
513	CENTESIMA	S	3.0112092	0.08728021	9.73153	2.7483902	5.2253017	10.79	20.0	5.23000
514	ARMIDA	F	3.0473654	0.03936019	3.87892	2.9274206	5.3196959	9.93	50.0	0.00000
515	ATHALIA	S	3.1263232	0.17495538	2.02722	2.5793562	5.5277810	11.80	18.0	0.00000
516	AMHERSTIA	M	2.6826828	0.27253464	12.98353	1.9515587	4.3939385	9.19	32.0	7.00000
517	EDITH	C	3.1360927	0.19972326	3.21388	2.5097420	5.5537119	10.33	41.0	0.00000
518	HALAWE		2.5362878	0.22188984	6.74137	1.9735115	4.0392227	12.52	9.5	0.00000
519	SYLVANIA	S	2.7887371	0.18566515	11.01868	2.2709658	4.6570549	10.18	23.0	17.96800
520	FRANZISKA	U	3.0103128	0.10735590	10.97071	2.6871378	5.2229686	11.54	11.5	0.00000
521	BRIXIA	C	2.7436447	0.27901164	10.56845	1.9781359	4.5445600	9.23	68.0	30.00000
522	HELGA	CX	3.6260209	0.07782478	4.45128	3.3438265	6.9047103	10.01	46.0	0.00000
523	ADA		2.9660034	0.18072397	4.32057	2.4299755	5.1080775	10.69	22.0	0.00000
524	FIDELIO	C	2.6350520	0.12889545	8.21083	2.2954056	4.2774382	10.87	32.0	0.00000
525	ADELAIDE	S	2.2448955	0.10197356	5.99697	2.0159755	3.3635211	13.73	4.6	0.00000
526	JENA	C	3.1194425	0.13829792	2.17303	2.6880300	5.5095415	10.83	24.0	0.00000
527	EURYANTHE		2.7259588	0.15182512	9.66659	2.3120899	4.5006886	11.38	16.0	0.00000
528	REZIA		3.3952823	0.00865435	12.69792	3.3658981	6.2562461	10.16	46.2	0.00000
529	PREZIOSA	SU	3.0178075	0.08905275	11.00730	2.7490633	5.2424860	11.08	11.0	0.00000
530	TURANDOT	CMEU	3.1866269	0.21995772	8.56942	2.4857037	5.6884875	10.21	26.1	0.00000
531	ZERLINA	X	2.7870176	0.19692621	33.96791	2.2381806	4.6527481	12.22	10.9	0.00000
532	HERCULINA	S	2.7725463	0.17612579	16.35407	2.2842293	4.6165566	6.76	110.0	9.40800
533	SARA	S	2.9802332	0.04307878	6.54801	2.8518484	5.1448812	10.79	17.0	0.00000
534	NASSOVIA	S	2.8846574	0.05667752	3.27812	2.7211623	4.8993831	10.84	15.0	9.00000
535	MONTAGUE	C	2.5692358	0.02551292	6.77907	2.5036871	4.1181860	10.48	39.0	0.00000
536	MERAPI	C	3.4953454	0.09029124	19.44207	3.1797464	6.5348430	8.99	76.0	0.00000
537	PAULY	C	3.0580170	0.23957975	9.92404	2.3253779	5.3476110	10.00	68.0	0.00000
538	FRIEDERIKE		3.1764879	0.14966303	6.48465	2.7010851	5.6613607	10.49	24.2	0.00000
539	PAMINA		2.7389910	0.21331765	6.79556	2.1547160	4.5330024	11.02	18.9	0.00000
540	ROSAMUNDE	S	2.2186567	0.09015285	5.57795	2.0186403	3.3047283	12.05	9.8	0.00000
541	DEBORAH		2.8137505	0.05247873	5.99644	2.6660883	4.7198520	11.30	16.6	0.00000
542	SUSANNA	SU	2.9080498	0.13895197	12.06153	2.5039706	4.9590988	10.42	20.0	0.00000
543	CHARLOTTE		3.0652840	0.14495748	8.47989	2.6209483	5.3666849	10.65	22.4	0.00000
544	JETTA		2.5917413	0.15223086	8.33601	2.1971984	4.1724148	11.26	16.9	0.00000
545	MESSALINA	C	3.1887019	0.18280689	11.13418	2.6057851	5.6940446	9.77	52.0	7.20000
546	HERODIAS	CU	2.5978713	0.11389632	14.88232	2.3019834	4.1872263	10.71	35.0	0.00000
547	PRAXEDIS	MEU	2.7700102	0.23876841	16.92855	2.1086192	4.6102242	10.49	23.7	0.00000
548	KRESSIDA	S	2.2813485	0.18570580	3.87349	1.8576888	3.4457791	12.47	7.9	0.00000
549	JESSONDA		2.6831918	0.25854161	3.94965	1.9894751	4.3951893	11.73	13.6	0.00000
550	SENTA	S	2.5914404	0.21779346	10.08403	2.0270417	4.1716881	10.50	26.0	0.00000
551	ORTRUD	C	2.9666543	0.11935570	0.40433	2.6125672	5.1097589	10.48	38.0	0.00000
552	SIGELINDE		3.1597586	0.06838768	7.63901	2.9436698	5.6166949	10.84	20.6	0.00000
553	KUNDRY		2.2303760	0.11045219	5.39018	1.9840261	3.3309424	13.45	4.9	0.00000
554	PERAGA	C	2.3747013	0.15287846	2.93685	2.0116606	3.6594300	9.81	50.0	13.63000
555	NORMA		3.1623530	0.17485537	2.64693	2.6093986	5.6236143	11.60	14.5	0.00000
556	PHYLLIS		2.4656487	0.10054265	5.22131	2.2177458	3.8716562	10.47	24.3	4.28300
557	VIOLETTA		2.4406991	0.10370524	2.49704	2.1875858	3.8130403	13.27	6.7	0.00000
558	CARMEN	M	2.9062269	0.04383210	8.37099	2.7788408	4.9544363	9.95	32.0	10.00000
559	NANON	C	2.7109709	0.06473082	9.30134	2.5354874	4.4636202	10.39	40.0	0.00000
560	DELILA	CMEU	2.7498846	0.16038312	8.46736	2.3088493	4.5600719	11.67	14.0	0.00000
561	INGWELDE		3.1758685	0.12814859	1.50390	2.7688854	5.6597042	12.21	17.6	0.00000
562	SALOME	S	3.0161519	0.10272216	11.12028	2.7063262	5.2381725	10.98	15.0	0.00000
563	SULEIKA	S	2.7115059	0.23640186	10.23030	2.0705009	4.4649420	9.62	27.0	5.69300
564	DUDU	C	2.7504184	0.27209672	18.04001	2.0020387	4.5613999	11.43	25.0	0.00000
565	MARBACHIA	SU	2.4453199	0.12827601	10.95735	2.1316440	3.8238738	11.96	9.8	0.00000
566	STEREOSKOPIA	C	3.3915279	0.09228421	4.91863	3.0785434	6.2458725	9.07	73.0	12.00000
567	ELEUTHERIA	CU	3.1300354	0.09675579	9.26380	2.8271863	5.5376291	10.23	42.0	0.00000
568	CHERUSKIA		2.8840425	0.16682410	18.37668	2.4029148	4.8978162	10.48	24.3	0.00000
569	MISA	C	2.6570401	0.18155485	1.28684	2.1746416	4.3310895	11.11	29.0	0.00000
570	KYTHERA	C	3.4307187	0.10229817	1.75971	3.0797625	6.3544455	9.74	55.0	0.00000

No.	Name	Type	A (A.U.)	E	I (deg.)	q (A.U.)	Period (years)	Mag.	Rad. (km)	Rot. (hours)
571	DULCINEA	S	2.4104002	0.24107020	5.23895	1.8293246	3.7422581	12.81	6.8	0.00000
572	REBEKKA	C	2.3999839	0.15666957	10.55481	2.0239794	3.7180266	11.97	19.0	0.00000
573	RECHA		3.0167072	0.10911149	9.81821	2.6875496	5.2396193	10.47	19.3	0.00000
574	REGINHILD	S	2.2525568	0.23987049	5.68516	1.7122350	3.3807547	13.74	4.4	0.00000
575	RENATE		2.5547357	0.12617536	15.05321	2.2323909	4.0833721	12.27	8.4	0.00000
576	EMANUELA		2.9906077	0.19087580	10.19142	2.4197731	5.1717696	11.00	19.1	0.00000
577	RHEA		3.1243670	0.13809837	5.27271	2.6928971	5.5225935	10.91	19.9	0.00000
578	HAPPELIA		2.7503555	0.19232996	6.15468	2.2213798	4.5612435	10.43	24.8	0.00000
579	SIDONIA	S	3.0142293	0.07420613	11.02195	2.7905550	5.2331648	8.90	39.9	13.00000
580	SELENE		3.2301183	0.08310271	3.64610	2.9616866	5.8053398	10.83	20.7	0.00000
581	TAUNTONIA		3.2139633	0.01929455	21.83342	3.1519513	5.7618418	10.79	21.0	0.00000
582	OLYMPIA	S	2.6119857	0.22459784	29.94350	2.0253394	4.2213969	10.28	22.0	0.00000
583	KLOTILDE	C	3.1747351	0.15891635	8.24278	2.6702178	5.6566753	10.19	44.0	0.00000
584	SEMIRAMIS	S	2.3738172	0.23410295	10.70957	1.8180997	3.6573870	9.70	26.0	5.06800
585	BILKIS	C	2.4287453	0.13295180	7.56922	2.1058390	3.7850614	11.33	26.0	0.00000
586	THEKLA	CU	3.0398321	0.06960228	1.61938	2.8282528	5.2999821	10.14	44.0	0.00000
587	HYPSIPYLE		2.3370628	0.16575135	24.99014	1.9496915	3.5727742	13.48	6.1	0.00000
588	ACHILLES	D	5.1748791	0.14934061	10.33331	4.4020596	11.7720022	9.64	58.0	0.00000
589	CROATIA	C	3.1380222	0.05365356	10.80352	2.9696560	5.5588379	10.14	45.0	0.00000
590	TOMYRIS		3.0034122	0.07792640	11.17008	2.7693672	5.2050200	10.98	15.2	0.00000
591	IRMGARD	MU	2.6787269	0.20796891	12.53676	2.1216350	4.3842230	11.63	14.0	0.00000
592	BATHSEBA		3.0185275	0.13729842	10.18348	2.6040885	5.2443628	10.70	21.9	0.00000
593	TITANIA	C	2.6987095	0.21658431	16.93937	2.1142113	4.4333720	10.29	41.0	9.89800
594	MIREILLE		2.6282148	0.35334736	32.57342	1.6995420	4.2608013	13.71	5.5	0.00000
595	POLYXENA		3.2067740	0.05414967	17.92390	3.0331283	5.7425203	9.16	44.6	0.00000
596	SCHEILA	U	2.9284925	0.16423853	14.67673	2.4475212	5.0114818	9.89	66.0	0.00000
597	BANDUSIA		2.6719511	0.14558351	12.80723	2.2829590	4.3675985	10.39	25.3	0.00000
598	OCTAVIA	C	2.7619839	0.25016445	12.20560	2.0710337	4.5902009	10.48	39.0	0.00000
599	LUISA	S	2.7693937	0.29617560	16.64068	1.9491668	4.6086855	9.38	33.0	9.57000
600	MUSA		2.6588118	0.05488661	10.21536	2.5128787	4.3354220	11.35	16.3	0.00000
601	NERTHUS	CEU	3.1287558	0.11252850	16.12231	2.7766817	5.5342340	10.59	21.6	0.00000
602	MARIANNA	C	3.0885470	0.24351512	15.22320	2.3364391	5.4278936	9.32	68.0	0.00000
603	TIMANDRA		2.5406129	0.17077684	7.99156	2.1067350	4.0495596	13.46	6.2	0.00000
604	TEKMESSA		3.1420689	0.20251147	4.42791	2.5057640	5.5695944	10.22	27.4	0.00000
605	JUVISIA		3.0041084	0.13513283	19.61347	2.5981548	5.2068305	10.52	23.8	0.00000
606	BRANGANE		2.5874102	0.21693178	8.64155	2.0261188	4.1619601	11.49	15.2	0.00000
607	JENNY		2.8521650	0.07719720	10.10928	2.6319859	4.8168368	10.78	21.1	0.00000
608	ADOLFINE		3.0235431	0.11916324	9.38658	2.6632481	5.2574391	11.74	10.7	0.00000
609	FULVIA		3.0867958	0.02817575	4.18103	2.9998231	5.4232779	11.18	17.6	0.00000
610	VALESKA		3.0759270	0.26488015	12.73933	2.2611749	5.3946590	13.20	6.9	0.00000
611	VALERIA	S	2.9773228	0.12386171	13.44962	2.6085465	5.1373472	10.28	21.0	0.00000
612	VERONIKA		3.1421268	0.27176511	20.76426	2.2882063	5.5697484	12.29	10.5	0.00000
613	GINEVRA	PF	2.9206507	0.05720711	7.68491	2.7535689	4.9913664	10.64	59.0	0.00000
614	PIA		2.6936910	0.10827871	7.01783	2.4020214	4.4210114	12.00	12.1	0.00000
615	ROSWITHA	C	2.6309068	0.10936490	2.77406	2.3431780	4.2673497	11.33	26.0	0.00000
616	ELLY	S	2.5525808	0.05888037	14.99217	2.4022839	4.0782070	11.79	11.0	0.00000
617	PATROCLUS	P	5.2321711	0.13940611	22.04409	4.5027742	11.9680367	9.17	82.0	0.00000
618	ELFRIEDE	C	3.1841545	0.08679062	17.01267	2.9077997	5.6818690	9.24	68.0	0.00000
619	TRIBERGA	S	2.5205896	0.07585030	13.75756	2.3294022	4.0017800	10.95	16.0	0.00000
620	DRAKONIA	CMEU	2.4352126	0.13423772	7.71567	2.1083152	3.8001902	12.26	10.1	0.00000
621	WERDANDA	CEU	3.1121747	0.15342049	2.32200	2.6347034	5.4902983	11.39	24.5	12.00000
622	ESTHER		2.4154196	0.24291383	8.64749	1.8286808	3.7539535	11.67	14.0	47.50000
623	CHIMAERA	C	2.4609079	0.11512066	14.16581	2.1776066	3.8604956	12.03	19.0	0.00000
624	HEKTOR	D	5.1722636	0.02304326	18.23892	5.0530777	11.7630777	8.53	116.0	6.92400
625	XENIA		2.6463339	0.22599582	12.06107	2.0482736	4.3049388	11.47	15.4	0.00000
626	NOTBURGA	C	2.5747478	0.24329774	25.35479	1.9483174	4.1314454	9.96	48.0	0.00000
627	CHARIS	CMEU	2.9010761	0.05881175	6.46930	2.7304585	4.9412708	11.02	17.9	0.00000
628	CHRISTINE	U	2.5820477	0.04241231	11.50681	2.4725370	4.1490278	10.27	27.0	20.00000
629	BERNARDINA		3.1192720	0.17146724	9.33337	2.5844190	5.5090899	10.80	20.9	0.00000
630	EUPHEMIA		2.6229258	0.11390240	13.86771	2.3241682	4.2479463	12.38	10.1	0.00000
631	PHILIPPINA	S	2.7915676	0.08433289	18.94299	2.5561466	4.6641469	9.81	27.0	5.92000
632	PYRRHA		2.6651120	0.19000208	2.22805	2.1587350	4.3508406	13.37	6.4	4.60000
633	ZELIMA	U	3.0154221	0.08854695	10.92362	2.7484157	5.2362719	10.77	22.0	10.00000
634	UTE		3.0503967	0.18123025	12.26254	2.4975727	5.3276353	11.01	19.0	0.00000
635	VUNDTIA	C	3.1423123	0.09116534	11.02806	2.8558424	5.5702415	9.98	47.0	0.00000
636	ERIKA		2.9119182	0.17103997	7.91544	2.4138637	4.9689970	10.74	21.5	14.63000
637	CHRYSOTHEMI		3.1552043	0.14413990	0.28533	2.7004135	5.6045566	11.86	21.1	0.00000
638	MOIRA		2.7334836	0.16029622	7.70676	2.2953165	4.5193367	10.85	20.5	0.00000
639	LATONA	S	3.0174649	0.11154247	8.56895	2.6808894	5.2415934	9.28	48.0	0.00000
640	BRAMBILLA	U	3.1684170	0.07385676	13.35153	2.9344079	5.6397972	10.30	26.0	0.00000
641	AGNES		2.2195003	0.12905537	1.71277	1.9330620	3.3066089	13.58	4.6	8.90000
642	CLARA	S	3.1983719	0.13148221	8.10003	2.7778428	5.7199655	11.19	14.0	0.00000
643	SCHEHEREZADE	EMP	3.3536649	0.08258137	13.65636	3.0767145	6.1415715	10.72	34.6	0.00000
644	COSIMA	SU	2.5985689	0.15677015	1.04291	2.1911907	4.1889129	11.71	11.0	0.00000
645	AGRIPPINA	S	3.1891167	0.17067054	7.04099	2.6448283	5.6951561	11.18	14.0	0.00000
646	KASTALIA		2.3247578	0.21284357	6.90943	1.8299479	3.5445943	14.20	4.4	0.00000
647	ADELGUNDE	CMEU	2.4411924	0.19338678	7.31045	1.9690981	3.8141961	12.45	9.5	0.00000
648	PIPPA	C	3.2131312	0.20335908	9.86718	2.5597119	5.7596049	10.43	39.0	0.00000
649	JOSEFA		2.5505438	0.27282184	12.63392	1.8546997	4.0733261	14.10	4.6	0.00000
650	AMALASUNTHA	CFPD	2.4590642	0.18385699	2.55319	2.0069480	3.8561580	13.47	10.0	0.00000
651	ANTIKLEIA	S	3.0239277	0.09644480	10.76572	2.7322855	5.2584424	11.10	14.7	0.00000
652	JUBILATRIX		2.5553327	0.12483232	15.78350	2.2363446	4.0848036	12.53	7.5	0.00000

No.	Name	Type	A (A.U.)	E	I (deg.)	q (A.U.)	Period (years)	Mag.	Rad. (km)	Rot. (hours)
653	BERENIKE	S	3.0115843	0.04710892	11.29519	2.8697116	5.2262783	10.27	21.0	0.00000
654	ZELINDA	C	2.2974269	0.23085691	18.13681	1.7670500	3.4822710	9.47	60.4	31.90000
655	BRISEIS		2.9902976	0.08488475	6.50527	2.7364669	5.1709652	10.70	21.9	0.00000
656	BEAGLE	U	3.1501572	0.13345782	0.51376	2.7297440	5.5911140	10.85	24.0	0.00000
657	GUNLOD		2.6131680	0.11430048	10.23139	2.3144817	4.2242637	11.70	13.8	0.00000
658	ASTERIA	SU	2.8541505	0.06388815	1.50852	2.6718042	4.8218679	11.61	12.0	0.00000
659	NESTOR	C	5.2430668	0.11188998	4.51197	4.6564202	12.0054407	9.69	55.0	0.00000
660	CRESCENTIA	S	2.5341742	0.10265840	15.25578	2.2740200	4.0341744	10.20	21.0	7.92000
661	CLOELIA	S	3.0155442	0.04097931	9.25190	2.8919692	5.2365894	10.71	22.0	0.00000
662	NEWTONIA		2.5559175	0.21304941	4.12204	2.0113809	4.0862060	11.54	14.9	0.00000
663	GERLINDE	C	3.0594940	0.15471362	17.85387	2.5861485	5.3514857	10.10	45.0	0.00000
664	JUDITH	C	3.1851861	0.23046222	8.54726	2.4511211	5.6846304	10.96	30.0	0.00000
665	SABINE		3.1618814	0.15366629	14.65198	2.6760068	5.6223564	9.62	36.1	0.00000
666	DESDEMONA		2.5923195	0.24049918	7.60789	1.9688687	4.1738110	11.87	12.8	0.00000
667	DENISE		3.1854591	0.19092602	25.37715	2.5772722	5.6853609	10.19	27.7	0.00000
668	DORA		2.7984717	0.23108792	6.83707	2.1517787	4.6814604	13.22	6.9	0.00000
669	KYPRIA	SU	3.0116773	0.07656055	10.79916	2.7811015	5.2265201	11.33	13.0	0.00000
670	OTTEGEBE		2.8047700	0.19192909	7.53420	2.2664530	4.6972742	11.10	18.2	0.00000
671	CARNEGIA		3.0947738	0.05996490	8.02592	2.9091959	5.4443164	11.43	15.7	0.00000
672	ASTARTE		2.5558007	0.13639244	11.14324	2.2072086	4.0859256	12.48	9.7	0.00000
673	EDDA	C	2.8153939	0.01139153	2.87532	2.7833223	4.7239876	11.25	27.0	0.00000
674	RACHELE	S	2.9202666	0.19624332	13.54353	2.3471837	4.9903817	8.57	60.0	50.00000
675	LUDMILLA		2.7693796	0.20317225	9.76779	2.2067184	4.6086497	9.20	43.8	7.71700
676	MELITTA	C	3.0578094	0.13155347	12.88268	2.6555438	5.3470664	10.52	37.0	0.00000
677	AALTJE		2.9550982	0.04961758	8.48623	2.8084733	5.0799317	10.81	20.8	0.00000
678	FREDEGUNDIS		2.5748315	0.21603833	6.08459	2.0185692	4.1316471	10.60	23.0	0.00000
679	PAX	U	2.5854557	0.31215829	24.39807	1.7783842	4.1572452	10.19	36.0	7.62500
680	GENOVEVA	CMEU	3.1432018	0.28950384	17.81953	2.2332330	5.5726070	10.27	25.4	0.00000
681	GORGO	SM	3.1112001	0.09038568	12.58286	2.8299923	5.4877200	11.78	12.0	0.00000
682	HAGAR		2.6531196	0.17216580	11.49718	2.1963432	4.3215070	13.47	6.1	0.00000
683	LANZIA		3.1177676	0.04957977	18.49237	2.9631894	5.5051050	9.53	37.6	4.32200
684	HILDBURG		2.4310782	0.03787159	5.50704	2.3390093	3.7905166	11.79	13.3	0.00000
685	HERMIA		2.2359722	0.19680035	3.64726	1.7959321	3.3434868	12.83	6.5	0.00000
686	GERSUIND	S	2.5896792	0.26576740	15.72148	1.9014269	4.1674361	10.78	17.0	0.00000
687	TINETTE		2.7225697	0.27048552	14.91283	1.9861540	4.4922976	12.95	7.8	0.00000
688	MELANIE		2.6978095	0.13906100	10.23438	2.3226495	4.4311547	11.57	14.7	0.00000
689	ZITA	C	2.3153284	0.22962894	5.74058	1.7836620	3.5230508	13.12	4.9	0.00000
690	WRATISLAVIA	CEU	3.1411564	0.18457565	11.26788	2.5613756	5.5671682	8.63	53.3	0.00000
691	LEHIGH	C	3.0090222	0.12613231	13.01184	2.6294875	5.2196107	10.28	43.0	0.00000
692	HIPPODAMIA	S	3.3690774	0.18276726	26.07114	2.7533202	6.1839571	10.13	23.0	0.00000
693	ZERBINETTA	CU	2.9426556	0.03076185	14.19828	2.8521340	5.0478811	10.42	41.0	0.00000
694	EKARD	C	2.6725535	0.32163566	15.84865	1.8129650	4.3690763	10.03	50.0	5.92500
695	BELLA	S	2.5382450	0.16003157	13.88803	2.1320455	4.0438986	9.78	22.0	14.22200
696	LEONORA	CMEU	3.1726599	0.24948654	13.01650	2.3811238	5.6511297	10.24	27.1	0.00000
697	GALILEA	C	2.8812783	0.15746532	15.14048	2.4275768	4.8907766	10.62	36.0	0.00000
698	ERNESTINA		2.8677022	0.10843186	11.52757	2.5567520	4.8562503	11.80	13.2	0.00000
699	HELA		2.6128900	0.40996405	15.28911	1.5416991	4.2235894	11.33	16.4	3.65600
700	AURAVICTRIX		2.2295506	0.10443448	6.78760	1.9967086	3.3290935	12.36	8.1	5.80000
701	ORIOLA	CU	3.0133102	0.03861923	7.12253	2.8969386	5.2307720	10.24	42.0	0.00000
702	ALAUDA	C	3.1909845	0.02945803	20.57447	3.0969844	5.7001600	8.17	108.0	8.36000
703	NOEMI		2.1747017	0.13767529	2.45555	1.8752991	3.2070045	13.59	4.6	0.00000
704	INTERAMNIA	F	3.0614548	0.14816850	17.30240	2.6078434	5.3566313	7.01	169.0	8.72700
705	ERMINIA	C	2.9231110	0.04974492	25.01111	2.7777009	4.9976740	9.36	64.0	0.00000
706	HIRUNDO		2.7269616	0.19561806	14.46423	2.1935186	4.5031719	12.00	12.1	0.00000
707	STEINA		2.1801367	0.10829412	4.27104	1.9440407	3.2190342	13.36	6.4	0.00000
708	RAPHAELA	SU	2.6692433	0.08661159	3.49428	2.4380560	4.3609614	11.76	11.0	0.00000
709	FRINGILLA	C	2.9165282	0.11131743	16.26389	2.5918677	4.9808021	10.01	48.0	52.40000
710	GERTRUD		3.1263888	0.13584664	1.75305	2.7016792	5.5279546	12.15	18.5	0.00000
711	MARMULLA		2.2368362	0.19591565	6.08415	1.7986050	3.3454249	12.54	7.4	0.00000
712	BOLIVIANA	C	2.5747950	0.18881567	12.80786	2.0886333	4.1315594	9.39	64.0	11.75000
713	LUSCINIA	C	3.3894436	0.16921924	10.16902	2.8158846	6.2401152	9.81	64.0	0.00000
714	ULULA	S	2.5345156	0.05612503	14.28550	2.3922658	4.0349898	10.21	27.0	0.00000
715	TRANSVAALIA		2.7691905	0.08605293	13.81241	2.5308936	4.6081781	11.13	18.0	0.00000
716	BERKELEY	U	2.8111899	0.08638743	8.50771	2.5683384	4.7134104	11.93	13.0	0.00000
717	WISIBADA	CMEU	3.1658392	0.24268728	1.68067	2.3975303	5.6329165	12.01	11.5	0.00000
718	ERIDA		3.0658431	0.19485509	6.91688	2.4684479	5.3681531	10.73	21.6	0.00000
719	ALBERT	LOST	2.5839057	0.54037249	10.82100	1.1876341	4.1535072	16.77	1.3	0.00000
720	BOHLINIA	S	2.8864839	0.01721449	2.36193	2.8367944	4.9040365	10.89	17.0	0.00000
721	TABORA	P	3.5449533	0.12038209	8.34115	3.1182044	6.6744547	10.31	47.0	0.00000
722	FRIEDA		2.1721365	0.14440192	5.64022	1.8584758	3.2013316	13.00	7.6	0.00000
723	HAMMONIA		2.9935760	0.05804455	4.98237	2.8198152	5.1794715	11.06	18.6	0.00000
724	HAPAG	LOST	2.4506347	0.25397739	11.76800	1.8282290	3.8363471	14.74	3.4	0.00000
725	AMANDA		2.5731933	0.21748720	3.78757	2.0135567	4.1277051	12.27	10.6	0.00000
726	JOELLA		2.5670164	0.28150481	15.41179	1.8443888	4.1128507	12.07	11.7	13.04000
727	NIPPONIA	U	2.5677793	0.10563209	15.02734	2.2965393	4.1146846	10.67	22.0	0.00000
728	LEONISIS		2.2537143	0.08829483	4.25824	2.0547230	3.3833609	13.78	4.2	0.00000
729	WATSONIA	U	2.7590430	0.09663949	18.04190	2.4924104	4.5828714	10.35	26.0	0.00000
730	ATHANASIA		2.2436178	0.17655414	4.23268	1.8474977	3.3606501	14.66	3.5	0.00000
731	SORGA	C	2.9880493	0.14400017	10.69472	2.5577695	5.1651344	10.49	38.0	0.00000
732	TJILAKI		2.4573803	0.04060751	10.97710	2.3575921	3.8521974	11.78	13.3	0.00000
733	MOCIA	C	3.3843641	0.07836318	20.36990	3.1191545	6.2260933	10.05	46.0	0.00000
734	BENDA		3.1542871	0.08389541	5.80917	2.8896568	5.6021123	11.06	18.6	0.00000
735	MARGHANNA	C	2.7308869	0.32180387	16.81466	1.8520770	4.5128989	10.52	37.0	0.00000

No.	Name	Type	A (A.U.)	E	I (deg.)	q (A.U.)	Period (years)	Mag.	Rad. (km)	Rot. (hours)
736	HARVARD	S	2.2020044	0.16521205	4.37804	1.8382066	3.2675879	12.82	6.8	6.70000
737	AREQUIPA	S	2.5926328	0.24128301	12.37747	1.9670745	4.1745677	9.94	23.0	14.13000
738	ALAGASTA	U	3.0320618	0.06151300	3.53292	2.8455505	5.2796731	11.16	17.0	0.00000
739	MANDEVILLE	EMP	2.7360704	0.14385682	20.69548	2.3424678	4.5257535	9.60	34.8	15.90000
740	CANTABIA	C	3.0484543	0.11262162	10.84349	2.7051325	5.3225474	9.93	49.0	0.00000
741	BOTOLPHIA	CMEU	2.7188032	0.06967105	8.41851	2.5293813	4.4829783	11.47	15.4	0.00000
742	EDISONA	S	3.0154235	0.11171557	11.22069	2.6785538	5.2362757	10.68	23.0	0.00000
743	EUGENISIS		2.7923295	0.05683232	4.82726	2.6336348	4.6660566	11.28	16.8	0.00000
744	AGUNTINA	U	3.1693830	0.12317048	7.71329	2.7790086	5.6423769	11.14	17.0	0.00000
745	MAURITIA		3.2704720	0.04933790	13.42779	3.1091139	5.9144678	10.99	19.2	0.00000
746	MARLU	CFPD	3.1086657	0.23753987	17.50657	2.3702335	5.4810157	10.71	42.0	0.00000
747	WINCHESTER	C	2.9978089	0.34259853	18.16811	1.9707640	5.1904607	8.65	102.0	9.40000
748	SIMEISA	P	3.9330266	0.18596570	2.26022	3.2016187	7.7999229	9.86	54.0	0.00000
749	MALZOVIA	S	2.2432945	0.17375223	5.38896	1.8535171	3.3599238	12.95	6.3	0.00000
750	OSKAR	F	2.4422896	0.13330334	3.94696	2.1167243	3.8167677	12.98	17.0	0.00000
751	FAINA	C	2.5512264	0.15341857	15.60394	2.1598208	4.0749612	9.62	56.0	0.00000
752	SULAMITIS		2.4626379	0.07420385	5.94862	2.2799008	3.8645670	11.24	17.1	0.00000
753	TIFLIS	SU	2.3285961	0.22113359	10.09993	1.8136653	3.5533767	11.56	12.0	9.84000
754	MALABAR	C	2.9885657	0.05215974	24.52336	2.8326828	5.1664734	10.16	44.0	0.00000
755	QUINTILLA	M	3.1663475	0.15515219	3.23459	2.6750817	5.6342731	10.84	19.0	0.00000
756	LILLIANA		3.1963055	0.14331761	20.36858	2.7382185	5.7144232	11.16	17.7	0.00000
757	PORTLANDIA		2.3734097	0.10785200	8.17732	2.1174326	3.6564453	11.32	16.1	0.00000
758	MANCUNIA	U	3.1914165	0.14798124	5.60264	2.7191467	5.7013173	9.17	43.2	0.00000
759	VINIFERA		2.6153092	0.20869526	19.97253	2.0695066	4.2294569	11.62	14.4	0.00000
760	MASSINGA	SX	3.1994390	0.20977506	12.59132	2.5282764	5.7228289	9.37	34.0	0.00000
761	BRENDELIA		2.8638930	0.06581983	2.16415	2.6753922	4.8465781	11.83	10.3	0.00000
762	PULCOVA	C	3.1722188	0.09169292	13.02914	2.8813486	5.6499510	9.27	66.0	0.00000
763	CUPIDO		2.2403448	0.16601537	4.08365	1.8684131	3.3532991	13.42	5.0	0.00000
764	GEDANIA	C	3.1791685	0.11717081	10.07779	2.8066626	5.6685281	10.48	38.0	0.00000
765	MATTIACA		2.5479119	0.27994433	5.57520	1.8346385	4.0670228	14.01	4.8	0.00000
766	MOGUNTIA	C	3.0243244	0.09446144	10.09742	2.7386422	5.2594767	11.03	20.0	0.00000
767	BONDIA	CFPD	3.1161845	0.18536273	2.42157	2.5385599	5.5009122	11.01	24.0	0.00000
768	STRUVEANA	EMP	3.1486709	0.20116986	16.27166	2.5152533	5.5871577	11.26	16.9	0.00000
769	TATJANA		3.1923447	0.16419601	7.37370	2.6681745	5.7038050	10.14	28.4	0.00000
770	BALI	U	2.2211487	0.15047193	4.38830	1.8869282	3.3102932	12.09	9.5	0.00000
771	LIBERA		2.6551037	0.24497005	14.91884	2.0046830	4.3263559	11.68	14.0	0.00000
772	TANETE	C	2.9992859	0.09611483	28.81318	2.7110102	5.1942973	9.25	66.0	12.00000
773	IRMINTRAUD	PD	2.8564527	0.08151770	16.68574	2.6236012	4.8277030	10.48	55.0	0.00000
774	ARMOR		3.0528603	0.16169620	5.53047	2.5592244	5.3340907	9.90	31.7	0.00000
775	LUMIERE		3.0106473	0.06880668	9.28430	2.8034947	5.2238393	11.54	12.0	0.00000
776	BERBERICIA	C	2.9301739	0.16553181	18.22393	2.4451368	5.0157986	8.59	92.0	7.67200
777	GUTEMBERGA		3.2386174	0.09687500	13.06603	2.9248762	5.8282671	11.10	18.2	0.00000
778	THEOBALDA	F	3.2111423	0.24244559	13.41232	2.4326150	5.7542582	10.46	33.0	11.65900
779	NINA		2.6652033	0.22703187	14.57306	2.0601172	4.3510647	9.65	35.6	11.16300
780	ARMENIA		3.1215317	0.08437886	19.03230	2.8581405	5.5150776	10.07	29.3	0.00000
781	KARTVELIA	CEU	3.2293003	0.09623993	19.17620	2.9185126	5.8031344	10.51	23.9	0.00000
782	MONTEFIORE	U	2.1806443	0.03844053	5.26100	2.0968192	3.2201581	12.55	7.5	0.00000
783	NORA	C	2.3420601	0.22994828	9.32889	1.8035074	3.5842397	12.03	19.0	0.00000
784	PICKERINGIA	CFPD	3.1181464	0.22365257	12.38339	2.4207649	5.5061083	10.22	44.0	0.00000
785	ZWETANA	F	2.5738783	0.20605876	12.69683	2.0435081	4.1293530	10.29	24.0	0.00000
786	BREDICHINA		3.1650119	0.17201488	14.57521	2.6205828	5.6307077	9.82	32.9	0.00000
787	MOSKVA		2.5397358	0.12722246	14.85270	2.2166243	4.0474625	11.32	16.5	0.00000
788	HOHENSTEINA		3.1269050	0.12583223	14.39316	2.7334394	5.5293236	9.46	38.8	0.00000
789	LENA		2.6848938	0.14683995	10.78074	2.2906442	4.3993721	12.18	11.1	0.00000
790	PRETORIA	P	3.4055572	0.15426186	20.55912	2.8802097	6.2846670	9.04	88.0	10.37000
791	ANI		3.1162198	0.19938037	16.39663	2.4949067	5.5010061	10.34	25.9	0.00000
792	METCALFIA		2.6205881	0.13343707	8.62104	2.2709045	4.2422686	11.07	18.5	9.17000
793	ARIZONA		2.7970574	0.12209559	15.79964	2.4555490	4.6779118	11.35	16.4	0.00000
794	IRENAEA		3.1485529	0.28473777	5.40492	2.2520409	5.5868430	12.28	10.6	0.00000
795	FINI		2.7486749	0.10044193	19.06786	2.4725926	4.5570631	10.93	19.7	0.00000
796	SARITA	CMEU	2.6319799	0.32339343	19.07737	1.7808149	4.2699604	10.16	26.9	7.75000
797	MONTANA	S	2.5361559	0.05695408	4.47920	2.3917117	4.0389080	11.55	12.0	5.00000
798	RUTH	SM	3.0172460	0.03575472	9.20609	2.9093654	5.2410235	10.48	27.0	0.00000
799	GUDULA		2.5414455	0.02348287	5.27641	2.4817648	4.0515499	11.41	15.8	0.00000
800	KRESSMANNIA	SX	2.1931398	0.20130804	4.26177	1.7516432	3.2478764	12.81	7.0	0.00000
801	HELWERTHIA	EMP	2.6062160	0.07567859	14.09432	2.4089811	4.2074175	12.13	11.4	0.00000
802	EPYAXA		2.1961920	0.07940822	5.20336	2.0217962	3.2546587	13.51	4.8	0.00000
803	PICKA		3.2034552	0.05942128	8.65251	3.0131018	5.7336082	10.75	21.4	0.00000
804	HISPANIA	C	2.8403866	0.13822506	15.34333	2.4477739	4.7870302	8.69	70.0	10.00000
805	HORMUTHIA	CF	3.2006550	0.18528414	15.71031	2.6076243	5.7260914	10.60	36.0	0.00000
806	GYLDENIA		3.2247846	0.07190042	14.13376	2.9929211	5.7909665	11.09	18.3	0.00000
807	CERASKIA	S	3.0184472	0.05955467	11.31035	2.8386846	5.2441530	11.65	11.0	0.00000
808	MERXIA		2.7438867	0.12926468	4.72232	2.3891990	4.5451608	10.68	22.1	0.00000
809	LUNDIA		2.2835581	0.19220926	7.14891	1.8446372	3.4507866	13.12	7.2	0.00000
810	ATOSSA		2.1790278	0.18071605	2.61226	1.7852426	3.2165787	14.11	3.6	0.00000
811	NAUHEIMA	S	2.8966701	0.07291391	3.13696	2.6854627	4.9300189	11.89	10.4	0.00000
812	ADELE		2.6601615	0.16492656	13.29617	2.2214303	4.3387241	12.41	10.0	0.00000
813	BAUMEIA		2.2234347	0.02526491	6.29657	2.1672599	3.3154051	13.30	6.6	0.00000
814	TAURIS	C	3.1534386	0.30823418	21.81573	2.1814408	5.5998521	9.90	49.0	35.80000
815	COPPELIA		2.6584983	0.07515118	13.89171	2.4587090	4.3346558	11.93	12.4	0.00000
816	JULIANA		3.0115569	0.10375969	14.34058	2.6990786	5.2262068	11.32	16.5	0.00000
817	ANNIKA		2.5892906	0.17785224	11.36307	2.1287794	4.1664982	11.88	12.7	0.00000
818	KAPTEYNIA		3.1675162	0.09617840	15.64871	2.8628695	5.6373925	10.42	24.9	0.00000

No.	Name	Type	A (A.U.)	E	I (deg.)	q (A.U.)	Period (years)	Mag.	Rad. (km)	Rot. (hours)
819	BARNARDIANA		2.1972208	0.14223403	4.89320	1.8847013	3.2569458	13.13	7.2	0.00000
820	ADRIANA		3.1248136	0.06232338	5.95152	2.9300647	5.5237770	11.46	15.5	0.00000
821	FANNY	C	2.7785301	0.20447889	5.36992	2.2103794	4.6315107	12.47	15.0	0.00000
822	LALAGE		2.2554405	0.15432061	0.71630	1.9073794	3.3872483	13.22	6.8	0.00000
823	SISIGAMBIS		2.2208128	0.09097424	3.64460	2.0187762	3.3095424	12.51	7.5	0.00000
824	ANASTASIA	S	2.7951231	0.13253513	8.10166	2.4246712	4.6730604	11.54	12.0	0.00000
825	TANINA	SX	2.2256980	0.07432214	3.40181	2.0602794	3.3204682	12.86	6.8	0.00000
826	HENRIKA		2.7123806	0.20493305	7.11327	2.1565242	4.4671030	12.72	8.6	0.00000
827	WOLFIANA		2.2748346	0.15637422	3.42258	1.9191092	3.4310319	13.64	5.7	0.00000
828	LINDEMANNIA		3.1863942	0.04599283	1.14336	3.0398428	5.6878643	11.28	15.8	0.00000
829	ACADEMIA		2.5791683	0.09985877	8.31395	2.3216157	4.1420903	11.91	12.6	0.00000
830	PETROPOLITANA	S	3.2096455	0.06435597	3.82366	3.0030856	5.7502351	10.35	25.0	0.00000
831	STATEIRA		2.2121711	0.14545122	4.83991	1.8904082	3.2902436	13.48	4.8	0.00000
832	KARIN		2.8655877	0.07838477	1.00233	2.6409693	4.8508801	12.23	8.6	0.00000
833	MONICA		3.0127864	0.11664332	9.77771	2.6613650	5.2294078	12.23	8.6	0.00000
834	BURNHAMIA	U	3.1593258	0.22168764	3.95084	2.4589424	5.6155410	10.41	24.0	0.00000
835	OLIVIA		3.2099514	0.10496563	3.69871	2.8730166	5.7510567	12.20	11.0	0.00000
836	JOLE		2.1903002	0.17616034	4.84265	1.8044562	3.2415707	14.27	3.3	0.00000
837	SCHWARZSCHILDA		2.2986424	0.04124012	6.73455	2.2038462	3.4850347	12.95	7.8	0.00000
838	SERAPHINA	EMP	2.8970470	0.13362557	10.39824	2.5099275	4.9309812	11.24	17.1	0.00000
839	VALBORG	S	2.6156507	0.15118875	12.56536	2.2201936	4.2302847	11.79	11.0	0.00000
840	ZENOBIA		3.1362443	0.08955022	9.93834	2.8553929	5.5541148	10.50	24.0	0.00000
841	ARABELLA		2.2547827	0.07046023	3.79305	2.0959101	3.3857665	13.81	5.2	3.39000
842	KERSTIN		3.2202618	0.13651343	14.62630	2.7806528	5.7787876	11.71	13.8	0.00000
843	NICOLAIA		2.2791340	0.20910661	7.99223	1.8025521	3.4407635	14.22	4.3	0.00000
844	LEONTINA		3.1970763	0.08200456	8.84067	2.9349015	5.7164907	10.77	21.2	0.00000
845	NAEMA		2.9418471	0.07014337	12.61444	2.7354960	5.0458016	11.11	18.2	0.00000
846	LIPPERTA	CEU	3.1204469	0.18713409	0.26339	2.5365050	5.5122032	11.31	24.9	0.00000
847	AGNIA	S	2.7823181	0.09262176	2.47836	2.5246148	4.6409850	11.42	13.0	0.00000
848	INNA		3.1020243	0.17413624	1.04495	2.5618494	5.4634600	11.84	21.3	0.00000
849	ARA	P	3.1651874	0.17962660	19.55896	2.5966356	5.6311765	9.00	76.0	4.11900
850	ALTONA		2.9953394	0.13138512	15.50752	2.6017964	5.1840487	10.61	22.9	0.00000
851	ZEISSIA		2.2279139	0.09084394	2.39377	2.0255215	3.3254285	12.90	6.3	0.00000
852	WLADILENA		2.3622551	0.27439910	23.01906	1.7140543	3.6306984	11.12	18.1	4.61100
853	NANSENIA	C	2.3126414	0.10526653	9.21900	2.0691977	3.5169194	12.76	14.0	0.00000
854	FROSTIA		2.3684492	0.17507349	6.08431	1.9537964	3.6449878	13.45	6.2	0.00000
855	NEWCOMBIA		2.3620424	0.18018214	10.89913	1.9364445	3.6302083	13.02	7.5	0.00000
856	BACKLUNDA		2.4370353	0.11666067	14.31896	2.1527290	3.8044574	11.91	12.6	0.00000
857	GLASENAPPIA	U	2.1902127	0.08827905	5.30192	1.9968629	3.2413764	12.22	10.0	0.00000
858	EL DJEZAIR	S	2.8091557	0.10435646	8.89645	2.5160022	4.7082958	11.22	14.0	0.00000
859	BOUZAREAH		3.2238133	0.10782201	13.60367	2.8762152	5.7883506	10.98	19.3	0.00000
860	URSINA	EMP	2.7945390	0.10838671	13.33153	2.4916482	4.6715956	10.75	21.4	0.00000
861	AIDA		3.1435876	0.09417468	8.04880	2.8475411	5.5736327	10.76	21.3	0.00000
862	FRANZIA		2.8032012	0.08063946	13.89035	2.5771525	4.6933331	11.23	17.2	0.00000
863	BENKOELA	A	3.1981549	0.04626128	25.44455	3.0502043	5.7193842	10.32	17.0	0.00000
864	AASE	SU	2.2080142	0.19041437	5.44847	1.7875766	3.2809739	14.20	3.6	0.00000
865	ZUBAIDA		2.4162500	0.19727933	13.29861	1.9395738	3.7558894	13.14	7.1	0.00000
866	FATME		3.1255145	0.06886571	8.65226	2.9102736	5.5256357	10.35	25.8	0.00000
867	KOVACIA		3.0649364	0.12940288	5.97914	2.6683247	5.3657713	12.09	11.6	0.00000
868	LOVA	C	2.7035642	0.14660597	5.83495	2.3072054	4.4453406	11.21	27.0	0.00000
869	MELLENA		2.6905475	0.21848059	7.83451	2.1027150	4.4132748	13.22	6.9	0.00000
870	MANTO	SM	2.3219559	0.26451752	6.19566	1.7077579	3.5381887	12.91	7.8	0.00000
871	AMNERIS		2.2222152	0.11992694	4.25480	1.9557116	3.3126774	13.69	4.4	0.00000
872	HOLDA	CMEU	2.7347579	0.07866821	7.37187	2.5196195	4.5224972	10.90	19.4	0.00000
873	MECHTHILD	EMP	2.6300869	0.14606155	5.27352	2.2459323	4.2653546	12.05	11.8	10.60000
874	ROTRAUT		3.1662378	0.06461386	11.12627	2.9616549	5.6339798	10.85	20.5	0.00000
875	NYMPHE		2.5542965	0.14720942	14.61459	2.1782801	4.0823193	12.80	6.6	0.00000
876	SCOTT	SU	3.0104835	0.11477849	11.36057	2.6649446	5.2234130	11.96	9.8	0.00000
877	WALKURE	CMEU	2.4865823	0.15919280	4.26337	2.0907362	3.9210668	11.66	13.3	0.00000
878	MILDRED	LOST	2.3633323	0.23106872	2.02100	1.8172401	3.6331823	16.50	1.5	0.00000
879	RICARDA		2.5310948	0.15472372	13.69558	2.1394744	4.0268235	12.67	7.0	0.00000
880	HERBA	CF	3.0019112	0.32141116	15.08424	2.0370634	5.2011185	12.96	12.0	0.00000
881	ATHENE		2.6099591	0.20927796	14.24081	2.0637522	4.2164850	13.50	6.0	0.00000
882	SWETLANA		3.1492796	0.24900132	6.08460	2.3651047	5.5887771	11.65	14.2	0.00000
883	MATTERANIA	S	2.2386217	0.19860449	4.71644	1.7940215	3.3494315	13.57	4.8	0.00000
884	PRIAMUS	D	5.1558661	0.12214905	8.91793	4.5260820	11.7071838	9.74	47.0	0.00000
885	ULRIKE		3.0993829	0.18124987	3.29727	2.5376201	5.4564834	11.91	12.6	0.00000
886	WASHINGTONIA		3.1764786	0.26917613	16.63627	2.3214464	5.6613359	9.60	36.4	0.00000
887	ALINDA	S	2.4931006	0.55806977	9.24838	1.1017766	3.9364953	14.96	2.0	73.97000
888	PARYSATIS	S	2.7086024	0.19521458	13.83084	2.1798437	4.4577723	10.62	19.0	0.00000
889	ERYNIA		2.4465353	0.20523578	8.07895	1.9444188	3.8267250	12.60	9.1	0.00000
890	WALTRAUT	U	3.0238020	0.05678944	10.85705	2.8520820	5.2581143	11.77	16.0	0.00000
891	GUNHILD		2.8591397	0.02987095	13.53470	2.7737343	4.8345165	11.22	17.3	0.00000
892	SEELIGERIA		3.2350681	0.08831969	21.32629	2.9493477	5.8186884	10.52	23.8	0.00000
893	LEOPOLDINA	CMEU	3.0567448	0.14214785	16.99173	2.6222351	5.3442750	10.40	23.7	0.00000
894	ERDA		3.1083064	0.12459742	12.72641	2.7210195	5.4800653	10.88	20.2	0.00000
895	HELIO	CMEU	3.2001293	0.14784171	26.06315	2.7270167	5.7246809	9.74	34.1	0.00000
896	SPHINX		2.2856884	0.16310780	8.18231	1.9128747	3.4556162	12.81	8.3	0.00000
897	LYSISTRATA	S	2.5423858	0.09138827	14.29576	2.3100417	4.0537987	12.14	23.0	0.00000
898	HILDEGARD		2.7327230	0.36860874	10.14940	1.7254175	4.5174508	13.37	6.4	0.00000
899	JOKASTE	CMEU	2.9074018	0.20057075	12.44152	2.3242619	4.9574409	11.09	17.3	0.00000
900	ROSALINDE		2.4740512	0.16008458	11.55946	2.0779939	3.8914640	12.87	8.1	0.00000
901	BRUNSIA		2.2246580	0.22078992	3.44446	1.7334759	3.3181412	12.75	6.7	0.00000

No.	Name	Type	A (A.U.)	E	I (deg.)	q (A.U.)	Period (years)	Mag.	Rad. (km)	Rot. (hours)
902	PROBITAS		2.4460115	0.17826597	6.36302	2.0099709	3.8254960	13.54	5.9	0.00000
903	NEALLEY		3.2434309	0.02481581	11.69293	3.1629424	5.8412657	10.75	21.4	0.00000
904	ROCKEFELLIA		2.9926603	0.08574614	15.16000	2.7360511	5.1770949	11.34	16.3	0.00000
905	UNIVERSITAS		2.2159805	0.15304789	5.32342	1.8768294	3.2987461	12.71	6.9	10.00000
906	REPSOLDA		2.8937230	0.08427227	11.79144	2.6498623	4.9224963	10.63	22.6	0.00000
907	RHODA	C	2.7982430	0.16219771	19.60040	2.3443744	4.6808872	10.54	37.0	0.00000
908	BUDA		2.4728730	0.14707085	13.37753	2.1091855	3.8886845	11.98	12.2	0.00000
909	ULLA	C	3.5379527	0.10352695	18.84070	3.1716790	6.6546926	9.64	56.0	0.00000
910	ANNELIESE		2.9296007	0.15238398	9.23525	2.4831765	5.0143270	11.17	17.7	0.00000
911	AGAMEMNON	D	5.2037778	0.06770517	21.84061	4.8514552	11.8707495	8.92	77.0	8.00000
912	MARITIMA		3.1176136	0.19110858	18.31957	2.5218110	5.5046973	10.23	27.2	0.00000
913	OTILA		2.1974192	0.17030054	5.80653	1.8231975	3.2573872	13.69	4.4	0.00000
914	PALISANA	C	2.4565034	0.21195902	25.29398	1.9358253	3.8501360	10.52	38.0	0.00000
915	COSETTE		2.2278783	0.13940696	5.54929	1.9172966	3.3253489	13.02	6.0	0.00000
916	AMERICA		2.3651886	0.23620071	11.12205	1.8065293	3.6374633	12.60	9.1	0.00000
917	LYKA		2.3814437	0.20062342	5.13241	1.9036703	3.6750264	12.55	9.4	0.00000
918	ITHA		2.8645773	0.18868834	12.08981	2.3240650	4.8483148	11.90	12.6	0.00000
919	ILSEBILL		2.7708364	0.08630617	8.15305	2.5316961	4.6122870	12.40	10.0	0.00000
920	ROGERIA	U	2.6211202	0.10501140	11.60600	2.3458726	4.2435608	12.25	11.0	0.00000
921	JOVITA		3.1905866	0.15900648	16.39977	2.6832626	5.6990933	11.10	18.2	0.00000
922	SCHLUTIA		2.6876202	0.19411565	7.28864	2.1659110	4.4060745	13.01	7.6	0.00000
923	HERLUGA		2.6135111	0.19707581	14.51449	2.0984511	4.2250953	12.65	8.9	0.00000
924	TONI	C	2.9390383	0.15359686	8.99590	2.4876113	5.0385766	10.37	40.0	0.00000
925	ALPHONSINA	S	2.6997552	0.08060923	21.10905	2.4821298	4.4359493	9.58	30.0	7.92000
926	IMHILDE		2.9822073	0.18225093	16.32875	2.4386971	5.1499944	11.61	14.4	0.00000
927	RATISBONA	C	3.2430294	0.08180031	14.46299	2.9777486	5.8401814	10.52	37.0	0.00000
928	HILDRUN		3.1342921	0.15004697	17.65856	2.6640010	5.5489292	10.72	21.7	0.00000
929	ALGUNDE		2.2383258	0.11383598	3.91357	1.9835238	3.3487670	13.45	4.9	0.00000
930	WESTPHALIA		2.4308689	0.14487228	15.29158	2.0787034	3.7900269	12.49	9.6	0.00000
931	WHITTEMORA	EMP	3.1874931	0.21920881	11.31227	2.4887664	5.6908073	9.99	30.4	0.00000
932	HOOVERIA	C	2.4199886	0.09031070	8.12223	2.2014377	3.7646101	10.94	30.0	0.00000
933	SUSI		2.3677881	0.16590115	5.54299	1.9749693	3.6434619	13.20	6.9	0.00000
934	THURINGIA		2.7498910	0.21602809	14.09375	2.1558373	4.5600877	11.43	15.7	0.00000
935	CLIVIA		2.2188494	0.14664479	4.02426	1.8934667	3.3051543	14.33	3.3	0.00000
936	KUNIGUNDE	SM	3.1369336	0.17093424	2.37105	2.6007242	5.5559454	11.02	18.0	0.00000
937	BETHGEA	S	2.2314911	0.21760213	3.69556	1.7459137	3.3334405	12.83	6.8	0.00000
938	CHLOSINDE		3.1649103	0.18150017	2.66099	2.5904787	5.6304369	12.34	16.9	0.00000
939	ISBERGA	SX	2.2473269	0.17672759	2.58674	1.8501621	3.3689871	13.33	5.5	0.00000
940	KORDULA	U	3.3651104	0.16801351	6.25440	2.7997265	6.1730385	10.13	26.0	0.00000
941	MURRAY	C	2.7850263	0.19343577	6.62595	2.2463026	4.6477628	12.52	15.0	0.00000
942	ROMILDA		3.1507907	0.17691340	10.58013	2.5933738	5.5928006	11.48	15.3	0.00000
943	BEGONIA	U	3.1394930	0.19827721	12.02512	2.5170031	5.5627465	10.81	21.0	0.00000
944	HIDALGO	MEU	5.8495355	0.65592128	42.39553	2.0127006	14.1475773	11.90	14.3	10.00000
945	BARCELONA	S	2.6378934	0.16175675	32.81249	2.2111962	4.2843590	11.10	15.0	7.38000
946	POESIA	C	3.1234431	0.13303488	1.43410	2.7079163	5.5201440	11.44	20.0	0.00000
947	MONTEROSA		2.7508800	0.25052908	6.71344	2.0617046	4.5625482	11.26	16.9	0.00000
948	JUCUNDA		3.0345895	0.16001266	8.66335	2.5490167	5.2862773	12.49	9.6	0.00000
949	HEL		2.9976258	0.19805545	10.69504	2.4039297	5.1899853	10.66	22.3	0.00000
950	AHRENSA		2.3708148	0.16060551	23.50025	1.9900489	3.6504502	12.41	10.0	0.00000
951	GASPRA		2.2096343	0.17333360	4.09920	1.8266304	3.2845857	12.90	6.3	20.00000
952	CAIA		2.9885759	0.24735554	10.03824	2.2493353	5.1665001	10.20	27.6	7.51000
953	PAINLEVA		2.7901082	0.18543261	8.67513	2.2727311	4.6604896	11.47	15.4	0.00000
954	LI	U	3.1465106	0.15362369	1.14853	2.6631322	5.5814085	10.84	31.1	0.00000
955	ALSTEDE		2.5899520	0.29472604	10.65218	1.8266258	4.1680951	12.63	9.0	0.00000
956	ELISA		2.2986071	0.20417868	5.95912	1.8292806	3.4849546	13.40	6.3	0.00000
957	CAMELIA		2.9223371	0.08103765	14.75849	2.6855178	4.9956894	10.92	19.8	0.00000
958	ASPLINDA	ERU	3.9702842	0.18920974	5.64485	3.2190678	7.9110184	11.03	16.0	0.00000
959	ARNE		3.2030106	0.19895139	4.46762	2.5657673	5.7324142	11.80	13.2	0.00000
960	BIRGIT		2.2482140	0.16511269	3.02398	1.8770055	3.3709824	14.19	4.4	0.00000
961	GUNNIE		2.6926973	0.09091608	10.99855	2.4478877	4.4185653	12.46	9.7	0.00000
962	ASLOG	S	2.9043527	0.09871152	2.60269	2.6176596	4.9496446	12.68	7.4	0.00000
963	IDUBERGA	S	2.2478292	0.13703068	7.98382	1.9398075	3.3701167	13.66	4.6	0.00000
964	SUBAMARA		3.0538380	0.11022694	9.04143	2.7172229	5.3366537	12.04	11.8	0.00000
965	ANGELICA		3.1440973	0.28756997	21.49392	2.2399492	5.5749884	11.29	16.7	0.00000
966	MUSCHI	S	2.7189016	0.12980455	14.40819	2.3659759	4.4832220	11.06	15.0	0.00000
967	HELIONAPE		2.2257755	0.16822770	5.41542	1.8513384	3.3206418	13.61	4.5	0.00000
968	PETUNIA	SU	2.8662288	0.13859788	11.55878	2.4689755	4.8525081	11.15	15.0	0.00000
969	LEOCADIA	F	2.4620638	0.20364907	2.30227	1.9606668	3.8632157	13.45	9.4	0.00000
970	PRIMULA		2.5617173	0.26800528	5.04525	1.8751636	4.1001220	13.43	6.2	0.00000
971	ALSATIA		2.6393690	0.16239145	13.78821	2.2107582	4.2879548	10.99	19.2	0.00000
972	COHNIA		3.0669532	0.22608601	8.36966	2.3735580	5.3710690	10.63	22.6	0.00000
973	ARALIA		3.2209845	0.10176753	15.76445	2.8931928	5.7807331	10.90	20.0	0.00000
974	LIOBA		2.5340178	0.11068455	5.46854	2.2535412	4.0338011	11.48	15.3	0.00000
975	PERSEVERANTIA	S	2.8375206	0.03302790	2.56120	2.7438033	4.7797871	11.52	12.0	0.00000
976	BENJAMINA	C	3.1953435	0.10807835	7.55223	2.8499959	5.7118435	10.32	42.0	0.00000
977	PHILIPPA	C	3.1160078	0.02538806	15.18303	3.0368986	5.5004449	10.73	34.0	0.00000
978	AIDAMINA	CMEU	3.1913383	0.24025416	21.66836	2.4246058	5.7011075	10.65	21.1	0.00000
979	ILSEWA		3.1532700	0.14361060	10.03536	2.7004271	5.5994034	10.91	19.9	0.00000
980	ANACOSTIA	S	2.7410095	0.20024680	15.92209	2.1921310	4.5380139	9.00	40.0	20.10000
981	MARTINA	U	3.1067922	0.19284967	2.07157	2.5076482	5.4760609	11.63	21.6	8.00000
982	FRANKLINA		3.0812240	0.22340003	13.50997	2.3928783	5.4086003	11.35	16.3	0.00000
983	GUNILA	CMEU	3.1544504	0.09997118	14.85795	2.8390963	5.6025476	10.60	22.3	0.00000
984	GRETIA		2.8051364	0.19545569	9.10286	2.2568567	4.6981945	10.55	23.5	5.76000

No.	Name	Type	A (A.U.)	E	I (deg.)	q (A.U.)	Period (years)	Mag.	Rad. (km)	Rot. (hours)
985	ROSINA		2.3000946	0.27682835	4.06181	1.6633632	3.4883380	14.11	4.6	0.00000
986	AMELIA		3.1510501	0.18556459	14.83494	2.5663269	5.5934911	10.54	23.6	0.00000
987	WALLIA		3.1620183	0.21714757	8.86412	2.4753938	5.6227212	10.52	23.8	10.00000
988	APPELLA		3.1661003	0.21734211	1.57655	2.4779732	5.6336126	12.39	10.1	0.00000
989	SCHWASSMANNIA		2.6581924	0.25245377	14.68752	1.9871217	4.3339071	13.32	6.6	0.00000
990	YERKES		2.6683583	0.21656707	8.78696	2.0904796	4.3587928	12.69	8.8	0.00000
991	MCDONALDA	CU	3.1508398	0.14556801	2.09540	2.6921782	5.5929308	12.05	18.0	0.00000
992	SWASEY		3.0216005	0.09349831	10.86570	2.7390859	5.2523732	11.97	12.2	0.00000
993	MOULTONA		2.8618555	0.04639023	1.77132	2.7290933	4.8414063	13.80	7.0	0.00000
994	OTTHILD		2.5296581	0.11500450	15.35338	2.2387359	4.0233955	11.41	15.8	0.00000
995	STERNBERGA		2.6154900	0.16671100	13.07350	2.1794591	4.2298951	11.17	17.7	0.00000
996	HILARITAS	U	3.0955124	0.13027310	0.66110	2.6922505	5.4462657	11.73	13.0	0.00000
997	PRISKA		2.6684492	0.18340902	10.50374	2.1790316	4.3590150	12.99	7.6	0.00000
998	BODEA		3.1319742	0.19869827	15.56007	2.5096564	5.5427756	12.11	11.5	0.00000
999	ZACHIA		2.6116941	0.21723063	9.76545	2.0443542	4.2206898	11.87	12.8	0.00000
1000	PIAZZIA		3.2021675	0.24115632	20.56446	2.4299445	5.7301507	11.29	16.7	0.00000
1001	GAUSSIA	CP	3.1982894	0.14644285	9.34602	2.7299230	5.7197447	10.60	40.0	0.00000
1002	OLBERSIA		2.7896037	0.15057033	10.74618	2.3695722	4.6592259	12.04	11.8	0.00000
1003	LILOFEE		3.1499014	0.14736496	1.83901	2.6857164	5.5904331	11.23	28.2	0.00000
1004	BELOPOLSKYA	U	3.4022818	0.08942430	2.97923	3.0980351	6.2756019	10.98	19.0	0.00000
1005	ARAGO		3.1659007	0.10986222	19.16950	2.8180878	5.6330800	10.81	20.8	0.00000
1006	LAGRANGEA		3.1704135	0.34056616	11.02522	2.0906780	5.6451287	12.72	8.6	0.00000
1007	PAWLOWIA		2.7069552	0.10932503	2.54410	2.4110172	4.4537063	12.60	9.1	0.00000
1008	LA PAZ		3.0928967	0.07707920	8.93868	2.8544989	5.4393644	11.63	14.3	0.00000
1009	SIRENE		2.6259568	0.45462853	15.76804	1.4321219	4.2553115	15.30	1.3	0.00000
1010	MARLENE		2.9315095	0.10303697	3.90970	2.6294556	5.0192280	11.74	13.6	0.00000
1011	LAODAMIA	S	2.3960941	0.34635055	5.47815	1.5662055	3.7089911	13.79	3.5	0.00000
1012	SAREMA	CEU	2.4810421	0.13254294	4.04294	2.1521976	3.9079697	13.25	6.3	0.00000
1013	TOMBECKA	C	2.6835868	0.20980503	11.88908	2.1205568	4.3961601	10.98	31.0	0.00000
1014	SEMPHYRA		2.8031912	0.19938350	2.27355	2.2442813	4.6933084	12.85	8.1	0.00000
1015	CHRISTA	C	3.2085099	0.07977101	9.46392	2.9525638	5.7471838	9.99	48.0	0.00000
1016	ANITRA		2.2192574	0.12722391	6.04026	1.9369147	3.3060658	13.28	6.7	0.00000
1017	JACQUELINE		2.6058338	0.07573762	7.94714	2.4084740	4.2064919	12.17	11.1	0.00000
1018	ARNOLDA		2.5383673	0.24979201	7.67679	1.9043033	4.0441909	12.33	10.4	0.00000
1019	STRACKEA	S	1.9116361	0.07152391	26.97907	1.7749084	2.6430650	13.90	4.2	0.00000
1020	ARCADIA		2.7913823	0.03931615	4.04800	2.6816359	4.6636825	12.11	11.5	0.00000
1021	FLAMMARIO		2.7384310	0.28404269	15.83535	1.9605997	4.5316119	9.97	28.9	0.00000
1022	OLYMPIADA		2.8082092	0.17319962	21.05885	2.3218284	4.7059159	11.24	17.1	0.00000
1023	THOMANA	U	3.1596069	0.11431680	10.06143	2.7984107	5.6162906	10.77	21.0	0.00000
1024	HALE		2.8659673	0.22730720	16.07483	2.2145121	4.8518438	11.66	14.1	0.00000
1025	RIEMA	E	1.9791516	0.03973124	26.87268	1.9005173	2.7843163	14.04	2.3	0.00000
1026	INGRID	LOST	2.2504308	0.17995262	5.39000	1.8454599	3.3759694	14.52	3.0	0.00000
1027	AESCULAPIA	CFPD	3.1527708	0.13021165	1.25413	2.7422433	5.5980735	11.84	20.0	0.00000
1028	LYDINA	C	3.3842561	0.13113672	9.47924	2.9404559	6.2257953	10.36	40.0	0.00000
1029	LA PLATA	U	2.8893106	0.02720058	2.43319	2.8107195	4.9112420	11.90	9.9	14.00000
1030	VITJA		3.1401143	0.10702413	14.69871	2.8040462	5.5643978	11.52	15.0	0.00000
1031	ARCTICA	C	3.0463943	0.06542651	17.60638	2.8470793	5.3171530	10.51	37.0	0.00000
1032	PAFURI		3.1402016	0.13493791	9.46061	2.7164693	5.5646296	10.98	19.3	0.00000
1033	SIMONA		3.0012455	0.11888708	10.65742	2.6444361	5.1993885	12.14	8.9	0.00000
1034	MOZARTIA		2.2920418	0.26387420	3.97936	1.6872309	3.4700344	13.66	5.6	0.00000
1035	AMATA		3.1323643	0.20591611	18.10667	2.4873600	5.5438104	11.69	13.9	0.00000
1036	GANYMED	S	2.6631849	0.53732580	26.44871	1.2321872	4.3461227	10.43	20.0	0.00000
1037	DAVIDWEILLA		2.2550933	0.19100516	5.89892	1.8243588	3.3864665	15.05	2.3	0.00000
1038	TUCKIA	CFPD	3.9616981	0.23296532	9.21965	3.0387597	7.8853698	11.59	25.0	0.00000
1039	SONNEBERGA		2.6790209	0.06115598	4.54252	2.5151827	4.3849449	12.30	10.5	0.00000
1040	KLUMPKEA		3.1165233	0.19000100	16.67711	2.5243807	5.5018096	11.09	18.3	0.00000
1041	ASTA		3.0690453	0.15061378	13.93003	2.6068046	5.3765655	11.10	18.2	0.00000
1042	AMAZONE		3.2256081	0.10719660	20.69361	2.8798339	5.7931852	11.23	17.2	0.00000
1043	BEATE	S	3.0925903	0.04449748	8.93312	2.9549778	5.4385557	11.00	16.0	0.00000
1044	TEUTONIA		2.5769143	0.13938399	4.26543	2.2177336	4.1366615	11.95	12.3	0.00000
1045	MICHELA		2.3587356	0.15759797	0.25941	1.9870036	3.6225872	14.02	4.8	0.00000
1046	EDWIN		2.9831212	0.05970436	7.91373	2.8050158	5.1523614	11.49	15.2	0.00000
1047	GEISHA		2.2414234	0.19229993	5.66230	1.8103977	3.3557208	13.19	5.5	0.00000
1048	FEODOSIA	C	2.7345483	0.17943013	15.80233	2.2438879	4.5219779	10.64	36.0	0.00000
1049	GOTHO		3.0948830	0.12990230	15.15291	2.6928506	5.4446044	11.75	13.5	0.00000
1050	META		2.6255894	0.17620231	12.50275	2.1629543	4.2544184	13.83	5.2	0.00000
1051	MEROPE		3.2238650	0.09370710	23.31235	2.9217660	5.7884898	10.94	19.6	0.00000
1052	BELGICA	S	2.2361999	0.14302829	4.69535	1.9163600	3.3439972	13.14	5.9	0.00000
1053	VIGDIS		2.6153584	0.09677992	8.33160	2.3622441	4.2295756	13.63	5.7	0.00000
1054	FORSYTIA		2.9222260	0.13433394	10.86012	2.5296719	4.9954047	11.57	14.7	0.00000
1055	TYNKA	S	2.1983185	0.20803896	5.27613	1.7409827	3.2593873	12.65	7.2	0.00000
1056	AZALEA		2.2295539	0.17771569	5.42969	1.8333273	3.3291011	12.66	7.0	0.00000
1057	WANDA		2.8963571	0.24407846	3.51942	2.1894186	4.9292192	12.19	11.0	0.00000
1058	GRUBBA	S	2.1967089	0.18726058	3.68985	1.7853519	3.2558079	12.82	6.8	0.00000
1059	MUSSORGSKIA		2.6425796	0.18559894	10.10643	2.1521194	4.2957807	12.27	10.6	0.00000
1060	MAGNOLIA		2.2371807	0.20212682	5.92370	1.7849866	3.3461978	14.27	3.3	0.00000
1061	PAEONIA	C	3.1214745	0.21918970	2.49281	2.4372795	5.5149260	11.98	19.0	0.00000
1062	LJUBA	CFPD	3.0034938	0.07238023	5.60920	2.7861001	5.2052317	11.20	26.0	0.00000
1063	AQUILEGIA		2.3141694	0.04042808	5.97527	2.2206120	3.5204058	12.50	9.6	0.00000
1064	AETHUSA		2.5462692	0.17081469	9.44767	2.1113291	4.0630903	12.25	10.7	0.00000
1065	AMUNDSENIA		2.3608720	0.29783636	8.37305	1.6577187	3.6275105	13.78	5.3	0.00000
1066	LOBELIA		2.4027176	0.20805675	4.81367	1.9028161	3.7243812	14.10	4.6	0.00000
1067	LUNARIA		2.8723145	0.19045147	10.54975	2.3252780	4.8679709	11.85	12.9	0.00000

No.	Name	Type	A (A.U.)	E	I (deg.)	q (A.U.)	Period (years)	Mag.	Rad. (km)	Rot. (hours)
1068	NOFRETETE		2.9083812	0.09493447	5.50305	2.6322756	4.9599466	12.33	10.4	0.00000
1069	PLANCKIA		3.1327467	0.10071424	13.55375	2.8172345	5.5448256	10.72	21.7	0.00000
1070	TUNICA		3.2273901	0.08769971	17.03635	2.9443488	5.7979865	11.98	12.2	0.00000
1071	BRITA		2.7991424	0.11178711	5.38150	2.4862342	4.6831436	11.18	17.6	0.00000
1072	MALVA		3.1635926	0.24193969	8.03109	2.3981938	5.6269212	11.68	14.0	10.00000
1073	GELLIVARA		3.1630845	0.20576753	1.61574	2.5122242	5.6255655	12.52	15.6	0.00000
1074	BELJAWSKYA	CFPD	3.1661184	0.16151583	0.81848	2.6547401	5.6336613	11.25	40.0	0.00000
1075	HELINA	U	3.0163705	0.10705603	11.53867	2.6934497	5.2387424	11.22	15.0	0.00000
1076	VIOLA	F	2.4757743	0.14230032	3.31801	2.1234708	3.8955305	12.78	11.0	7.33600
1077	CAMPANULA		2.3921354	0.19777630	5.39810	1.9190278	3.6998036	13.91	5.0	0.00000
1078	MENTHA		2.2690787	0.13902552	7.37390	1.9536188	3.4180179	12.60	9.1	0.00000
1079	MIMOSA	SU	2.8741367	0.04733850	1.18025	2.7380793	4.8726039	12.32	8.2	0.00000
1080	ORCHIS		2.4195590	0.25624704	4.59446	1.7995543	3.7636077	13.48	6.1	0.00000
1081	RESEDA		3.0849519	0.16051626	4.23149	2.5897670	5.4184189	12.11	11.5	0.00000
1082	PIROLA	C	3.1334636	0.16888805	1.84557	2.6042590	5.5467296	11.39	25.0	0.00000
1083	SALVIA		2.3282320	0.18316980	5.13852	1.9017702	3.5525434	13.91	5.0	0.00000
1084	TAMARIWA		2.6873803	0.13207448	3.89237	2.3324461	4.4054847	11.74	13.6	0.00000
1085	AMARYLLIS		3.1806831	0.05998844	6.64394	2.9898789	5.6725798	10.77	21.2	0.00000
1086	NATA		3.1644080	0.05191257	8.35504	3.0001354	5.6290965	10.68	22.1	0.00000
1087	ARABIS	C	3.0155213	0.08771023	10.05551	2.7510293	5.2365303	10.81	28.0	0.00000
1088	MITAKA	SR	2.2015195	0.19594041	7.65121	1.7701529	3.2665086	12.61	7.7	0.00000
1089	TAMA		2.2144747	0.12764733	3.72898	1.9318029	3.2953844	12.83	6.5	0.00000
1090	SUMIDA		2.3585134	0.22221725	21.52757	1.8344111	3.6220758	13.97	4.9	0.00000
1091	SPIRAEA		3.4237189	0.04836880	1.15961	3.2581177	6.3350077	11.85	21.2	0.00000
1092	LILIUM		2.9032042	0.08264364	5.39344	2.6632729	4.9467092	11.45	15.5	0.00000
1093	FREDA	C	3.1468277	0.25994995	25.17005	2.3288100	5.5822520	9.77	52.0	0.00000
1094	SIBERIA		2.5445714	0.13480496	13.99880	2.2015505	4.0590272	13.28	6.7	0.00000
1095	TULIPA		3.0268483	0.01882212	10.02471	2.9698765	5.2660618	11.28	13.3	0.00000
1096	REUNERTA		2.6028042	0.18993132	9.47942	2.1084502	4.1991582	11.28	16.8	0.00000
1097	VICIA		2.6405685	0.29520565	1.52370	1.8610578	4.2908783	12.78	8.4	0.00000
1098	HAKONE		2.6883633	0.11633739	13.39572	2.3756061	4.4079022	11.68	14.0	0.00000
1099	FIGNERIA		3.1841972	0.27540508	11.70604	2.3072531	5.6819830	11.67	14.0	0.00000
1100	ARNICA		2.8977015	0.06610208	1.03190	2.7061572	4.9326515	12.42	7.9	0.00000
1101	CLEMATIS		3.2423007	0.07348877	21.31942	3.0040281	5.8382130	11.98	12.2	0.00000
1102	PEPITA	C	3.0653386	0.11981080	15.80047	2.6980779	5.3668280	10.73	34.0	0.00000
1103	SEQUOIA	E	1.9337559	0.09435729	17.89696	1.7512919	2.6890724	13.50	2.9	0.00000
1104	SYRINGA		2.6264584	0.34812152	6.42129	1.7121317	4.2565308	13.50	6.0	0.00000
1105	FRAGAR1A		3.0104327	0.10224012	10.97334	2.7026458	5.2232809	10.78	16.7	0.00000
1106	CYDONIA		2.5969644	0.12259856	13.10365	2.2785802	4.1850338	12.87	8.1	0.00000
1107	LICTORIA		3.1816440	0.12726182	7.06826	2.7767422	5.6751504	10.32	26.1	0.00000
1108	DEMETER		2.4259338	0.25903574	24.92428	1.7975303	3.7784915	12.34	10.3	0.00000
1109	TATA		3.2095611	0.11908419	4.17407	2.8273530	5.7500081	10.92	18.1	0.00000
1110	JAROSLAWA		2.2188144	0.24109495	5.85289	1.6838694	3.3050759	13.19	7.0	0.00000
1111	REINMUTHIA		2.9943955	0.10351457	3.89005	2.6844318	5.1815982	11.70	12.8	0.00000
1112	POLONIA	C	3.0222814	0.10880878	8.99781	2.6934307	5.2541490	11.14	29.2	0.00000
1113	KATJA		3.1103675	0.14339805	13.28465	2.6643469	5.4855170	10.60	23.0	0.00000
1114	LORRAINE		3.0879803	0.08343832	10.74478	2.8303244	5.4263997	10.79	21.0	0.00000
1115	SABAUDA		3.0985422	0.17236292	15.34959	2.5644684	5.4542637	10.39	25.3	0.00000
1116	CATRIONA		2.9267557	0.22748893	16.55777	2.2609510	5.0070243	10.73	21.6	0.00000
1117	REGINITA		2.2477021	0.19739537	4.33907	1.8040162	3.3698313	13.18	5.5	0.00000
1118	HANSKYA		3.2049179	0.06019961	14.02549	3.0119832	5.7375350	10.87	20.3	0.00000
1119	EUBOEA		2.6142402	0.15232301	7.85110	2.2160313	4.2268634	12.59	9.2	0.00000
1120	CANNONIA		2.2165399	0.15515170	4.05165	1.8726399	3.2999952	13.33	5.2	0.00000
1121	NATASCHA		2.5464034	0.15846364	6.14723	2.1428912	4.0634117	12.52	9.5	0.00000
1122	NEITH		2.6040714	0.26009864	4.74111	1.9267559	4.2022252	12.74	8.6	20.00000
1123	SHAPLEYA		2.2254970	0.15627991	6.41731	1.8776965	3.3200185	12.65	8.9	0.00000
1124	STROOBANTIA	MP	2.9276025	0.03071515	7.79376	2.8376808	5.0091977	11.88	12.7	0.00000
1125	CHINA		3.1364925	0.20954323	3.03974	2.4792619	5.5547738	14.20	4.4	0.00000
1126	OTERO		2.2724719	0.14645563	6.50154	1.9396555	3.4256878	13.70	5.5	0.00000
1127	MIMI	C	2.5968554	0.26155797	14.77799	1.9176271	4.1847701	11.89	40.0	0.00000
1128	ASTRID		2.7871032	0.04334175	1.01940	2.6663053	4.6529622	11.89	12.7	0.00000
1129	NEUJMINA	U	3.0190356	0.08806052	8.60584	2.7531776	5.2456865	11.11	16.0	0.00000
1130	SKULD		2.2285447	0.19762106	2.16647	1.7881374	3.3268411	13.15	5.6	0.00000
1131	PORZIA		2.2282574	0.28583851	3.23335	1.5913357	3.3261976	15.32	2.6	0.00000
1132	HOLLANDIA		2.6868150	0.27375922	7.23487	1.9512748	4.4040947	11.82	13.1	0.00000
1133	LUGDUNA		2.1862681	0.18716645	5.37372	1.7770720	3.2326233	13.22	5.4	0.00000
1134	KEPLER		2.6809361	0.46732295	15.18873	1.4280730	4.3896475	15.29	2.6	0.00000
1135	COLCHIS		2.6651568	0.11564576	4.55335	2.3569427	4.3509507	11.45	15.5	0.00000
1136	MERCEDES		2.5634599	0.25944293	8.95630	1.8983883	4.1043062	12.08	11.6	0.00000
1137	RAISSA		2.4233327	0.09648631	4.32560	2.1895144	3.7724161	11.98	12.2	0.00000
1138	ATTICA		3.1490543	0.06390623	14.00652	2.9478099	5.5881777	12.20	11.0	0.00000
1139	ATAMI	S	1.9476762	0.25546652	13.09298	1.4501101	2.7181606	14.25	3.5	0.00000
1140	CRIMEA	S	2.7731133	0.11069882	14.13443	2.4661329	4.6179729	11.47	13.0	0.00000
1141	BOHMIA		2.2700839	0.16519822	4.27807	1.8950702	3.4202898	14.51	3.0	0.00000
1142	AETOLIA		3.1757145	0.09925184	2.10086	2.8605192	5.6592932	11.45	15.5	0.00000
1143	ODYSSEUS	D	5.2452035	0.09302093	3.13910	4.7572894	12.0127792	9.45	87.0	0.00000
1144	ODA	D	3.7541118	0.08017425	9.66545	3.4531288	7.2737908	11.20	28.0	0.00000
1145	ROBELMONTE		2.4249475	0.11512119	6.22610	2.1457846	3.7761874	12.15	11.2	0.00000
1146	BIARMIA	CMEU	3.0495143	0.24927239	17.15934	2.2893546	5.3253236	10.65	21.5	0.00000
1147	STAVROPOLIS		2.2716858	0.23037443	3.87450	1.7483476	3.4239106	13.09	7.3	0.00000
1148	RARAJU	S	3.0159791	0.10985463	10.83865	2.6846597	5.2377224	11.13	17.0	0.00000
1149	VOLGA		2.8997607	0.09410880	11.75333	2.6268678	4.9379110	11.41	15.8	0.00000
1150	ACHAIA		2.1913221	0.20428944	2.38767	1.7436581	3.2438390	14.42	3.1	0.00000

No.	Name	Type	A (A.U.)	E	I (deg.)	q (A.U.)	Period (years)	Mag.	Rad. (km)	Rot. (hours)
1151	ITHAKA		2.4063790	0.27579072	6.56487	1.7427220	3.7328973	14.81	3.3	0.00000
1152	PAWONA		2.4261594	0.04417738	5.07219	2.3189781	3.7790184	12.23	10.8	0.00000
1153	WALLENBERGIA		2.1961620	0.16075087	3.33504	1.8431270	3.2545917	13.31	5.2	0.00000
1154	ASTRONOMIA		3.3975320	0.05442440	4.53010	3.2126231	6.2624650	11.44	25.6	0.00000
1155	AENNA		2.4642849	0.16235697	6.61207	2.0641911	3.8684442	12.80	8.3	0.00000
1156	KIRA		2.2367556	0.04729158	1.39924	2.1309760	3.3452439	13.89	5.0	0.00000
1157	ARABIA		3.1953766	0.13371404	9.55204	2.7681098	5.7119327	11.19	17.5	0.00000
1158	LUDA		2.5630138	0.11284240	14.88370	2.2737973	4.1032352	12.05	9.3	0.00000
1159	GRANADA		2.3800149	0.05754953	13.04954	2.2430463	3.6717198	12.62	9.1	0.00000
1160	ILLYRIA		2.5599284	0.11932800	14.99930	2.2544572	4.0958281	12.77	6.7	0.00000
1161	THESSALIA		3.1641154	0.09760579	9.36020	2.8552794	5.6283164	12.70	8.7	0.00000
1162	LARISSA	P	3.9366927	0.11301494	1.89900	3.4917874	7.8108311	10.43	30.0	0.00000
1163	SAGA		3.2104781	0.06200218	9.01454	3.0114214	5.7524729	11.72	13.7	0.00000
1164	KOBOLDA		2.3058200	0.19686776	25.18372	1.8518783	3.5013709	14.03	4.7	0.00000
1165	IMPRINETTA		3.1330602	0.20618147	12.79155	2.4870813	5.5456586	11.71	13.8	0.00000
1166	SAKUNTALA		2.5401573	0.20247047	18.82439	2.0258503	4.0484700	12.58	9.2	0.00000
1167	DUBIAGO	D	3.4001763	0.07708725	5.77076	3.1380661	6.2697783	10.95	31.0	0.00000
1168	BRANDIA		2.5529044	0.21737953	12.75270	1.9979553	4.0789824	13.01	7.6	11.48000
1169	ALWINE		2.3175719	0.15547100	4.05032	1.9572567	3.5281725	14.29	4.2	0.00000
1170	SIVA	S	2.3246698	0.30077189	22.20091	1.6254745	3.5443933	13.11	6.0	0.00000
1171	RUSTHAWELIA	P	3.1655397	0.20913607	3.05198	2.5035112	5.6321168	10.80	29.0	0.00000
1172	ANEAS	D	5.1605630	0.10393589	16.71823	4.6241951	11.7231846	9.26	81.0	0.00000
1173	ANCHISES	C	5.3022041	0.13662861	6.91126	4.5777712	12.2091284	9.82	58.0	0.00000
1174	MARMARA		3.0213099	0.11394650	10.10131	2.6770420	5.2516150	12.84	6.5	0.00000
1175	MARGO		3.2201912	0.04805298	16.26910	3.0654514	5.7785983	11.53	15.0	0.00000
1176	LUCIDOR		2.6907935	0.14403036	6.64067	2.3032374	4.4138803	12.18	11.1	0.00000
1177	GONNESSIA	X	3.3501668	0.01951952	15.09373	3.2847731	6.1319642	10.11	47.3	0.00000
1178	IRMELA	C	2.6783087	0.18614762	6.96894	2.1797478	4.3831964	12.89	10.0	0.00000
1179	MALLY	LOST	2.6165960	0.17541827	8.74100	2.1575973	4.2325788	15.00	3.0	0.00000
1180	RITA	P	3.9912059	0.16501549	7.20316	3.3325951	7.9736323	10.13	50.0	0.00000
1181	LILITH		2.6657763	0.19491385	5.57590	2.1461797	4.3524675	12.63	9.0	0.00000
1182	ILONA		2.2591760	0.11780500	9.39244	1.9930339	3.3956673	12.50	9.6	0.00000
1183	JUTTA		2.3843422	0.12829395	2.79921	2.0784457	3.6817381	12.98	7.7	0.00000
1184	GAEA		2.6676586	0.07034237	11.31381	2.4800093	4.3570786	12.32	10.4	0.00000
1185	NIKKO		2.2370789	0.10605001	5.70122	1.9998367	3.3459692	13.20	5.5	0.00000
1186	TURNERA	U	3.0198951	0.10416168	10.76265	2.7053375	5.2479267	10.25	20.0	0.00000
1187	AFRA		2.6411624	0.22065939	10.71687	2.0583651	4.2923255	12.56	9.3	0.00000
1188	GOTHLANDIA		2.1904657	0.18022470	4.82018	1.7956897	3.2419381	13.11	5.7	0.00000
1189	TERENTIA		2.9304280	0.11337195	9.85747	2.5981996	5.0164509	11.10	18.2	0.00000
1190	PELAGIA		2.4299562	0.13305378	3.16805	2.1066413	3.7878926	13.17	7.0	0.00000
1191	ALFATERNA		2.8915100	0.04851655	18.46105	2.7512240	4.9168510	11.64	14.2	0.00000
1192	PRISMA		2.3646111	0.26081330	23.86358	1.7478890	3.6361318	13.58	5.8	6.55800
1193	AFRICA		2.6467292	0.12342824	14.14578	2.3200481	4.3059034	13.22	6.9	0.00000
1194	ALETTA		2.9118462	0.09266925	10.85892	2.6420076	4.9688129	11.70	13.8	0.00000
1195	ORANGIA		2.2583439	0.20091237	7.18600	1.8046148	3.3937914	14.51	3.8	0.00000
1196	SHEBA		2.6538527	0.17791015	17.69023	2.1817052	4.3232980	11.48	15.3	0.00000
1197	RHODESIA		2.8869636	0.23552030	12.88556	2.2070251	4.9052596	11.26	16.9	0.00000
1198	ATLANTIS		2.2508583	0.33506334	2.72443	1.4966781	3.3769312	16.69	1.4	0.00000
1199	GELDONIA	C	3.0186136	0.03465473	8.77821	2.9140043	5.2445869	11.40	20.0	0.00000
1200	IMPERATRIX		3.0553823	0.11579955	4.60362	2.7015705	5.3407021	11.75	13.5	0.00000
1201	STRENUA		2.6989520	0.03886772	6.99295	2.5940497	4.4339695	12.56	9.3	0.00000
1202	MARINA	CFPD	3.9749556	0.18188924	3.35970	3.2519538	7.9249845	11.31	29.0	0.00000
1203	NANNA		2.8881352	0.24529293	5.94496	2.1796961	4.9082451	13.09	7.3	0.00000
1204	RENZIA		2.2628746	0.29326731	1.87651	1.5992475	3.4040096	13.32	6.6	7.90000
1205	EBELLA		2.5344505	0.27704799	8.90006	1.8322860	4.0348344	15.19	2.8	0.00000
1206	NUMEROWIA		2.8654985	0.05513477	13.02530	2.7075100	4.8506536	11.38	16.0	0.00000
1207	OSTENIA		3.0194864	0.09227923	10.37224	2.7408507	5.2468619	12.04	9.4	8.40000
1208	TROILUS	C	5.1967735	0.09116829	33.64291	4.7229924	11.8467894	9.79	62.0	0.00000
1209	PUMMA		3.1649137	0.13241029	6.93459	2.7458465	5.6304460	11.48	15.3	0.00000
1210	MOROSOVIA	SM	3.0092139	0.06122494	11.27007	2.8249750	5.2201090	11.09	17.0	0.00000
1211	BRESSOLE		2.9279404	0.16041686	12.77341	2.4582493	5.0100646	12.01	12.0	0.00000
1212	FRANCETTE	P	3.9479299	0.18761200	7.59126	3.2072508	7.8442988	10.52	47.0	16.00000
1213	ALGERIA		3.1311116	0.14217561	13.05163	2.6859438	5.5404859	12.14	11.3	0.00000
1214	RICHILDE		2.7117581	0.11657039	9.83879	2.3956473	4.4655652	12.04	11.8	0.00000
1215	BOYER	S	2.5794420	0.13235441	15.89140	2.2380414	4.1427493	11.81	8.0	0.00000
1216	ASKANIA		2.2323611	0.17882068	7.60149	1.8331687	3.3353903	13.09	6.0	0.00000
1217	MAXIMILIANA		2.3534429	0.15299894	5.15548	1.9933687	3.6104016	14.48	3.8	0.00000
1218	ASTER		2.2629497	0.10971430	3.15940	2.0146718	3.4041789	14.14	4.5	0.00000
1219	BRITTA		2.2132962	0.12492484	4.41516	1.9368005	3.2927542	13.15	5.6	5.56000
1220	CROCUS		3.0027113	0.07453334	11.36722	2.7789092	5.2031980	12.24	8.5	0.00000
1221	AMOR		1.9211660	0.43450823	11.89642	1.0863979	2.6628332	19.06	0.5	0.00000
1222	TINA		2.7921770	0.24827580	19.71201	2.0989468	4.6656742	13.16	7.1	0.00000
1223	NECKAR	S	2.8695023	0.05780561	2.55039	2.7036290	4.8608236	11.61	13.0	8.60000
1224	FANTASIA	SU	2.3035307	0.19913109	7.87854	1.8448262	3.4961579	12.53	7.8	12.00000
1225	ARIANE		2.2333298	0.07480240	3.07611	2.0662713	3.3375616	13.60	4.6	0.00000
1226	GOLIA		2.5823631	0.11349305	9.83123	2.2892828	4.1497884	13.21	6.9	0.00000
1227	GERANIUM		3.2172873	0.19710211	16.25946	2.5831530	5.7707829	11.45	15.5	0.00000
1228	SCABIOSA		2.7717433	0.03879797	3.28870	2.6642053	4.6145515	12.71	8.7	0.00000
1229	TILIA		3.2382765	0.14899901	0.97649	2.7557764	5.8273473	12.91	13.0	0.00000
1230	RICEIA		2.5732789	0.17741424	10.48398	2.1167426	4.1279106	14.64	3.6	0.00000
1231	AURICULA		2.6685810	0.08689574	11.47581	2.4366927	4.3593383	12.68	8.8	0.00000
1232	CORTUSA		3.1835382	0.13174988	10.17282	2.7641075	5.6802192	11.28	16.8	0.00000
1233	KOBRESIA		2.5550940	0.05886723	5.59636	2.4046826	4.0842314	12.32	10.4	0.00000

No.	Name	Type	A (A.U.)	E	I (deg.)	q (A.U.)	Period (years)	Mag.	Rad. (km)	Rot. (hours)
1234	ELYNA		3.0136230	0.09017054	8.51488	2.7418828	5.2315860	11.72	10.8	0.00000
1235	SCHORRIA	C	1.9100677	0.15452097	25.00274	1.6149222	2.6398129	15.30	4.2	0.00000
1236	THAIS	EMPD	2.4304719	0.24447520	13.14913	1.8362819	3.7890987	12.81	8.3	0.00000
1237	GENEVIEVE		2.6116889	0.07744627	9.71456	2.4094231	4.2206774	12.00	12.1	0.00000
1238	PREDAPPIA		2.6670945	0.14155915	12.14906	2.2895429	4.3556962	13.01	7.6	0.00000
1239	QUETELETA		2.6621368	0.23138075	1.66681	2.0461695	4.3435574	13.67	5.6	0.00000
1240	CENTENARIA		2.8704097	0.17061143	10.14462	2.3806851	4.8631296	10.90	20.0	14.00000
1241	DYSONA	C	3.1809053	0.10884650	23.55097	2.8346751	5.6731744	10.47	39.0	0.00000
1242	ZAMBESIA		2.7351203	0.19139940	10.16837	2.2116201	4.5233965	11.32	16.5	0.00000
1243	PAMELA		3.0992866	0.03387289	13.27116	2.9943049	5.4562292	10.94	19.6	0.00000
1244	DEIRA		2.3429580	0.09771328	8.70008	2.1140199	3.5863008	12.39	10.1	0.00000
1245	CALVINIA	S	2.8920133	0.08021756	2.88837	2.6600232	4.9181352	11.04	15.1	4.85500
1246	CHAKA		2.6234691	0.30558020	16.03034	1.8217890	4.2492661	12.72	8.6	0.00000
1247	MEMORIA	C	3.1484671	0.15512279	1.75416	2.6600680	5.5866146	11.47	24.0	0.00000
1248	JUGURTHA		2.7242727	0.01539507	9.13605	2.6823323	4.4965129	10.90	20.0	0.00000
1249	RUTHERFORDIA		2.2243416	0.07672308	4.87053	2.0536833	3.3174336	12.96	6.1	0.00000
1250	GALANTHUS		2.5548604	0.26833084	15.13555	1.8693125	4.0836711	14.15	4.5	3.92000
1251	HEDERA	EMP	2.7168903	0.15589778	6.05174	2.2933331	4.4782481	11.58	13.6	0.00000
1252	CELESTIA	S	2.6957560	0.20524988	33.86798	2.1424525	4.4260964	12.04	9.8	0.00000
1253	FRISIA		3.1536806	0.21725062	1.35006	2.4685414	5.6004968	13.22	11.3	0.00000
1254	ERFORDIA		3.1333213	0.04089763	7.08282	3.0051758	5.5463514	11.54	14.9	0.00000
1255	SCHILOWA		3.1593399	0.15761861	8.54056	2.6613691	5.6155787	11.59	14.6	0.00000
1256	NORMANNIA	D	3.9203618	0.08006347	4.10844	3.6064839	7.7622781	10.81	40.0	0.00000
1257	MORA		2.4880521	0.08156050	3.91558	2.2851255	3.9245441	12.82	8.3	0.00000
1258	SICILIA		3.1884260	0.02927014	7.73558	3.0951004	5.6933060	11.61	14.4	0.00000
1259	OGYALLA		3.0953746	0.13613296	2.38664	2.6739919	5.4459019	11.89	20.8	12.00000
1260	WALHALLA		2.6150415	0.03827786	8.01621	2.5149434	4.2288070	12.87	8.1	0.00000
1261	LEGIA		3.1526260	0.16592190	2.42853	2.6295364	5.5976882	11.83	13.0	0.00000
1262	SNIADECKIA		3.0010991	0.00466418	13.13120	2.9871013	5.1990080	11.29	16.7	0.00000
1263	VARSAVIA	C	2.6655664	0.18916593	29.22738	2.1613319	4.3519535	11.46	25.0	0.00000
1264	LETABA		2.8612869	0.15560593	25.06030	2.4160535	4.8399634	10.84	20.6	0.00000
1265	SCHWEIKARDA		3.0256677	0.07011046	9.52267	2.8135366	5.2629814	11.03	14.9	0.00000
1266	TONE	CX	3.3716559	0.03296458	17.28308	3.2605107	6.1910582	10.41	40.0	0.00000
1267	GEERTRUIDA		2.4656973	0.18272381	4.78121	2.0151558	3.8717711	13.53	6.0	5.50000
1268	LIBYA	F	3.9438808	0.10577008	4.41548	3.5267363	7.8322344	10.01	49.0	0.00000
1269	ROLLANDIA	D	3.8953102	0.09688030	2.76120	3.5179315	7.6879945	9.72	55.0	0.00000
1270	DATURA		2.2342663	0.20797811	5.99254	1.7695879	3.3396614	13.79	4.2	0.00000
1271	ISERGINA		3.1336658	0.13749214	6.66699	2.7028115	5.5472665	11.60	14.5	0.00000
1272	GEFION		2.7832880	0.15156777	8.42493	2.3614314	4.6434121	13.51	6.0	0.00000
1273	HELMA		2.3942263	0.16208884	5.40875	2.0061488	3.7046549	14.05	4.7	0.00000
1274	DELPORTIA	S	2.2288713	0.11395572	4.39761	1.9748787	3.3275723	12.99	6.3	0.00000
1275	CIMBRIA	EMP	2.6813743	0.16657026	12.85000	2.2347372	4.3907247	11.73	12.8	0.00000
1276	UCCLIA		3.1630037	0.11066412	23.39231	2.8129728	5.6253500	11.83	13.0	0.00000
1277	DOLORES		2.6997654	0.23711163	6.98244	2.0596197	4.4359746	12.21	10.9	0.00000
1278	KENYA		2.4049859	0.26078656	10.88700	1.7777979	3.7296562	12.07	11.7	0.00000
1279	UGANDA		2.3699880	0.20960397	5.72129	1.8732291	3.6485410	13.65	5.6	0.00000
1280	BAILLAUDA		3.4100680	0.06972134	6.42603	3.1723137	6.2971578	11.02	31.1	0.00000
1281	JEANNE		2.5606542	0.20185186	7.44049	2.0437815	4.0975699	12.59	9.2	0.00000
1282	UTOPIA		3.1213467	0.11921303	18.01967	2.7492416	5.5145874	11.15	17.8	0.00000
1283	KOMSOMOLIA		3.1971469	0.21097414	8.87424	2.5226314	5.7166796	11.92	12.5	0.00000
1284	LATVIA	S	2.6449478	0.17016017	10.89769	2.1948829	4.3015566	11.40	14.0	25.00000
1285	JULIETTA		2.9916196	0.04709493	5.70160	2.8507295	5.1743946	11.34	16.3	0.00000
1286	BANACHIEWICZA	S	3.0214067	0.09628901	9.73879	2.7304783	5.2518673	11.80	11.0	0.00000
1287	LORCIA		3.0112519	0.05786078	9.82580	2.8370185	5.2254133	12.13	9.0	0.00000
1288	SANTA		2.8840082	0.06335337	7.55876	2.7012966	4.8977289	12.69	8.8	0.00000
1289	KUTAISSI	U	2.8608732	0.06042675	1.60666	2.6880000	4.8389139	11.47	11.0	0.00000
1290	ALBERTINE		2.3668406	0.15305705	5.59315	2.0045791	3.6412754	13.65	5.6	0.00000
1291	PHRYNE		3.0106733	0.09688044	9.10362	2.7189980	5.2239075	11.36	12.8	0.00000
1292	LUCE		2.5429304	0.05823172	2.15395	2.3948512	4.0551014	12.54	9.4	0.00000
1293	SONJA		2.2271917	0.27489045	5.36129	1.6149579	3.3238113	15.13	2.9	0.00000
1294	ANTWERPIA		2.6877830	0.23293050	8.70353	2.0617163	4.4064746	11.59	14.6	0.00000
1295	DEFLOTTE		3.3975258	0.11069540	2.86593	3.0214353	6.2624478	11.64	23.4	0.00000
1296	ANDREE		2.4182522	0.13981250	4.11487	2.0801504	3.7605588	12.62	9.1	0.00000
1297	QUADEA		3.0214744	0.06830045	9.01163	2.8151062	5.2520442	12.38	8.0	0.00000
1298	NOCTURNA		3.1224706	0.15276229	5.49268	2.6454749	5.5175662	11.98	12.2	0.00000
1299	MERTONA		2.8030975	0.18770333	7.87564	2.2769468	4.6930728	13.00	7.6	0.00000
1300	MARCELLE		2.7809887	0.00891695	9.53899	2.7561908	4.6376595	12.29	10.5	0.00000
1301	YVONNE	X	2.7625577	0.27240431	34.05976	2.0100250	4.5916314	11.78	13.3	0.00000
1302	WERRA	CFPD	3.1228495	0.16365436	2.59560	2.6117816	5.5185704	11.88	19.0	0.00000
1303	LUTHERA		3.2091401	0.12536895	19.56344	2.8068135	5.7488766	10.43	24.8	0.00000
1304	AROSA		3.2139821	0.10158951	18.80498	2.8874753	5.7618933	10.33	26.0	0.00000
1305	PONGOLA		3.0186749	0.06763742	2.31972	2.8144994	5.2447462	11.51	15.1	0.00000
1306	SCYTHIA	S	3.1428235	0.10894927	14.94790	2.8004150	5.5716004	10.75	17.0	0.00000
1307	CIMMERIA		2.2503281	0.09714659	3.94623	2.0317163	3.3757381	13.17	5.6	0.00000
1308	HALLERIA		2.9084222	0.01266261	5.57574	2.8715942	4.9600515	11.76	13.5	0.00000
1309	HYPERBOREA		3.2017994	0.15317281	10.30559	2.7113707	5.7291627	11.30	16.6	0.00000
1310	VILLIGERA	S	2.3937328	0.35659781	21.05304	1.5401329	3.7035096	12.69	7.3	0.00000
1311	KNOPFIA		2.4263349	0.04668789	2.81907	2.3130543	3.7794282	13.77	5.3	0.00000
1312	VASSAR		3.0917318	0.21791039	21.90633	2.4180112	5.4362912	12.09	11.6	0.00000
1313	BERNA		2.6567714	0.20522460	12.55162	2.1115365	4.3304329	13.00	7.8	0.00000
1314	PAULA	S	2.2953484	0.17396913	5.23421	1.8960288	3.4775465	13.82	4.2	0.00000
1315	BRONISLAWA		3.2015445	0.08235156	7.07323	2.9378924	5.7284789	11.08	18.4	0.00000
1316	KASAN		2.4106648	0.31933430	23.96012	1.6408569	3.7428744	14.79	3.3	0.00000

No.	Name	Type	A (A.U.)	E	I (deg.)	q (A.U.)	Period (years)	Mag.	Rad. (km)	Rot. (hours)
1317	SILVRETTA	C	3.1795905	0.25201082	20.71295	2.3782992	5.6696568	10.94	31.0	7.04800
1318	NERINA		2.3075373	0.20360810	24.67430	1.8377039	3.5052831	13.13	7.2	0.00000
1319	DISA		2.9859433	0.20531093	2.78141	2.3728964	5.1596746	11.71	13.8	0.00000
1320	IMPALA		2.9914329	0.22983053	19.81083	2.3039103	5.1739101	11.89	12.7	0.00000
1321	MAJUBA		2.9383795	0.17070773	9.51951	2.4367754	5.0368824	11.37	16.1	0.00000
1322	COPPERNICUS		2.4260545	0.23092976	23.28854	1.8658063	3.7787733	14.08	4.6	0.00000
1323	TUGELA		3.2018051	0.17552625	18.67808	2.6398044	5.7291784	11.33	16.4	0.00000
1324	KNYSNA		2.1852136	0.16341703	4.51533	1.8281125	3.2302849	13.56	4.6	0.00000
1325	INANDA		2.5395026	0.25903767	7.44732	1.8816758	4.0469046	13.21	6.9	0.00000
1326	LOSAKA	U	2.6651852	0.22667988	15.96671	2.0610414	4.3510199	11.98	12.0	0.00000
1327	NAMAQUA		2.7786322	0.16375689	5.81804	2.3236120	4.6317658	13.25	6.8	0.00000
1328	DEVOTA	U	3.4873431	0.15519941	5.77502	2.9461095	6.5124140	11.32	16.0	0.00000
1329	ELIANE	SU	2.6154587	0.17475633	14.50358	2.1583908	4.2298193	11.37	13.0	0.00000
1330	SPIRIDONIA	P	3.1711822	0.06910269	15.97967	2.9520450	5.6471820	11.12	28.0	0.00000
1331	SOLVEJG	CU	3.1141248	0.17432682	3.07649	2.5712492	5.4954596	11.16	27.0	15.00000
1332	MARCONIA		3.0600412	0.13231128	2.47164	2.6551633	5.3529220	11.32	16.5	0.00000
1333	CEVENOLA		2.6332674	0.13350368	14.60827	2.2817163	4.2730937	12.74	8.6	0.00000
1334	LUNDMARKA		2.9134011	0.09448381	11.44266	2.6381319	4.9727936	11.09	18.3	0.00000
1335	DEMOULINA		2.2408862	0.15384988	2.54776	1.8961262	3.3545146	14.89	2.5	0.00000
1336	ZEELANDIA	S	2.8500311	0.06029467	3.19446	2.6781895	4.8114328	12.13	9.4	0.00000
1337	GERARDA		2.9125073	0.09768305	17.96997	2.6280048	4.9705052	12.01	12.0	0.00000
1338	DUPONTA		2.2641008	0.11197848	4.81505	2.0105700	3.4067762	13.96	3.9	0.00000
1339	DESAGNEAUXA	U	3.0231192	0.04968727	8.67518	2.8729086	5.2563334	11.86	10.1	0.00000
1340	YVETTE		3.1734345	0.14305344	0.42036	2.7194638	5.6531992	12.56	15.3	0.00000
1341	EDMEE	CMEU	2.7421215	0.07685637	13.07616	2.5313721	4.5407758	11.54	14.1	0.00000
1342	BRABANTIA		2.2883286	0.20325138	20.93812	1.8232228	3.4616058	13.35	6.5	0.00000
1343	NICOLE		2.5704732	0.11209002	6.02377	2.2823489	4.1211615	12.50	9.6	0.00000
1344	CAUBETA		2.2477019	0.11956959	5.65409	1.9789451	3.3698306	14.06	3.7	0.00000
1345	POTOMAC	C	3.9820883	0.17907572	11.39357	3.2689929	7.9463253	10.65	35.0	0.00000
1346	GOTHA		2.6272511	0.17731737	13.88335	2.1613939	4.2584581	12.37	10.2	0.00000
1347	PATRIA		2.5715809	0.06792659	11.89535	2.3969021	4.1238256	12.12	11.4	0.00000
1348	MICHEL		2.7914143	0.13872188	6.58641	2.4041841	4.6637626	12.28	10.6	0.00000
1349	BECHUANA		3.0161903	0.15123618	10.00590	2.5600333	5.2382731	11.64	14.2	0.00000
1350	ROSSELIA		2.8562562	0.08935385	2.93852	2.6010387	4.8272047	11.54	11.8	6.00000
1351	UZBEKISTANI		3.1950328	0.06392279	9.72766	2.9907975	5.7110109	11.01	19.0	0.00000
1352	WAWEL		2.7788489	0.06453502	3.75115	2.5995159	4.6323075	12.34	10.3	0.00000
1353	MAARTJE		3.0110168	0.09295151	9.18324	2.7311382	5.2248015	11.05	14.8	0.00000
1354	BOTHA		3.1379786	0.20816810	5.95042	2.4847517	5.5587220	12.06	11.7	0.00000
1355	MAGOEBA		1.8536798	0.04463827	22.82316	1.7709347	2.5237834	13.80	3.0	0.00000
1356	NYANZA		3.0808380	0.04980507	7.94719	2.9273965	5.4075842	11.33	16.4	0.00000
1357	KHAMA		3.1792576	0.16108908	13.98526	2.6671138	5.6687665	12.03	11.5	0.00000
1358	GAIKA		2.4758165	0.17026912	2.16949	2.0542614	3.8956299	12.84	8.2	0.00000
1359	PRIESKA	C	3.1191628	0.06404560	11.07933	2.9193943	5.5088010	11.52	24.0	0.00000
1360	TARKA		2.6339271	0.21479098	22.79381	2.0681832	4.2746997	12.42	9.9	0.00000
1361	LEUSCHNERIA		3.0862117	0.12435659	21.53616	2.7024207	5.4217381	11.93	12.4	0.00000
1362	GRIQUA	CX	3.2355173	0.36442843	24.15206	2.0564027	5.8199005	12.00	16.0	7.00000
1363	HERBERTA		2.9015212	0.06806565	1.09622	2.7040274	4.9424081	12.66	7.0	0.00000
1364	SAFARA	S	3.0138381	0.06843435	11.47469	2.8075881	5.2321458	12.02	9.6	0.00000
1365	HENYEY		2.2482574	0.12364428	5.07485	1.9702733	3.3710802	13.27	5.3	0.00000
1366	PICCOLO		2.8718522	0.14176810	9.48553	2.4647152	4.8667960	11.50	15.2	0.00000
1367	NONGOMA		2.3435752	0.13144442	22.46775	2.0355253	3.5877182	14.20	4.4	0.00000
1368	NUMIDIA		2.5238531	0.06344759	14.84667	2.3637207	4.0095539	11.92	12.5	0.00000
1369	OSTANINA		3.1036541	0.22412166	14.27979	2.4080579	5.4677668	11.38	16.0	0.00000
1370	HELLA		2.2503860	0.17099622	4.80507	1.8655784	3.3758683	14.86	2.6	0.00000
1371	RESI		3.2145345	0.09310406	16.49107	2.9152484	5.7633786	12.26	10.7	0.00000
1372	HAREMARI		2.7680581	0.14665297	16.44975	2.3621142	4.6053514	12.70	8.7	0.00000
1373	CINCINNATI	X	3.3991158	0.32329348	39.04241	2.3002038	6.2668452	14.23	7.1	0.00000
1374	ISORA		2.2504549	0.27781326	5.29980	1.6252487	3.3760235	14.70	3.5	0.00000
1375	ALFREDA		2.4480095	0.06858098	5.83098	2.2801228	3.8301842	12.91	7.9	0.00000
1376	MICHELLE		2.2281494	0.21486451	3.54870	1.7493991	3.3259554	13.51	4.8	0.00000
1377	ROBERBAUXA		2.2596931	0.09395338	6.02619	2.0473874	3.3968329	14.18	4.4	0.00000
1378	LEONCE		2.3740394	0.15007636	3.60023	2.0177522	3.6579001	13.30	6.6	0.00000
1379	LOMONOSOWA		2.5265989	0.08695669	15.58002	2.3068943	4.0160995	12.01	12.0	0.00000
1380	VOLODIA		3.1460907	0.10737635	10.49269	2.8082750	5.5802913	13.08	7.3	0.00000
1381	DANUBIA		2.4881301	0.18046713	4.68294	2.0391045	3.9247289	12.94	7.8	0.00000
1382	GERTI		2.2205501	0.13186079	1.56197	1.9277467	3.3089550	13.33	5.2	0.00000
1383	LIMBURGIA		3.0868387	0.18217027	0.04879	2.5245085	5.4233909	12.84	13.5	0.00000
1384	KNIERTJE		2.6765399	0.18193579	11.85361	2.1895814	4.3788548	12.76	8.5	0.00000
1385	GELRIA		2.7423606	0.10797893	6.92468	2.4462433	4.5413694	12.01	12.0	0.00000
1386	STORERIA		2.3644433	0.28674918	11.78551	1.6864412	3.6357448	14.68	3.5	0.00000
1387	KAMA		2.2585206	0.20905524	5.52704	1.7863649	3.3941891	14.33	3.3	0.00000
1388	APHRODITE	SM	3.0192568	0.08922561	11.19470	2.7498617	5.2462630	12.01	11.0	0.00000
1389	ONNIE		2.8664126	0.01570110	2.03840	2.8214068	4.8529749	12.67	7.0	0.00000
1390	ABASTUMANI	P	3.4418170	0.02931081	19.95761	3.3409345	6.3853054	10.15	45.0	0.00000
1391	CARELIA	S	2.5459197	0.16900951	7.58323	2.1156349	4.0622535	12.98	6.3	0.00000
1392	PIERRE	MEU	2.6085236	0.20197760	12.24231	2.0816603	4.2130065	12.75	8.4	0.00000
1393	SOFALA		2.4353735	0.10733921	5.84161	2.1739624	3.8005667	13.13	7.2	0.00000
1394	ALGOA		2.4393246	0.07822448	2.66414	2.2485096	3.8098195	12.78	8.4	0.00000
1395	ARIBEDA		3.2056174	0.04112966	8.69083	3.0737715	5.7394137	12.67	8.9	0.00000
1396	OUTENIQUA		2.2482364	0.16385812	4.49759	1.8798447	3.3710330	12.92	6.2	0.00000
1397	UMTATA		2.6818044	0.25339115	3.51695	2.0022590	4.3917809	12.77	8.5	0.00000
1398	DONNERA		3.1519868	0.11549464	11.85604	2.7879493	5.5959859	11.35	16.3	0.00000
1399	TENERIFFA		2.2160285	0.16578710	6.51113	1.8486395	3.2988534	15.17	2.2	0.00000

No.	Name	Type	A (A.U.)	E	I (deg.)	q (A.U.)	Period (years)	Mag.	Rad. (km)	Rot. (hours)
1400	TIRELA		3.1076267	0.24747995	15.56243	2.3385513	5.4782677	12.88	8.0	0.00000
1401	LAVONNE	S	2.2269535	0.17935415	7.28708	1.8275402	3.3232782	13.40	5.2	0.00000
1402	ERI		2.6836531	0.15441459	14.28458	2.2692578	4.3963223	14.52	3.8	0.00000
1403	IDELSONIA		2.7158692	0.29448405	10.18190	1.9160891	4.4757237	13.56	5.9	0.00000
1404	AJAX		5.2666259	0.11427477	18.03952	4.6647835	12.0864477	10.17	46.0	0.00000
1405	SIBELIUS		2.2510900	0.14642256	7.03416	1.9214796	3.3774529	14.31	3.3	0.00000
1406	KOMPPA		2.6961436	0.09648015	12.43262	2.4360192	4.4270511	12.42	9.9	0.00000
1407	LINDELOF		2.7613649	0.28483787	5.78485	1.9748237	4.5886583	12.29	10.5	0.00000
1408	TRUSANDA		3.1120355	0.08449122	8.31304	2.8490958	5.4899302	12.00	12.1	0.00000
1409	ISKO		2.6759527	0.05645753	6.70204	2.5248749	4.3774142	11.65	14.2	0.00000
1410	MARGRET		3.0158503	0.11014943	10.36406	2.6836562	5.2373872	12.35	8.1	0.00000
1411	BRAUNA		3.0025010	0.06079063	8.04653	2.8199770	5.2026515	11.96	12.3	0.00000
1412	LAGRULA		2.2148097	0.11217218	4.71798	1.9663697	3.2961321	13.59	4.6	0.00000
1413	ROUCARIE		3.0193198	0.06524626	10.22486	2.8223205	5.2464271	12.43	7.8	0.00000
1414	JEROME		2.7920845	0.15625814	8.83777	2.3557985	4.6654425	13.74	5.4	0.00000
1415	MALAUTRA	S	2.2235377	0.08746277	3.42530	2.0290608	3.3156352	13.45	5.0	0.00000
1416	RENAUXA	U	3.0140338	0.11101742	10.06143	2.6794236	5.2326555	11.60	16.0	4.30000
1417	WALINSKIA		2.9711230	0.07796524	8.27415	2.7394786	5.1213083	12.25	10.7	0.00000
1418	FAYETA	S	2.2417333	0.20352612	7.19334	1.7854819	3.3564167	13.14	5.9	0.00000
1419	DANZIG		2.2929981	0.14682175	5.72507	1.9563360	3.4722061	12.58	9.2	0.00000
1420	RADCLIFFE		2.7488246	0.07817921	3.48044	2.5339236	4.5574355	12.82	8.3	0.00000
1421	ESPERANTO		3.0916817	0.07571057	9.80514	2.8576088	5.4361596	11.43	15.7	0.00000
1422	STROMGRENIA	S	2.2468846	0.16748469	2.67560	1.8705659	3.3679931	13.70	4.6	0.00000
1423	JOSE		2.8606970	0.07822183	2.91033	2.6369281	4.8384671	12.50	7.6	0.00000
1424	SUNDMANIA		3.1845412	0.07616450	9.21481	2.9419923	5.6829042	10.56	23.4	0.00000
1425	TUORLA		2.6114008	0.10050521	12.94397	2.3489416	4.2199793	12.76	8.5	0.00000
1426	RIVIERA		2.5784879	0.16565613	9.03597	2.1513455	4.1404510	12.05	11.8	0.00000
1427	RUVUMA		2.7528214	0.20852353	9.34998	2.1787934	4.5673790	11.81	13.2	0.00000
1428	MOMBASA		2.8102865	0.13827363	17.33768	2.4216981	4.7111392	11.43	15.7	0.00000
1429	PEMBA		2.5494096	0.34175846	7.72432	1.6781274	4.0706096	13.16	7.1	28.00000
1430	SOMALIA		2.5610433	0.19702825	3.29262	2.0564454	4.0985041	13.16	7.1	0.00000
1431	LUANDA		2.6203446	0.18140964	14.01855	2.1449890	4.2416773	12.51	9.5	0.00000
1432	ETHIOPIA		2.3818486	0.22571056	8.26938	1.8442402	3.6759639	13.28	6.7	0.00000
1433	GERAMTINA		2.7966518	0.17128488	8.23731	2.3176277	4.6768951	12.76	8.5	0.00000
1434	MARGOT	SU	3.0143836	0.06908870	10.82663	2.8061235	5.2335668	11.40	13.0	0.00000
1435	GARLENA		2.6458037	0.24769293	4.04273	1.9904567	4.3036447	14.70	3.5	0.00000
1436	SALONTA		3.1403544	0.07543910	13.87099	2.9034488	5.5650358	11.78	13.3	0.00000
1437	DIOMEDES	C	5.1002812	0.04529598	20.58966	4.8692589	11.5183735	9.28	86.0	18.00000
1438	WENDELINE		3.1760454	0.22555360	2.03372	2.4596770	5.6601777	12.70	8.7	0.00000
1439	VOGTIA	EMP	3.9918504	0.11332390	4.19863	3.5394783	7.9755635	11.40	25.5	0.00000
1440	ROSTIA		3.1404333	0.20342930	2.28906	2.5015771	5.5652461	12.80	13.7	0.00000
1441	BOLYAI		2.6326110	0.23702380	13.92521	2.0086195	4.2714963	14.13	4.5	0.00000
1442	CORVINA		2.8742099	0.07783413	1.24959	2.6504984	4.8727903	12.53	7.5	0.00000
1443	RUPPINA		2.9403696	0.06467905	1.92323	2.7501893	5.0420008	12.31	10.4	0.00000
1444	PANNONIA		3.1575420	0.13554277	17.73645	2.7295599	5.6107855	12.14	11.3	0.00000
1445	KONKOLYA	SM	3.1097698	0.18966195	2.28495	2.5199647	5.4839358	11.87	14.0	0.00000
1446	SILLANPAA		2.2456231	0.10040487	5.25728	2.0201516	3.3651569	13.87	4.0	0.00000
1447	UTRA		2.5354404	0.03929802	4.79362	2.4358027	4.0371990	12.83	8.2	0.00000
1448	LINDBLADIA		2.3716512	0.18727857	5.81590	1.9274917	3.6523819	14.29	4.2	0.00000
1449	1938DO	S	2.2220008	0.14270687	6.63967	1.9049060	3.3121982	13.73	4.5	0.00000
1450	RAIMONDA		2.6119254	0.16951008	4.85825	2.1691778	4.2212505	12.57	9.3	0.00000
1451	GRANO		2.2028046	0.11816082	5.11222	1.9425193	3.2693689	13.76	4.2	0.00000
1452	HUNNIA		3.1183643	0.19319864	14.19777	2.5159006	5.5066857	13.00	7.6	0.00000
1453	FENNIA	RU	1.8972480	0.02830168	23.67293	1.8435527	2.6132813	13.84	3.2	0.00000
1454	KALEVALA		2.3640840	0.14321601	5.09785	2.0255094	3.6349158	14.24	4.3	0.00000
1455	MITCHELLA		2.2462025	0.12564018	7.75262	1.9639891	3.3664589	14.41	3.1	0.00000
1456	SALDANHA	C	3.2108209	0.20962447	10.67286	2.5377543	5.7533937	11.87	20.0	0.00000
1457	ANKARA		2.6990943	0.15549380	6.07732	2.2794018	4.4343204	12.36	10.2	0.00000
1458	MINEURA		2.6252439	0.18353967	12.50614	2.1434073	4.2535787	12.45	9.0	0.00000
1459	MAGNYA		3.1397517	0.23595434	16.94885	2.3989134	5.5634341	11.86	12.9	0.00000
1460	HALTIA		2.5409231	0.19001676	6.70504	2.0581050	4.0503006	13.70	5.5	0.00000
1461	JEAN-JACQUES	M	3.1269796	0.04004918	15.28605	3.0017467	5.5295219	11.00	16.0	0.00000
1462	ZAMENHOF	SM	3.1530893	0.10113470	0.97023	2.8342025	5.5989218	12.09	13.0	0.00000
1463	NORDENMARKIA		3.1475258	0.20087145	7.30239	2.5152776	5.5841093	12.02	11.9	0.00000
1464	ARMISTICIA		3.0019112	0.04755186	11.55497	2.8591647	5.2011185	12.22	8.6	0.00000
1465	AUTONOMA		3.0294077	0.17392507	9.92911	2.5025177	5.2727432	12.14	11.3	0.00000
1466	MUNDLERIA		2.3780515	0.15564232	13.14031	2.0079260	3.6671772	14.02	4.8	0.00000
1467	MASHONA	C	3.3648901	0.14657681	22.08496	2.8716753	6.1724324	9.60	58.0	0.00000
1468	ZOMBA		2.1958890	0.27099276	9.95225	1.6008191	3.2539854	14.51	3.8	0.00000
1469	LINZIA		3.1231217	0.06548563	13.40082	2.9186022	5.5192919	10.84	20.6	0.00000
1470	CARLA		3.1639113	0.05265050	3.20898	2.9973297	5.6277714	12.20	11.0	0.00000
1471	TORNIO		2.7153192	0.11948477	13.61216	2.3908799	4.4743638	12.36	10.2	0.00000
1472	MUONIO		2.2344480	0.19852658	4.56686	1.7908506	3.3400683	13.67	4.4	0.00000
1473	OUNAS		2.5759087	0.23549940	13.67300	1.9692837	4.1342397	13.52	6.0	0.00000
1474	BEIRA	U	2.7372599	0.48786584	26.71487	1.4018443	4.5287051	13.50	5.6	0.00000
1475	YALTA		2.3489311	0.16781950	4.50501	1.9547347	3.6000242	14.09	4.6	0.00000
1476	COX		2.2816381	0.18923420	6.32318	1.8498743	3.4464355	14.79	2.6	0.00000
1477	BONSDORFFIA		3.1684771	0.29322261	15.75745	2.2394080	5.6399579	12.59	8.9	0.00000
1478	VIHURI		2.4662964	0.09264018	7.86395	2.2378182	3.8731821	13.49	6.1	0.00000
1479	INKERI		2.6746435	0.19553968	7.29575	2.1516445	4.3742023	12.31	9.9	0.00000
1480	AUNUS		2.2019415	0.10901601	4.86280	1.9618945	3.2674477	14.35	4.1	0.00000
1481	TUBINGIA		3.0165677	0.04126250	3.52510	2.8920965	5.2392559	11.83	13.0	0.00000
1482	SEBASTIANA		2.8727963	0.03748484	2.97661	2.7651100	4.8691959	11.95	9.7	0.00000

No.	Name	Type	A (A.U.)	E	I (deg.)	q (A.U.)	Period (years)	Mag.	Rad. (km)	Rot. (hours)
1483	NAKOILA		2.7156892	0.17996857	4.49404	2.2269504	4.4752784	12.54	9.4	0.00000
1484	POSTREMA		2.7376232	0.20511284	17.25983	2.1761014	4.5296068	12.22	10.9	0.00000
1485	ISA		3.0244498	0.11868037	8.93765	2.6655071	5.2598042	12.47	7.7	0.00000
1486	MARILYN		2.1985860	0.12419118	0.07528	1.9255410	3.2599819	14.51	3.8	0.00000
1487	BODA		3.1451478	0.11018948	2.46701	2.7985857	5.5777831	11.97	20.1	0.00000
1488	AURA		3.0380239	0.11929132	10.54695	2.6756141	5.2952538	12.00	12.1	0.00000
1489	ATTILA		3.2021363	0.14504378	2.41306	2.7376862	5.7300668	13.02	7.5	0.00000
1490	LIMPOPO		2.3525715	0.15462840	10.02199	1.9887972	3.6083968	12.66	8.9	0.00000
1491	BALDUINUS		3.2054951	0.16649829	3.76535	2.6717856	5.7390852	12.57	9.3	0.00000
1492	OPPOLZER		2.1730509	0.11640406	6.05927	1.9200989	3.2033534	14.33	4.1	0.00000
1493	SIGRID	F	2.4296558	0.20323832	2.59028	1.9358567	3.7871904	12.68	14.0	0.00000
1494	SAVO		2.1897433	0.13131678	2.45422	1.9021932	3.2403343	13.88	5.1	0.00000
1495	HELSINKI		2.6381121	0.15452819	12.74276	2.2304494	4.2848916	13.42	6.3	0.00000
1496	TURKU		2.2061508	0.16162059	2.49999	1.8495913	3.2768214	13.52	4.7	0.00000
1497	TAMPERE		2.8947673	0.07883475	1.05751	2.6665590	4.9251618	12.90	6.3	0.00000
1498	LAHTI		3.0936403	0.24392654	12.65074	2.3390193	5.4413261	13.00	7.6	0.00000
1499	PORI		2.6713650	0.18563394	12.18236	2.1754692	4.3661623	12.64	9.0	0.00000
1500	JYVASKYLA	SX	2.2427280	0.18952110	7.44241	1.8176838	3.3586514	14.24	3.6	0.00000
1501	BAADE		2.5442369	0.24073163	7.34395	1.9317586	4.0582266	13.50	6.0	0.00000
1502	ARENDA		2.7305009	0.08954120	4.08361	2.4860086	4.5119419	12.70	8.7	0.00000
1503	KUOPIO		2.6271086	0.10248624	12.34956	2.3578660	4.2581115	11.71	13.8	0.00000
1504	LAPPEENRANTA	S	2.3988812	0.15692329	11.03877	2.0224409	3.7154644	12.88	9.1	0.00000
1505	KORANNA		2.6592071	0.13446768	14.48756	2.3016298	4.3363891	12.47	9.7	0.00000
1506	XOSA		2.5692470	0.26407263	12.55454	1.8907793	4.1182132	13.11	7.2	0.00000
1507	VAASA		2.3311374	0.24446109	9.23765	1.7612652	3.5591955	14.59	3.7	0.00000
1508	KEMI	F	2.7700021	0.41907457	28.68331	1.6091647	4.6102042	13.15	11.0	0.00000
1509	ESCLANGONA		1.8664896	0.03209850	22.31829	1.8065782	2.5499892	14.00	2.8	0.00000
1510	CHARLOIS		2.6702456	0.15020224	11.84237	2.2691686	4.3634181	12.45	9.8	0.00000
1511	DALERA		2.3576827	0.10917050	4.06987	2.1002934	3.6201625	14.07	4.6	0.00000
1512	OULU	P	3.9360168	0.15847887	6.52065	3.3122413	7.8088202	10.57	45.0	0.00000
1513	MATRA		2.1923094	0.09881915	3.97906	1.9756671	3.2460315	14.36	4.1	30.00000
1514	RICOUXA		2.2401264	0.19989254	4.53594	1.7923418	3.3528087	13.54	4.7	0.00000
1515	PERROTIN		2.5717037	0.23240981	10.65053	1.9740146	4.1241212	13.85	5.1	0.00000
1516	HENRY		2.6235712	0.18616413	8.73919	2.1351564	4.2495141	12.91	7.9	0.00000
1517	BEOGRAD		2.7166405	0.04401733	5.28089	2.5970612	4.4776306	12.12	11.4	0.00000
1518	ROVANIEMI		2.2257531	0.14291431	6.71557	1.9076610	3.3205914	13.47	4.8	0.00000
1519	KAJAANI		3.1369071	0.23258451	12.58454	2.4073112	5.5558753	12.33	10.4	0.00000
1520	IMATRA	CFPD	3.1031749	0.10788235	15.26069	2.7683971	5.4665008	11.43	26.0	0.00000
1521	SEINAJOKI		2.8521514	0.13753442	15.06663	2.4598825	4.8168025	13.19	7.0	0.00000
1522	KOKKOLA		2.3678913	0.07119573	5.35226	2.1993077	3.6437006	13.64	5.7	0.00000
1523	PIEKSAMAKI		2.2420480	0.09419326	5.14403	2.0308621	3.3571239	13.31	5.2	5.33000
1524	JOENSUU		3.1179826	0.10940398	12.65361	2.7768631	5.5056748	11.82	13.1	0.00000
1525	SAVONLINNA		2.6980107	0.26232773	5.88296	1.9902476	4.4316502	13.57	5.8	0.00000
1526	MIKKELI		2.3154464	0.18692875	6.20912	1.8826228	3.5233197	14.73	3.4	0.00000
1527	MALMQUISTA		2.2269452	0.19803178	5.19073	1.7859393	3.3232598	13.57	4.6	0.00000
1528	CONRADA		2.4158230	0.14010596	8.53089	2.0773518	3.7548940	13.52	6.0	0.00000
1529	OTERMA	D	4.0088339	0.19341961	9.03125	3.2334468	8.0265169	11.08	30.0	0.00000
1530	RANTASEPPA		2.2483420	0.19872051	4.41766	1.8015504	3.3712704	14.46	3.1	0.00000
1531	HARTMUT		2.6267741	0.15253595	12.42204	2.2260964	4.2572980	13.02	7.5	0.00000
1532	INARI	SU	3.0032451	0.04852975	8.79112	2.8574984	5.2045860	11.85	10.0	0.00000
1533	SAIMAA	U	3.0120027	0.04183027	10.69486	2.8860097	5.2273674	12.06	9.3	0.00000
1534	NASI		2.7281442	0.25270113	9.83047	2.0387392	4.5061016	12.96	7.7	0.00000
1535	PAIJANNE		3.1502132	0.20231256	6.16931	2.5128856	5.5912628	12.80	8.3	0.00000
1536	PIELINEN		2.2043724	0.19512905	1.53085	1.7742352	3.2728601	14.69	2.8	0.00000
1537	TRANSYLVANIA		3.0504117	0.30096102	3.85909	2.1323566	5.3276744	13.06	7.4	0.00000
1538	DETRE		2.3624988	0.21799113	9.45112	1.8474948	3.6312602	15.48	2.4	0.00000
1539	BORRELLY		3.1583686	0.17500161	1.72202	2.6056490	5.6129889	12.19	18.2	0.00000
1540	KEVOLA		2.8494658	0.08249359	11.96873	2.6144030	4.8100009	11.79	13.3	0.00000
1541	ESTONIA		2.7671945	0.06877910	4.89398	2.5768695	4.6031971	12.56	9.3	0.00000
1542	SCHALEN		3.0959983	0.10898339	2.73867	2.7585857	5.4475479	11.50	15.2	0.00000
1543	BOURGEOIS		2.6308272	0.32428950	11.09745	1.7776777	4.2671556	13.54	5.9	0.00000
1544	VINTERHANSENIA		2.3737025	0.10372436	3.33407	2.1274917	3.6571221	12.81	8.3	0.00000
1545	THERNOE		2.7707775	0.24068671	2.95314	2.1038880	4.6121397	12.70	8.7	0.00000
1546	IZSAK		3.1852503	0.11667247	16.13679	2.8136194	5.6848025	11.65	14.2	0.00000
1547	NELE	U	2.6442897	0.25426668	11.69964	1.9719349	4.2999516	11.80	13.0	0.00000
1548	PALOMAA		2.7878180	0.08211077	16.51820	2.5589080	4.6547527	12.84	8.2	0.00000
1549	MIKKO		2.2306917	0.08340785	5.54787	2.0446343	3.3316493	13.60	4.6	0.00000
1550	TITO		2.5465751	0.30811426	8.83694	1.7619390	4.0638223	13.48	6.1	0.00000
1551	ARGELANDER		2.3951011	0.06568123	3.76297	2.2377880	3.7066858	13.58	5.8	0.00000
1552	BESSEL		3.0070157	0.10313536	9.86563	2.6968861	5.2143908	12.63	7.1	0.00000
1553	BAUERSFELDA		2.9046211	0.10082728	3.23332	2.6117561	4.9503312	12.70	8.7	0.00000
1554	YUGOSLAVIA		2.6184368	0.20245682	12.16078	2.0883164	4.2370458	12.64	9.0	0.00000
1555	DEJAN		2.6892564	0.27599794	6.05019	1.9470272	4.4100990	12.62	9.1	0.00000
1556	WINGOLFIA		3.4150286	0.12838416	15.71926	2.9765930	6.3109031	11.33	27.0	0.00000
1557	ROEHLA		3.0074878	0.10915443	10.32222	2.6792071	5.2156186	12.34	8.1	0.00000
1558	JARNEFELT		3.2238331	0.05120257	10.48591	3.0587645	5.7884035	11.51	15.1	0.00000
1559	KUSTAANHEIMO		2.3895218	0.13676579	3.20164	2.0627170	3.6937416	13.06	7.4	0.00000
1560	STRATTONIA		2.6847987	0.21337418	6.27663	2.1119318	4.3991380	12.70	8.7	0.00000
1561	FRICKE		3.1810820	0.14361337	4.37068	2.7242360	5.6736465	11.98	12.2	0.00000
1562	GONDOLATSCH		2.2264619	0.07833141	4.88663	2.0520601	3.3221781	13.37	5.1	8.20000
1563	NOEL		2.1915939	0.08545937	5.98314	2.0043015	3.2444427	13.76	4.2	0.00000
1564	SRBIJA		3.1562552	0.21411523	11.06474	2.4804530	5.6073565	12.09	11.6	0.00000
1565	LEMAITRE		2.3927584	0.35051185	21.41470	1.5540682	3.7012486	13.68	5.6	0.00000

No.	Name	Type	A (A.U.)	E	I (deg.)	q (A.U.)	Period (years)	Mag.	Rad. (km)	Rot. (hours)
1566	ICARUS	U	1.0778852	0.82675558	22.89607	0.1867375	1.1190737	17.55	0.7	2.27300
1567	ALIKOSKI	C	3.2133865	0.08475595	17.25878	2.9410329	5.7602916	10.54	36.0	0.00000
1568	AISLEEN		2.3523624	0.25417253	24.89087	1.7544565	3.6079154	13.12	7.2	0.00000
1569	EVITA		3.1552792	0.12224638	12.23237	2.7695577	5.6047554	12.43	9.9	0.00000
1570	BRUNONIA		2.8462479	0.05882678	1.65688	2.6788125	4.8018556	12.52	7.5	0.00000
1571	CESCO		3.1463304	0.12012739	14.52352	2.7683699	5.5809288	13.13	7.2	0.00000
1572	POSNANIA		3.1073029	0.20659202	13.29466	2.4653587	5.4774113	11.14	17.9	0.00000
1573	VAISALA		2.3698199	0.23226485	24.57442	1.8193940	3.6481526	13.90	5.0	0.00000
1574	MEYER		3.5317230	0.04963398	14.40901	3.3564296	6.6371245	11.50	24.9	0.00000
1575	WINIFRED		2.3741572	0.18001096	24.80538	1.9467829	3.6581728	13.90	5.0	0.00000
1576	FABIOLA	CFPD	3.1290565	0.18534093	0.94059	2.5491142	5.5350313	11.91	15.0	6.70000
1577	REISS		2.2307167	0.16595264	4.35835	1.8605233	3.3317058	15.24	2.1	0.00000
1578	KIRKWOOD	D	3.9383905	0.23102129	0.81396	3.0285387	7.8158851	11.50	28.0	0.00000
1579	HERRICK	C	3.4174631	0.14931604	8.75118	2.9071810	6.3176527	11.02	30.0	0.00000
1580	BETULIA	U	2.1967196	0.48941329	52.01724	1.1216158	3.2558317	14.95	0.5	6.13000
1581	ABANDERADA	U	3.1593447	0.11696046	2.53836	2.7898264	5.6155915	11.29	27.5	0.00000
1582	MARTIR		3.1523063	0.12820350	11.60401	2.7481694	5.5968361	12.99	7.6	0.00000
1583	ANTILOCHUS	D	5.1087103	0.05405273	28.55044	4.8325706	11.5469398	9.62	79.0	0.00000
1584	FUJI	U	2.3748474	0.19592978	26.67629	1.9095441	3.6597681	12.02	12.9	10.00000
1585	UNION		2.9266007	0.30910525	26.19734	2.0219729	5.0066266	11.69	13.9	0.00000
1586	THIELE		2.4322197	0.10202837	4.06060	2.1840644	3.7931869	13.51	6.0	0.00000
1587	KAHRSTEDT		2.5459993	0.15110096	7.84623	2.1612964	4.0624447	12.81	8.3	0.00000
1588	DESCAMISADA		3.0311062	0.06535687	11.27574	2.8330026	5.2771778	12.06	11.7	0.00000
1589	FANATICA		2.4176483	0.09242793	5.25561	2.1941900	3.7591505	13.18	7.0	0.00000
1590	TSIOLKOVSKAJA		2.2301853	0.15649292	4.34829	1.8811771	3.3305151	12.92	6.2	6.70000
1591	BAIZE		2.3904319	0.17778312	24.79953	1.9654534	3.6958518	13.18	7.0	0.00000
1592	MATHIEU		2.7632430	0.30690175	13.52532	1.9151988	4.5933399	12.69	8.8	0.00000
1593	FAGNES	SM	2.2253227	0.28059852	9.97601	1.6009005	3.3196287	14.51	3.3	47.00000
1594	DANJON		2.2688341	0.19518678	8.95062	1.8259878	3.4174654	13.38	6.4	0.00000
1595	TANGA	U	2.6450801	0.10874345	4.15919	2.3574450	4.3018794	12.87	7.6	0.00000
1596	ITZIGSOHN		2.8897040	0.12962370	13.27590	2.5151298	4.9122448	11.76	13.5	0.00000
1597	LAUGIER		2.8438032	0.09181660	11.83053	2.5826950	4.7956700	13.27	6.7	0.00000
1598	PALOQUE		2.3321199	0.08053749	7.53451	2.1442969	3.5614457	14.32	4.1	0.00000
1599	GIOMUS		3.1362472	0.13529199	6.09245	2.7119379	5.5541220	12.09	11.6	0.00000
1600	VYSSOTSKY	SREU	1.8490194	0.03761161	21.16867	1.7794749	2.5142715	14.17	2.6	0.00000
1601	PATRY		2.2344973	0.12970883	4.94345	1.9446632	3.3401787	13.64	4.5	0.00000
1602	INDIANA	S	2.2443123	0.10474261	4.16364	2.0092371	3.3622105	13.70	4.6	0.00000
1603	NEVA		2.7542970	0.09282474	8.56041	2.4986303	4.5710521	12.02	11.9	0.00000
1604	TOMBAUGH	U	3.0218909	0.10383640	9.40293	2.7081084	5.2531300	11.62	11.2	8.20000
1605	MILANKOVITCH		3.0126255	0.07150342	10.57089	2.7972124	5.2289886	11.19	13.8	0.00000
1606	JEKHOVSKY		2.6870892	0.31935322	7.68972	1.8289585	4.4047685	12.73	8.6	0.00000
1607	MAVIS		2.5466549	0.30976146	8.59883	1.7577994	4.0640135	12.68	8.8	0.00000
1608	MUNOZ		2.2144022	0.16925190	3.94414	1.8396103	3.2952223	13.68	4.4	0.00000
1609	BRENDA		2.5824847	0.24773967	18.73506	1.9427009	4.1500816	11.93	12.4	0.00000
1610	MIRNAYA		2.2022913	0.19868150	2.20071	1.7647367	3.2682261	14.67	2.8	0.00000
1611	BEYER		3.1861410	0.14824609	4.25118	2.7138081	5.6871872	11.84	13.0	0.00000
1612	HIROSE		3.0997002	0.09351732	16.85053	2.8098245	5.4573212	12.10	11.5	0.00000
1613	SMILEY		2.7385609	0.26107112	7.95312	2.0236018	4.5319347	12.83	8.2	0.00000
1614	GOLDSCHMIDT		2.9944775	0.07196038	14.09336	2.7789936	5.1818109	11.68	14.0	0.00000
1615	BARDWELL	SM	3.1101787	0.19386981	1.68163	2.5072088	5.4850173	11.88	14.0	20.00000
1616	FILIPOFF	SM	2.9101052	0.02062673	8.49066	2.8500793	4.9643574	12.29	13.0	0.00000
1617	ALSCHMITT		3.1916795	0.13961568	13.24690	2.7460709	5.7020221	12.03	11.9	0.00000
1618	DAWN		2.8685329	0.02685847	3.22458	2.7914886	4.8583608	12.87	6.4	0.00000
1619	UETA	S	2.2405400	0.17611282	6.21265	1.8459522	3.3537374	12.77	7.0	0.00000
1620	GEOGRAPHOS	S	1.2446955	0.33538482	13.32094	0.8272436	1.3886563	16.78	1.0	5.22500
1621	DRUZHBA	S	2.2307615	0.11910444	3.17249	1.9650679	3.3318059	12.80	6.9	0.00000
1622	CHACORNAC		2.2345476	0.16262622	6.46099	1.8711516	3.3402920	13.42	5.0	0.00000
1623	VIVIAN	SM	3.1249523	0.17372692	2.49340	2.5820639	5.5241456	11.81	15.0	0.00000
1624	RABE		3.1749218	0.11435474	1.98603	2.8118544	5.6571741	11.90	20.7	0.00000
1625	THE NORC	C	3.1840868	0.23320220	15.44123	2.4415507	5.6816878	11.55	23.2	0.00000
1626	SADEYA		2.3625412	0.27536872	25.30626	1.7119713	3.6313584	13.60	5.8	0.00000
1627	IVAR	S	1.8630213	0.39662695	8.44335	1.1240969	2.5428851	14.22	3.1	0.00000
1628	STROBEL		3.0165973	0.06104379	19.38622	2.8324528	5.2393327	11.56	14.8	0.00000
1629	PECKER		2.2382343	0.15369697	9.70618	1.8942245	3.3485618	14.00	4.8	0.00000
1630	MILET		3.0415459	0.16322310	4.52686	2.5450952	5.3044648	12.52	9.5	0.00000
1631	KOPFF		2.2352486	0.21364091	7.48329	1.7577080	3.3418636	13.57	4.6	0.00000
1632	SIEBOHME		2.6565299	0.13642935	5.71635	2.2941010	4.3298421	12.61	9.1	0.00000
1633	CHIMAY		3.1660185	0.14930432	2.67524	2.6933184	5.6333947	11.54	14.9	0.00000
1634	NDOLA		2.2458892	0.16266645	7.60462	1.8805584	3.3657551	14.54	3.0	0.00000
1635	BOHRMANN		2.8542569	0.05813130	1.81024	2.6883352	4.8221374	12.72	6.8	0.00000
1636	PORTER	S	2.2346337	0.12705724	4.43750	1.9507073	3.3404849	13.38	5.2	0.00000
1637	SWINGS		3.0672541	0.05350260	14.09368	2.9031479	5.3718591	11.26	16.9	0.00000
1638	RUANDA		2.7505858	0.18879807	0.28709	2.2312803	4.5618162	12.76	8.5	0.00000
1639	BOWER	C	2.5735667	0.15097269	8.40433	2.1850283	4.1286030	11.92	19.0	0.00000
1640	NEMO		2.2891471	0.34191474	7.10328	1.5064540	3.4634633	14.60	3.6	0.00000
1641	TANA		3.0210338	0.09691373	9.31341	2.7282541	5.2508955	12.50	7.6	0.00000
1642	HILL		2.7508855	0.06766784	10.81383	2.5647390	4.5625615	12.50	9.6	0.00000
1643	BROWN		2.4885364	0.20053899	3.52688	1.9894878	3.9256899	13.69	5.5	0.00000
1644	RAFITA	S	2.5482781	0.15383737	6.99262	2.1562579	4.0678997	12.11	9.3	5.15600
1645	WATERFIELD	CMEU	3.0547493	0.11953512	1.01683	2.6895995	5.3390422	12.58	9.2	0.00000
1646	ROSSELAND		2.3608351	0.11854157	8.38392	2.0809779	3.6274252	13.80	5.3	69.20000
1647	MENELAUS		5.2492046	0.02471780	5.63745	5.1194558	12.0265284	11.50	24.9	0.00000
1648	SHAJNA		2.2347176	0.20746610	4.56780	1.7710894	3.3406727	13.58	5.8	0.00000

No.	Name	Type	A (A.U.)	E	I (deg.)	q (A.U.)	Period (years)	Mag.	Rad. (km)	Rot. (hours)
1649	FABRE		3.0182047	0.04848612	10.82245	2.8718638	5.2435217	12.70	6.9	0.00000
1650	HECKMANN	F	2.4365487	0.16201915	2.73851	2.0417812	3.8033180	12.55	18.0	0.00000
1651	BEHRENS		2.1792340	0.06665280	5.07632	2.0339820	3.2170353	13.37	5.1	0.00000
1652	HERGE		2.2517443	0.14868025	3.19145	1.9169544	3.3789253	13.70	4.4	0.00000
1653	YAKHONTOVIA		2.6081936	0.32689464	4.05816	1.7555891	4.2122073	12.68	8.8	0.00000
1654	BOJEVA		3.0185928	0.08564074	10.44146	2.7600782	5.2445326	12.04	9.4	0.00000
1655	COMAS SOLA	CF	2.7821052	0.23367965	9.59323	2.1319838	4.6404519	12.77	13.0	0.00000
1656	SUOMI	S	1.8774574	0.12377776	25.07008	1.6450700	2.5724986	14.20	4.2	0.00000
1657	ROEMERA		2.3479252	0.23677124	23.41653	1.7920041	3.5977120	11.74	14.0	0.00000
1658	INNES	S	2.5628088	0.17982078	9.06996	2.1019626	4.1027427	12.73	7.3	0.00000
1659	PUNKAHARJU		2.7830184	0.25991777	16.54118	2.0596623	4.6427374	11.18	17.6	0.00000
1660	WOOD		2.3933556	0.30402511	20.56038	1.6657155	3.7026343	14.13	4.5	0.00000
1661	GRANULE		2.1834559	0.09111551	3.03175	1.9845093	3.2263885	14.00	3.8	0.00000
1662	HOFFMAN		2.7412472	0.17354777	4.24637	2.2655098	4.5386043	12.97	7.7	0.00000
1663	VAN DEN BOS		2.2392147	0.17989165	5.36442	1.8363986	3.3507619	14.78	2.6	0.00000
1664	FELIX		2.3378932	0.22597781	6.11340	1.8095812	3.5746787	13.70	5.5	0.00000
1665	GABY		2.4152944	0.20508435	10.81131	1.9199553	3.7536614	12.28	10.6	0.00000
1666	VAN GENT		2.1852293	0.18222912	2.68563	1.7870170	3.2303200	13.40	5.0	0.00000
1667	PELS		2.1893649	0.15617193	4.61976	1.8474474	3.2394941	13.50	4.8	0.00000
1668	HANNA		2.8056152	0.21520375	4.73050	2.2018361	4.6993971	13.46	6.2	0.00000
1669	DAGMAR	CU	3.1346633	0.11770904	0.94335	2.7656851	5.5499153	11.97	19.0	0.00000
1670	MINNAERT		2.9014864	0.10356078	10.51510	2.6010063	4.9423194	12.17	11.1	0.00000
1671	CHAIKA		2.5870144	0.25988418	3.96773	1.9146903	4.1610055	12.94	7.8	0.00000
1672	GEZELLE		3.2019348	0.25536299	1.04633	2.3842793	5.7295265	13.05	7.4	0.00000
1673	VAN HOUTEN		3.1057582	0.18014748	3.57930	2.5462637	5.4733276	12.10	11.5	0.00000
1674	GROENEVELD		3.1949818	0.14074980	2.68200	2.7452888	5.7108741	12.07	19.2	8.10000
1675	SIMONIDA		2.2338221	0.12516643	6.79548	1.9542226	3.3386655	13.06	5.8	5.30000
1676	KARIBA		2.2357984	0.18658476	6.13201	1.8186324	3.3430965	14.13	4.5	0.00000
1677	TYCHO BRAHE		2.5313134	0.10972927	14.79759	2.2535541	4.0273452	13.34	6.5	0.00000
1678	HVEEN		3.1628432	0.10501696	10.17901	2.8306911	5.6249223	11.98	12.2	0.00000
1679	NEVANLINNA		3.1266296	0.13758153	17.99885	2.6964631	5.5285935	11.54	14.9	0.00000
1680	PER BRAHE		2.7237928	0.18337476	4.26295	2.2243178	4.4953251	12.40	10.0	0.00000
1681	STEINMETZ	S	2.6960640	0.20538358	7.21387	2.1423366	4.4268546	12.71	7.2	0.00000
1682	KAREL		2.2389164	0.19182357	4.02801	1.8094395	3.3500929	14.50	3.0	0.00000
1683	CASTAFIORE		2.7362514	0.17788173	12.48157	2.2495222	4.5262027	12.84	8.2	0.00000
1684	IGUASSU		3.0868378	0.13658366	3.66363	2.6652260	5.4233880	12.70	8.7	0.00000
1685	TORO	S	1.3671771	0.43589413	9.37420	0.7712326	1.5985903	15.10	3.8	10.19600
1686	DE SITTER	SU	3.1566048	0.16838595	0.62184	2.6250768	5.6082878	11.92	8.5	0.00000
1687	GLARONA	CFPD	3.1466849	0.18572199	2.64217	2.5622764	5.5818725	11.24	22.0	6.50000
1688	WILKENS		2.6220908	0.23713449	11.74855	2.0003026	4.2459178	13.30	6.6	0.00000
1689	FLORIS-JAN		2.4502041	0.20589009	6.38200	1.9457314	3.8353357	12.80	8.3	0.00000
1690	MAYRHOFER		3.0414848	0.09019651	13.01167	2.7671535	5.3043051	11.80	13.2	0.00000
1691	OORT	C	3.1779253	0.15754117	1.05944	2.6772711	5.6652036	11.66	20.0	0.00000
1692	SUBBOTINA		2.7889709	0.13374865	2.42250	2.4159498	4.6576405	12.38	10.1	0.00000
1693	HERTZSPRUNG	CU	2.8021374	0.26891270	11.93080	2.0486071	4.6906619	12.01	19.0	0.00000
1694	KAISER	C	2.3961177	0.25716108	11.08323	1.7799295	3.7090459	13.66	9.0	0.00000
1695	WALBECK		2.7837181	0.29049331	16.64620	1.9750665	4.6444883	13.00	7.6	0.00000
1696	NURMELA		2.2616813	0.09917633	6.04230	2.0373759	3.4013169	14.30	3.3	0.00000
1697	KOSKENNIEMI		2.3741510	0.11736472	5.66707	2.0955093	3.6581583	13.20	6.9	0.00000
1698	CHRISTOPHE		3.1522098	0.12021517	1.52749	2.7732663	5.5965791	12.40	16.5	0.00000
1699	HONKASALO		2.2115827	0.16551232	1.97108	1.8455385	3.2889311	14.32	3.3	0.00000
1700	ZVEZDARA	EMP	2.3611765	0.22455075	4.51816	1.8309726	3.6282125	13.62	5.7	0.00000
1701	OKAVANGO		3.1829581	0.17144869	16.29182	2.6372442	5.6786671	11.50	15.2	0.00000
1702	KALAHARI	MU	2.8576336	0.14144932	9.95347	2.4534233	4.8306971	12.30	13.9	0.00000
1703	BARRY		2.2147145	0.17183816	4.52366	1.8341421	3.2959199	14.40	3.2	0.00000
1704	WACHMANN		2.2224369	0.08784454	0.96826	2.0272081	3.3131735	13.82	4.1	0.00000
1705	TAPIO		2.2986948	0.24571256	7.69600	1.7338767	3.4851542	14.20	4.4	0.00000
1706	DIECKVOSS		2.1255307	0.11380022	1.87069	1.8836448	3.0988536	13.88	5.1	0.00000
1707	CHANTAL	SU	2.2193563	0.17018872	4.03692	1.8416469	3.3062873	13.66	4.6	0.00000
1708	POLIT		2.9164970	0.30416927	6.06601	2.0293882	4.9807220	12.90	8.0	0.00000
1709	UKRAINA		2.3786173	0.21424793	7.57655	1.8690035	3.6684859	13.90	5.0	0.00000
1710	GOTHARD		2.3214371	0.26832542	8.46168	1.6985365	3.5370026	14.54	3.7	0.00000
1711	SANDRINE		3.0138364	0.10897656	11.10021	2.6853988	5.2321415	12.04	9.4	0.00000
1712	ANGOLA		3.1838493	0.13392876	19.22302	2.7574403	5.6810522	11.00	19.1	0.00000
1713	BANCILHON		2.2286911	0.18397024	3.74849	1.8186783	3.3271689	14.40	3.2	0.00000
1714	SY		2.5650458	0.15879852	7.99022	2.1577203	4.1081161	12.74	8.6	0.00000
1715	SALLI		2.4015114	0.23647183	11.46379	1.8336217	3.7215769	13.40	6.3	0.00000
1716	PETER		2.7332499	0.09215908	5.72319	2.4813559	4.5187569	13.00	7.6	0.00000
1717	ARLON	U	2.1956499	0.12899600	6.18966	1.9124198	3.2534535	13.55	5.9	0.00000
1718	NAMIBIA		2.3656130	0.27653113	7.69042	1.7114474	3.6384428	14.90	3.2	0.00000
1719	JENS		2.6555698	0.22332221	14.27945	2.0625222	4.3274951	12.60	12.1	0.00000
1720	NIELS		2.1882730	0.10363630	0.73189	1.9614884	3.2370710	14.26	4.3	0.00000
1721	WELLS		3.1557426	0.05130684	16.08261	2.9938314	5.6059909	11.90	12.6	0.00000
1722	GOFFIN		2.5141683	0.04863101	5.47742	2.3919017	3.9864976	13.00	7.6	0.00000
1723	KLEMOLA	C	3.0145688	0.03926581	10.92931	2.8961992	5.2340488	11.06	30.0	0.00000
1724	VLADIMIR	FU	2.7107687	0.05849018	12.23001	2.5522153	4.4631214	12.00	19.0	0.00000
1725	CRAO		2.9032247	0.08965069	3.17262	2.6429486	4.9467616	12.18	8.8	0.00000
1726	HOFFMEISTER		2.7866426	0.04405327	3.47523	2.6638820	4.6518092	13.00	7.6	0.00000
1727	METTE	RA	1.8539528	0.10224461	22.89979	1.6643960	2.5243409	14.20	2.4	0.00000
1728	GOETHE LINK		2.5623240	0.09170553	7.20825	2.3273449	4.1015792	12.70	8.7	0.00000
1729	BERYL		2.2295587	0.10021791	2.44268	2.0061171	3.3291118	13.49	6.1	0.00000
1730	MARCELINE		2.7816050	0.22619528	9.53240	2.1524191	4.6392007	12.80	11.1	0.00000
1731	SMUTS		3.1794457	0.10886592	5.89808	2.8333125	5.6692700	11.00	19.1	0.00000

No.	Name	Type	A (A.U.)	E	I (deg.)	q (A.U.)	Period (years)	Mag.	Rad. (km)	Rot. (hours)
1732	HEIKE		3.0091085	0.11637694	10.79771	2.6589177	5.2198348	11.89	10.0	0.00000
1733	SILKE		2.1926870	0.08356377	4.43275	2.0094578	3.2468708	14.13	3.6	0.00000
1734	ZHONGOLOVICH		2.7796302	0.22922501	8.33698	2.1424694	4.6342616	12.55	9.4	0.00000
1735	ITA		3.1482596	0.11366056	15.63775	2.7904267	5.5860629	10.70	21.9	0.00000
1736	FLOIRAC		2.2282858	0.16946538	4.55211	1.8506684	3.3262608	13.30	5.2	0.00000
1737	SEVERNY		3.0158515	0.04294697	9.40236	2.8863299	5.2373900	12.10	9.1	0.00000
1738	OOSTERHOFF		2.1834378	0.20263705	4.87724	1.7409925	3.2263484	13.70	4.4	0.00000
1739	MEYERMANN		2.2610734	0.12330985	3.40643	1.9822608	3.3999457	13.70	5.5	0.00000
1740	PAAVO NURMI		2.4665756	0.18895523	2.00229	2.0005033	3.8738399	14.40	4.0	0.00000
1741	GICLAS		2.8830087	0.07029673	2.89585	2.6803427	4.8951831	12.60	7.2	0.00000
1742	SCHAIFERS		2.8901770	0.09368156	2.48939	2.6194208	4.9134512	12.30	8.3	0.00000
1743	SCHMIDT		2.4710438	0.13894507	6.36964	2.1277044	3.8843708	13.48	6.1	0.00000
1744	HARRIET		2.2292032	0.12068725	4.40753	1.9601668	3.3283155	14.86	2.6	0.00000
1745	FERGUSON		2.8465226	0.05456217	3.26005	2.6912100	4.8025508	13.17	5.6	0.00000
1746	BROUWER	D	3.9681706	0.20003851	8.38195	3.1743836	7.9047022	10.90	32.0	0.00000
1747	WRIGHT		1.7089694	0.11051382	21.41228	1.5201046	2.2340939	14.82	4.1	0.00000
1748	MAUDERLI	D	3.9217870	0.23386978	3.29703	3.0045996	7.7665119	11.70	25.0	0.00000
1749	TELAMON		5.2373185	0.11305731	6.07375	4.6452012	11.9857025	11.20	28.6	0.00000
1750	ECKERT	S	1.9266099	0.17267968	19.07874	1.5939234	2.6741803	14.66	2.9	0.00000
1751	HERGET		2.7908247	0.17294107	8.11323	2.3081765	4.6622849	13.50	8.0	0.00000
1752	VAN HERK		2.2383580	0.20069855	3.50070	1.7891229	3.3488395	14.70	4.6	0.00000
1753	MIEKE		3.0159645	0.07570184	11.38281	2.7876503	5.2376842	12.30	13.9	8.80000
1754	CUNNINGHAM	EMU	3.9622166	0.16510457	12.11131	3.3080366	7.8869181	10.70	33.2	0.00000
1755	LORBACH	U	3.0910580	0.04357390	10.70062	2.9563687	5.4345145	12.01	13.0	0.00000
1756	GIACOBINI		2.5489321	0.22757661	5.12212	1.9688547	4.0694656	13.90	5.0	0.00000
1757	PORVOO		2.3514929	0.12575699	3.97552	2.0557761	3.6059151	14.10	6.1	4.89000
1758	NAANTALI		3.0113926	0.03383375	10.82260	2.9095058	5.2257791	12.00	9.5	0.00000
1759	KIENLE		2.6480925	0.31475845	4.56951	1.8145831	4.3092308	14.10	4.6	29.25000
1760	SANDRA		3.1570027	0.11775878	8.42427	2.7852378	5.6093483	12.60	9.1	0.00000
1761	EDMONDSON		3.1582985	0.24295048	2.46928	2.3909883	5.6128025	12.60	9.1	0.00000
1762	RUSSELL		2.8741241	0.07773321	2.27636	2.6507092	4.8725719	12.80	6.6	0.00000
1763	WILLIAMS		2.1883168	0.20368505	4.23808	1.7425894	3.2371683	14.20	3.5	0.00000
1764	COGSHALL		3.0856340	0.12931515	2.23334	2.6866148	5.4202161	12.50	15.7	0.00000
1765	WRUBEL	CMEU	3.1676118	0.18542530	19.94861	2.5802565	5.6376481	10.94	19.2	0.00000
1766	SLIPHER		2.7488482	0.08588535	5.21869	2.5127625	4.5574946	13.50	6.0	0.00000
1767	LAMPLAND	C	3.0224047	0.09388342	9.82095	2.7386510	5.2544699	13.22	11.0	0.00000
1768	APPENZELLA		2.4510403	0.17977329	3.27482	2.0104086	3.8372993	14.20	4.4	0.00000
1769	CARLOSTORRES		2.1782477	0.14175013	1.58840	1.8694808	3.2148514	14.00	4.8	0.00000
1770	SCHLESINGER		2.4569688	0.06075937	5.30101	2.3076849	3.8512299	14.10	4.6	0.00000
1771	MAKOVER		3.1209137	0.17695341	11.25278	2.5686574	5.5134397	11.20	17.4	0.00000
1772	GAGARIN		2.5303795	0.10116654	5.74654	2.2743897	4.0251169	13.20	6.9	0.00000
1773	RUMPELSTILZ		2.4358001	0.12882619	5.40066	2.1220052	3.8015654	13.30	6.6	0.00000
1774	KULIKOV		2.8748970	0.07044291	1.85247	2.6723809	4.8745375	13.50	4.8	0.00000
1775	ZIMMERWALD		2.6035204	0.18132597	12.58941	2.1314344	4.2008915	13.30	6.6	0.00000
1776	KUIPER	CFPD	3.1007929	0.02650961	9.48014	3.0185921	5.4602075	12.10	22.0	0.00000
1777	GEHRELS		2.6258781	0.01883958	3.15913	2.5764077	4.2551203	12.92	7.9	0.00000
1778	ALFVEN		3.1437871	0.12784509	2.46042	2.7418692	5.5741634	12.93	12.9	0.00000
1779	PARANA		2.1757109	0.16128314	0.89573	1.8248053	3.2092366	15.40	2.5	0.00000
1780	KIPPES		3.0134201	0.05726404	8.99468	2.8408594	5.2310576	11.90	10.0	0.00000
1781	VAN BIESBROECK		2.3949564	0.10720021	6.93854	2.1382165	3.7063498	14.00	4.8	0.00000
1782	SCHNELLER	CFPD	3.1268988	0.14899217	1.54164	2.6610153	5.5293069	12.70	11.0	0.00000
1783	ALBITSKIJ		2.6615369	0.13409764	11.47702	2.3046310	4.3420892	12.60	9.1	0.00000
1784	BENGUELA		2.4060559	0.13016015	1.47408	2.0928833	3.7321455	13.50	6.0	0.00000
1785	WURM		2.2370176	0.06818164	3.77194	2.0844941	3.3458321	13.90	4.0	0.00000
1786	RAAHE		3.0206802	0.10576844	10.43531	2.7011876	5.2499733	12.10	9.1	0.00000
1787	CHINY		3.0085096	0.04797816	8.92764	2.8641667	5.2182765	12.40	7.9	0.00000
1788	KIESS		3.1034946	0.16866399	0.67138	2.5800469	5.4673452	12.80	13.7	0.00000
1789	DOBROVOLSKY		2.2136812	0.18855810	1.97834	1.7962737	3.2936132	14.30	3.3	5.80000
1790	VOLKOV		2.2376370	0.10098655	5.11159	2.0116658	3.3472216	14.00	3.8	0.00000
1791	PATSAYEV		2.7460127	0.14281216	5.37089	2.3538487	4.5504446	13.10	7.3	0.00000
1792	RENI	C	2.7784452	0.27783528	8.99987	2.0064950	4.6312981	13.06	12.0	0.00000
1793	ZOYA		2.2244525	0.09738010	1.50752	2.0078351	3.3176818	13.70	4.4	7.00000
1794	FINSEN	C	3.1170311	0.16801469	14.56859	2.5995581	5.5031543	11.90	20.0	0.00000
1795	WOLTJER		2.7852862	0.18999924	7.55179	2.2560840	4.6484132	13.03	7.5	0.00000
1796	RIGA		3.3476965	0.06475094	22.75292	3.1309299	6.1251836	11.60	23.8	0.00000
1797	SCHAUMASSE		2.2371504	0.02445855	3.14131	2.1824329	3.3461297	13.90	5.0	0.00000
1798	WATTS		2.1995232	0.12253190	6.19616	1.9300114	3.2620666	13.70	5.5	0.00000
1799	KOUSSEVITSKY		3.0232778	0.12374454	11.51697	2.6491635	5.2567468	12.40	7.9	0.00000
1800	AGUILAR		2.3569198	0.13562460	5.79118	2.0372634	3.6184053	14.00	4.8	0.00000
1801	TITICACA		3.0186512	0.07439303	10.98988	2.7940848	5.2446852	12.30	8.3	0.00000
1802	ZHANG HENG		2.8429255	0.03677727	2.68338	2.7383704	4.7934504	13.10	9.6	0.00000
1803	ZWICKY		2.3503594	0.24767490	21.55842	1.7682344	3.6033082	13.30	6.6	0.00000
1804	CHEBOTAREV		2.4100962	0.02002916	3.64312	2.3618240	3.7415502	13.40	6.3	0.00000
1805	DIRIKIS		3.1286805	0.12727343	2.52205	2.7304826	5.5340338	12.50	15.7	0.00000
1806	DERICE		2.2364173	0.10597830	3.83969	1.9994056	3.3444850	14.00	4.8	0.00000
1807	SLOVAKIA		2.2266505	0.17795074	3.48360	1.8304164	3.3226001	14.00	4.8	0.00000
1808	BELLEROPHON		2.7464051	0.18031763	2.04265	2.2511799	4.5514202	13.33	6.5	4.00000
1809	PROMETHEUS		2.9251635	0.10133224	3.26177	2.6287501	5.0029387	12.80	8.3	0.00000
1810	EPIMETHEUS		2.2236252	0.09283619	4.03151	2.0171924	3.3158312	13.87	5.1	0.00000
1811	BRUWER		3.1456988	0.09058718	8.49589	2.8607388	5.5792484	12.24	10.8	0.00000
1812	GILGAMESH		3.0132089	0.07915759	10.26569	2.7746904	5.2305079	12.72	6.8	0.00000
1813	IMHOTEP		2.6846280	0.07965799	8.09450	2.4707758	4.3987184	13.62	5.7	0.00000
1814	BACH		2.2258468	0.13121496	4.34780	1.9337823	3.3208013	14.20	4.4	0.00000

No.	Name	Type	A (A.U.)	E	I (deg.)	q (A.U.)	Period (years)	Mag.	Rad. (km)	Rot. (hours)
1815	BEETHOVEN		3.1772430	0.17410760	2.71885	2.6240609	5.6633792	12.40	10.0	0.00000
1816	LIBERIA		2.3385394	0.21947223	26.14205	1.8252949	3.5761609	14.70	3.5	0.00000
1817	KATANGA		2.3723702	0.19056129	25.68105	1.9202884	3.6540434	13.30	6.6	0.00000
1818	BRAHMS		2.1641018	0.17810296	2.97731	1.7786689	3.1835856	15.20	2.8	0.00000
1819	LAPUTA		3.1594877	0.21387506	23.61522	2.4837520	5.6159730	12.00	12.1	0.00000
1820	LOHMANN		2.1980803	0.20998435	4.99784	1.7365178	3.2588570	14.60	3.6	0.00000
1821	ACONCAGUA		2.3788314	0.20107080	2.10444	1.9005179	3.6689813	14.80	3.3	0.00000
1822	WATERMAN		2.1698079	0.15338632	0.95473	1.8369890	3.1961851	14.60	3.6	0.00000
1823	GLIESE		2.2259631	0.13526373	2.88846	1.9248708	3.3210614	14.20	5.8	0.00000
1824	HAWORTH		2.8844712	0.04193549	1.93448	2.7635095	4.8989081	12.80	6.6	0.00000
1825	KLARE		2.6768117	0.11501821	4.04049	2.3689296	4.3795223	12.90	8.0	0.00000
1826	MILLER		3.0021963	0.08086345	9.20890	2.7594285	5.2018600	12.30	8.3	0.00000
1827	ATKINSON		2.7086565	0.18096364	4.51925	2.2184882	4.4579058	13.49	6.1	0.00000
1828	KASHIRINA		3.0566561	0.11518433	14.31813	2.7045772	5.3440423	12.20	11.0	0.00000
1829	DAWSON		2.2512541	0.12005581	6.32853	1.9809780	3.3778222	13.70	5.5	0.00000
1830	POGSON	S	2.1879969	0.05606458	3.95666	2.0653276	3.2364585	13.61	4.8	0.00000
1831	NICHOLSON		2.2391167	0.12856761	5.63763	1.9512388	3.3505423	13.20	6.9	0.00000
1832	MRKOS		3.2051189	0.11727171	15.02082	2.8292489	5.7380748	11.60	19.2	0.00000
1833	SHMAKOVA		2.6349521	0.11235546	10.00463	2.3389008	4.2771950	13.05	7.4	0.00000
1834	1969QP		3.0196970	0.07781754	9.44296	2.7847116	5.2474103	12.70	6.9	0.00000
1835	GAJDARIYA		2.8326726	0.08786319	0.98983	2.5837848	4.7675424	12.70	6.9	0.00000
1836	KOMAROV		2.7828269	0.19239151	7.00956	2.2474346	4.6422582	12.60	9.1	0.00000
1837	OSITA		2.2057602	0.08627031	3.84134	2.0154686	3.2759514	14.90	3.2	0.00000
1838	URSA		3.2071269	0.02965508	22.07812	3.1120193	5.7434678	11.90	12.6	0.00000
1839	RAGAZZA		2.8004327	0.16569664	10.16896	2.3364103	4.6863823	12.70	8.7	0.00000
1840	HUS		2.9177046	0.01567095	2.41405	2.8719814	4.9838161	12.90	10.6	0.00000
1841	MASARYK		3.4340267	0.08519980	2.63066	3.1414483	6.3636384	11.70	22.7	0.00000
1842	HYNEK	S	2.2669344	0.17952314	5.35496	1.8599674	3.4131742	13.50	4.9	0.00000
1843	JARMILA		2.6530771	0.16797511	8.42343	2.2074263	4.3214035	12.60	9.1	0.00000
1844	SUSILVA		3.0145433	0.04837436	11.78423	2.8687167	5.2339826	12.50	7.6	0.00000
1845	HELEWALDA		2.9682751	0.05611200	10.71305	2.8017192	5.1139469	12.90	8.0	0.00000
1846	BENGT		2.3384838	0.14114176	3.18727	2.0084262	3.5760336	14.64	3.6	0.00000
1847	STOBBE		2.6101346	0.01941709	11.14621	2.5594532	4.2169099	12.00	12.1	0.00000
1848	DELVAUX		2.8704891	0.04119536	1.44666	2.7522385	4.8633313	11.80	10.4	0.00000
1849	KRESAK		3.0515912	0.01631848	10.77958	3.0017939	5.3307648	12.20	11.0	0.00000
1850	KOHOUTEK		2.2510235	0.12469046	4.05059	1.9703424	3.3773034	14.20	4.4	0.00000
1851	LACROUTE		3.1126211	0.18414211	1.66766	2.5394566	5.4914799	13.10	11.9	0.00000
1852	CARPENTER		3.0143745	0.06631405	11.20143	2.8144791	5.2335429	11.80	10.4	0.00000
1853	MCELROY		3.0644317	0.04991427	15.76432	2.9114728	5.3644466	11.60	14.5	0.00000
1854	SKVORTSOV		2.5396800	0.13612652	4.91048	2.1939621	4.0473289	13.60	5.8	0.00000
1855	KOROLEV		2.2477927	0.08393115	3.07833	2.0591331	3.3700349	13.80	5.3	0.00000
1856	RUZENA		2.2367105	0.08016550	4.74213	2.0574036	3.3451431	13.40	6.3	0.00000
1857	PARCHOMENKO		2.2439811	0.13418929	4.39446	1.9428627	3.3614664	13.70	5.5	0.00000
1858	LOBACHEVSKIJ		2.6981592	0.07840187	1.65601	2.4866185	4.4320159	12.80	8.3	0.00000
1859	KOVALEVSKAYA		3.2050645	0.10844831	7.72379	2.8574805	5.7379289	11.20	17.4	0.00000
1860	BARBAROSSA		2.5640793	0.20611674	9.95268	2.0355797	4.1057944	12.60	9.1	0.00000
1861	KOMENSKY		3.0156269	0.07015201	10.47934	2.8040745	5.2368050	12.90	6.3	0.00000
1862	APOLLO	S	1.4711586	0.56006277	6.34982	0.6472174	1.7843878	17.00	0.7	3.06500
1863	ANTINOUS		2.2594247	0.60650402	18.41831	0.8890746	3.3962281	16.50	1.5	0.00000
1864	DAEDALUS	SU	1.4609457	0.61489320	22.15929	0.5626202	1.7658390	15.96	1.6	8.57000
1865	CERBERUS	S	1.0800560	0.46707243	16.09410	0.5755916	1.1224562	17.50	0.8	6.80000
1866	SISYPHUS	U	1.8935918	0.53944683	41.14318	0.8720997	2.6057308	14.50	3.8	0.00000
1867	DEIPHOBOS	D	5.1691647	0.04504657	26.85278	4.9363117	11.7525082	9.42	70.0	0.00000
1868	THERSITES	CFPD	5.2610359	0.11008479	16.80346	4.6818757	12.0672112	10.75	52.0	0.00000
1869	PHILOCTETES		5.3108206	0.06386228	3.95938	4.9716592	12.2389011	12.30	13.9	0.00000
1870	GLAUKOS		5.2424116	0.03253635	6.58042	5.0718427	12.0031900	11.90	20.7	0.00000
1871	ASTYANAX		5.3385363	0.03520115	8.56669	5.1506138	12.3348341	12.30	17.3	0.00000
1872	HELENOS		5.1726894	0.04583840	14.73885	4.9355817	11.7645302	11.50	24.9	0.00000
1873	AGENOR		5.2663283	0.09167741	21.84358	4.7835250	12.0854235	11.70	22.7	0.00000
1874	KACIVELIA		3.1293249	0.30270168	4.87535	2.1820729	5.5357437	12.10	11.5	0.00000
1875	1969QQ		3.1262879	0.17836092	13.42096	2.5686803	5.5276875	13.50	6.0	0.00000
1876	NAPOLITANIA		1.9642235	0.04824370	23.11127	1.8694621	2.7528741	16.00	1.1	0.00000
1877	MARSDEN	CFPD	3.9466646	0.21338572	17.54474	3.1045029	7.8405285	12.40	22.0	0.00000
1878	HUGHES		2.8492901	0.01217764	1.77143	2.8145926	4.8095560	12.40	7.9	0.00000
1879	BROEDERSTROOM		2.2455535	0.14798611	1.72221	1.9132427	3.3650000	14.20	5.8	0.00000
1880	MCCROSKY		2.6758776	0.07624406	4.84836	2.4718578	4.3772297	12.70	8.7	0.00000
1881	SHAO		3.1624410	0.11188726	9.88209	2.8086040	5.6238489	12.10	11.5	0.00000
1882	RAUMA		3.0071957	0.09250975	9.49217	2.7290008	5.2148585	12.10	9.1	0.00000
1883	RIMITO		2.4132373	0.26227117	25.47420	1.7803148	3.7488673	14.30	4.2	0.00000
1884	SKIP		2.4245992	0.26562893	21.76786	1.7805554	3.7753735	14.30	4.2	0.00000
1885	HERERO		2.2494760	0.24756144	5.66341	1.6925924	3.3738208	14.70	3.5	0.00000
1886	LOWELL		2.6264095	0.15775277	14.92748	2.2120862	4.2564120	12.70	8.7	0.00000
1887	VIRTON		3.0047290	0.11709671	9.64135	2.6528852	5.2084436	11.80	10.4	0.00000
1888	ZU CHONG-ZHI		2.5491855	0.16366529	5.86393	2.1319723	4.0700727	13.10	7.3	0.00000
1889	PAKHMUTOVA		3.0844858	0.11745868	13.20054	2.7221861	5.4171910	12.00	12.1	0.00000
1890	KONOSHENKOVA		3.2125463	0.13709483	9.90179	2.7721229	5.7580323	12.50	9.6	0.00000
1891	GONDOLA		2.7054732	0.06880130	11.50977	2.5193331	4.4500499	13.00	7.6	0.00000
1892	LUCIENNE		2.4607191	0.09083927	13.99784	2.2371891	3.8600509	13.20	6.9	0.00000
1893	JAKOBA		2.7082369	0.05363757	10.04302	2.5629735	4.4568701	12.40	10.0	0.00000
1894	HAFFNER		2.8857536	0.07287147	0.90538	2.6754646	4.9021759	13.40	5.0	0.00000
1895	LARINK		3.1684868	0.16863278	1.83052	2.6341760	5.6399837	13.00	12.5	0.00000
1896	BEER		2.3676009	0.22137986	2.22128	1.8434618	3.6430302	14.80	3.3	0.00000
1897	HIND		2.2833655	0.14177449	4.05744	1.9596426	3.4503503	14.40	4.0	0.00000

No.	Name	Type	A (A.U.)	E	I (deg.)	q (A.U.)	Period (years)	Mag.	Rad. (km)	Rot. (hours)
1898	COWELL		3.1114504	0.17367497	1.02314	2.5710695	5.4883823	13.30	10.9	0.00000
1899	CROMMELIN		2.2649188	0.10716728	7.27817	2.0221934	3.4086227	14.00	4.8	0.00000
1900	KATYUSHA		2.2090394	0.13505119	6.54377	1.9107060	3.2832594	13.50	6.0	0.00000
1901	MORAVIA		3.2520328	0.05919721	23.94410	3.0595214	5.8645182	12.50	9.6	0.00000
1902	SHAPOSHNIKOV	C	3.9754615	0.22536227	12.49770	3.0795424	7.9264975	10.60	36.0	0.00000
1903	ADZHIMUSHKAJ		3.0010779	0.05108126	10.98579	2.8477788	5.1989527	12.00	9.5	0.00000
1904	MASSEVITCH		2.7428875	0.07443818	12.82626	2.5387120	4.5426788	12.80	8.3	0.00000
1905	AMBARTSUMIAN		2.2239008	0.16281734	2.61584	1.8618112	3.3164475	14.00	4.8	0.00000
1906	NAEF	X	2.3729224	0.13463309	6.46958	2.0534487	3.6553195	13.80	5.3	0.00000
1907	RUDNEVA		2.5457847	0.04556800	3.21210	2.4297786	4.0619311	13.20	6.9	0.00000
1908	POBEDA		2.8916049	0.03373096	4.78476	2.7940681	4.9170928	12.50	9.6	0.00000
1909	ALEKHIN		2.4250879	0.22203997	1.77412	1.8866215	3.7765152	13.40	6.3	0.00000
1910	MIKHAILOV		3.0438976	0.05122764	10.35247	2.8879659	5.3106184	11.70	13.8	0.00000
1911	SCHUBART	C	3.9808073	0.16125935	1.65536	3.3388648	7.9424906	11.30	41.0	0.00000
1912	ANUBIS		2.9028351	0.09194544	3.16103	2.6359327	4.9457660	13.14	5.6	0.00000
1913	SEKANINA		2.8787916	0.07913785	1.57651	2.6509700	4.8844461	12.30	8.3	0.00000
1914	HARTBEESPOORTDAM		2.4056485	0.14943977	5.68799	2.0461490	3.7311976	13.60	5.8	0.00000
1915	QUETZALCOATL	SU	2.5311143	0.57686192	20.50367	1.0710109	4.0268703	20.10	0.2	4.90000
1916	BOREAS	S	2.2734425	0.44980657	12.84554	1.2508332	3.4278829	16.10	1.5	0.00000
1917	CUYO		2.1489129	0.50507063	23.98923	1.0635601	3.1501284	16.50	1.5	0.00000
1918	AIQUILLON		3.1980393	0.11712587	9.22453	2.8234663	5.7190738	12.30	10.5	0.00000
1919	CLEMENCE		1.9360728	0.09514294	19.33563	1.7518691	2.6939065	15.00	1.7	0.00000
1920	SARMIENTO		1.9300973	0.10573095	22.79615	1.7260263	2.6814444	15.60	1.3	0.00000
1921	PALA	X	3.2404256	0.40977538	19.61179	1.9125791	5.8331490	15.60	2.3	0.00000
1922	ZULU	X	3.2490613	0.47587177	35.37846	1.7029248	5.8564830	12.90	8.0	0.00000
1923	OSIRIS		2.4347210	0.06577125	4.95801	2.2745862	3.7990391	14.64	3.6	0.00000
1924	HORUS		2.3388729	0.13222110	2.73373	2.0296247	3.5769258	14.34	4.1	0.00000
1925	FRANKLIN-ADAMS		2.5517342	0.17796566	7.71904	2.0976131	4.0761781	13.50	6.0	0.00000
1926	DEMIDDLAER		2.6568134	0.10735550	13.74168	2.3715899	4.3305354	13.20	6.9	0.00000
1927	SUVANTO		2.6502371	0.14886625	13.36772	2.2557063	4.3144665	12.90	8.0	0.00000
1928	SUMMA		2.4777648	0.19885220	4.55731	1.9850559	3.9002292	13.90	5.0	0.00000
1929	KOLLAA		2.3626535	0.07656966	7.77364	2.1817460	3.6316171	13.50	6.0	0.00000
1930	LUCIFER		2.8973093	0.14452885	14.07137	2.4785645	4.9316506	12.30	10.5	0.00000
1931	1969QB		2.5392332	0.27224779	8.20147	1.8479326	4.0462608	14.40	6.2	0.00000
1932	JANSKY		2.3715982	0.15779157	1.89277	1.9973800	3.6522598	14.60	3.6	0.00000
1933	TINCHEN		2.3531256	0.12278327	6.88827	2.0642011	3.6096711	14.40	4.0	0.00000
1934	JEFFERS		2.3892682	0.29938498	23.16091	1.6739571	3.6931531	14.00	4.8	0.00000
1935	LUCERNA		2.6279178	0.22707577	9.55560	2.0311811	4.2600789	14.40	4.0	0.00000
1936	LUGANO		2.6739314	0.13903624	10.23984	2.3021581	4.3724556	12.50	9.6	0.00000
1937	LOCARNO		2.3778820	0.15626629	12.46131	2.0062990	3.6667848	13.30	6.6	0.00000
1938	LAUSANNA		2.2356365	0.16050997	3.33607	1.8767945	3.3427334	14.00	4.8	0.00000
1939	LORETTA		3.1217608	0.12595733	0.90945	2.7285523	5.5156851	12.00	19.8	0.00000
1940	WHIPPLE		3.0602193	0.06830019	6.55716	2.8512056	5.3533893	12.50	9.6	0.00000
1941	WILD	CFPD	3.9934976	0.27763435	3.94531	2.8847654	7.9805012	12.50	13.0	0.00000
1942	JABLUNKA		2.3185012	0.18449935	24.35307	1.8907392	3.5302949	14.20	4.4	0.00000
1943	ANTEROS	S	1.4301506	0.25594398	8.70271	1.0641121	1.7103014	16.50	2.0	0.00000
1944	GUNTER		2.2396207	0.23634858	5.48446	1.7102895	3.3516736	14.80	3.3	0.00000
1945	WESSELINK		2.5558705	0.17749229	4.21496	2.1022232	4.0860934	13.50	6.0	0.00000
1946	1931PH		2.2933800	0.23527101	8.15952	1.7538143	3.4730742	14.00	4.8	10.22300
1947	ISO-HEIKKIL		3.1511581	0.04068936	11.87011	3.0229394	5.5937786	11.50	15.2	0.00000
1948	KAMPALA		2.5339868	0.16844189	5.82179	2.1071572	4.0337267	13.50	6.0	0.00000
1949	MESSINA		2.3841181	0.22993964	4.64740	1.8359149	3.6812186	14.60	3.6	0.00000
1950	WEMPE		2.1782537	0.08444192	4.22303	1.9943178	3.2148645	14.00	4.8	0.00000
1951	LICK		1.3905056	0.06171719	39.09571	1.3046875	1.6396800	17.20	1.1	0.00000
1952	HESBURGH	C	3.1099133	0.14859524	14.23650	2.6477950	5.4843154	11.34	26.0	0.00000
1953	RUPERTWILDT		3.1213205	0.17052154	2.46109	2.5890682	5.5145178	12.90	13.1	0.00000
1954	KUKARKIN		2.9375944	0.31069899	14.84972	2.0248868	5.0348639	13.20	6.9	0.00000
1955	MCMATH		2.8531890	0.06554613	0.99775	2.6661737	4.8194318	12.60	7.2	0.00000
1956	ARTEK		3.2236903	0.09075382	1.46626	2.9311280	5.7880192	13.00	12.5	0.00000
1957	ANGARA		3.0077150	0.05345072	11.19848	2.8469505	5.2162099	12.00	9.5	0.00000
1958	CHANDRA		3.1076522	0.16164076	10.56491	2.6053288	5.4783354	12.10	11.5	0.00000
1959	KARBYSHEV		2.3156610	0.13350032	6.20182	2.0065196	3.5238099	14.00	4.8	0.00000
1960	GUISAN		2.5241313	0.12679411	8.45935	2.2040863	4.0102177	12.61	9.1	0.00000
1961	DUFOUR		3.2037776	0.10698755	6.62967	2.8610132	5.7344732	12.30	10.5	0.00000
1962	DUNANT		3.2053454	0.21801138	1.59274	2.5065436	5.7386832	13.30	10.9	0.00000
1963	BEZOVEC		2.4235165	0.20894460	25.00467	1.9171357	3.7728448	12.00	12.1	0.00000
1964	LUYTEN		2.4657137	0.19320183	2.36887	1.9893333	3.8718095	14.46	3.9	0.00000
1965	VAN DE KAMP		2.5682762	0.10757001	2.22583	2.2920065	4.1158786	13.42	6.3	0.00000
1966	TRISTAN		2.4477885	0.08873491	2.48862	2.2305841	3.8296652	15.14	2.8	0.00000
1967	MENZEL		2.2333214	0.13846759	3.90149	1.9240789	3.3375430	14.00	4.8	0.00000
1968	MEHLTRETTER		2.7391870	0.11263972	4.60168	2.4306457	4.5334883	12.80	8.3	0.00000
1969	ALAIN		3.1087151	0.14368577	3.27304	2.6620369	5.4811459	12.60	9.1	0.00000
1970	1949BF		2.7810383	0.15980205	7.08755	2.3366227	4.6377835	12.70	8.7	0.00000
1971	HAGIHARA		2.9981966	0.08279067	8.68804	2.7499740	5.1914678	13.40	6.3	0.00000
1972	YI XING		2.4195275	0.16906680	4.12998	2.0104656	3.7635341	14.50	3.8	0.00000
1973	COLOCOLO		3.1747582	0.09597004	10.62681	2.8700767	5.6567373	13.00	7.6	0.00000
1974	CAUPOLICAN		3.1647561	0.09815699	10.22658	2.8541131	5.6300254	13.10	7.3	0.00000
1975	PIKELNER		2.8033850	0.11504352	6.30121	2.4808738	4.6937952	13.20	6.9	0.00000
1976	KAVERIN		2.3818672	0.07507404	2.37401	2.2030506	3.6760066	13.80	5.3	0.00000
1977	SHURA		2.7810490	0.07396153	7.75222	2.5753582	4.6378098	12.40	10.0	0.00000
1978	PATRICE		2.1946507	0.21375296	4.34473	1.7255375	3.2512329	14.20	4.4	0.00000
1979	SAKHAROV	X	2.3740544	0.09954218	6.05506	2.1377358	3.6579351	14.66	3.5	0.00000
1980	TEZCATLIPOC	U	1.7095205	0.36531571	26.84633	1.0850056	2.2351744	15.11	3.1	0.00000

No.	Name	Type	A (A.U.)	E	I (deg.)	q (A.U.)	Period (years)	Mag.	Rad. (km)	Rot. (hours)
1981	MIDAS		1.7758766	0.64979732	39.84247	0.6219168	2.3665690	18.00	0.8	0.00000
1982	CLINE		2.3096132	0.24966937	6.84129	1.7329735	3.5100143	13.50	6.0	0.00000
1983	BOK		2.6223404	0.09770045	9.39704	2.3661366	4.2465243	14.00	4.8	0.00000
1984	FEDYNSKIJ		3.0142763	0.08307804	4.77515	2.7638562	5.2332873	12.30	10.5	0.00000
1985	HOPMANN		3.1247499	0.14790250	17.26912	2.6625917	5.5236087	12.30	10.5	0.00000
1986	1935SV173SA		3.1061206	0.18959630	2.20002	2.5172117	5.4742861	13.10	11.9	0.00000
1987	KAPLAN		2.3822572	0.22662853	23.64258	1.8423698	3.6769099	12.90	8.0	0.00000
1988	DELORES		2.1535437	0.10246420	4.25442	1.9328825	3.1603162	14.70	3.5	0.00000
1989	TATRY		2.3510349	0.07548478	7.77403	2.1735673	3.6048615	13.30	6.6	0.00000
1990	PILCHER		2.1748719	0.05094720	3.13193	2.0640683	3.2073810	14.00	4.8	0.00000
1991	DARWIN		2.2493057	0.20678978	5.91324	1.7841724	3.3734381	14.60	3.6	0.00000
1992	GALVARINO		2.9932077	0.04604384	10.55208	2.8553889	5.1785154	13.20	6.9	0.00000
1993	GUACOLDA		3.0569127	0.06761809	11.46392	2.8502100	5.3447146	13.50	6.0	0.00000
1994	SHANE		2.6805341	0.20566244	10.22145	2.1292489	4.3886609	13.50	6.0	0.00000
1995	HAJEK		2.5277288	0.05753008	10.80698	2.3823082	4.0187936	13.70	5.5	0.00000
1996	ADAMS		2.5575843	0.13909250	15.11734	2.2018435	4.0902033	13.20	5.5	0.00000
1997	LEVERRIER		2.2094786	0.20645477	6.06536	1.7533211	3.2842383	14.40	4.0	0.00000
1998	TITIUS		2.4182973	0.06585100	7.63829	2.2590499	3.7606637	12.80	8.3	0.00000
1999	HIRAYAMA		3.1180899	0.11445445	12.51315	2.7612104	5.5059586	12.00	12.1	0.00000
2000	HERSCHEL	S	2.3810692	0.29929432	22.73994	1.6684287	3.6741595	12.33	8.6	0.00000
2001	EINSTEIN	E	1.9334466	0.09861696	22.68791	1.7427762	2.6884272	14.00	2.2	0.00000
2002	EULER		2.4168370	0.07040404	8.52197	2.2466819	3.7572582	13.30	6.6	0.00000
2003	1941BH73AG1		3.0623331	0.12420440	1.88140	2.6819777	5.3589368	12.91	7.9	0.00000
2004	LEXELL		2.1725309	0.07915215	2.49846	2.0005703	3.2022035	13.90	5.0	0.00000
2005	HENCKE		2.6226871	0.16582626	12.21492	2.1877766	4.2473660	13.50	6.0	0.00000
2006	POLONSKAYA		2.3244524	0.19260056	4.91808	1.8767616	3.5438962	14.10	4.6	0.00000
2007	MCCUSKEY		2.3825269	0.11788174	3.05210	2.1016705	3.6775341	13.00	7.6	0.00000
2008	KONSTITUTSIYA		3.2121966	0.10246099	20.66452	2.8830717	5.7570920	11.20	17.4	0.00000
2009	VOLOSHINA	CFPD	3.1172090	0.14082229	2.86298	2.6782365	5.5036254	12.10	14.0	0.00000
2010	CHEBYSHEV	CU	3.0963647	0.18065009	2.43012	2.5370061	5.4485154	12.20	14.0	0.00000
2011	VETERANIYA		2.3872101	0.14863165	6.18641	2.0323951	3.6883829	14.00	4.8	0.00000
2012	GUO SHOU-JING		2.3280816	0.17860191	2.90730	1.9122818	3.5521991	14.30	4.2	0.00000
2013	TUCAPEL		2.2903068	0.22510175	7.51043	1.7747546	3.4660952	13.20	6.9	0.00000
2014	VASILEVSKIS		2.4021144	0.28385293	21.42630	1.7202671	3.7229784	12.90	8.0	0.00000
2015	KACHUEVSKAY		2.3353758	0.10358637	11.90916	2.0934627	3.5689063	13.40	6.3	0.00000
2016	HEINEMANN		3.1434162	0.17861532	0.92228	2.5819540	5.5731769	12.50	15.7	0.00000
2017	WESSON		2.2525916	0.18510008	4.85961	1.8356367	3.3808329	14.70	3.5	0.00000
2018	SCHUSTER		2.1834600	0.19237670	2.55664	1.7634132	3.2263975	15.60	2.3	0.00000
2019	1935SX1		2.2404013	0.16606776	4.04728	1.8683428	3.3534257	13.50	6.0	0.00000
2020	UKKO		3.0250893	0.06148773	11.13370	2.8390834	5.2614722	11.70	10.9	0.00000
2021	POINCARE		2.3089893	0.21944219	5.48422	1.8022995	3.5085919	14.70	3.5	0.00000
2022	WEST		2.7065358	0.11641177	5.66400	2.3914633	4.4526715	12.30	10.5	0.00000
2023	ASAPH		2.8777406	0.27935532	22.32340	2.0738285	4.8817720	12.70	8.7	0.00000
2024	MCLAUGHLIN		2.3256512	0.13804381	7.31155	2.0046093	3.5466378	14.40	4.0	0.00000
2025	1953LG		3.1699197	0.09928024	6.99676	2.8552094	5.6438098	11.80	13.2	0.00000
2026	COTTRELL		2.4449737	0.11906633	2.46609	2.1538596	3.8230615	14.50	3.8	0.00000
2027	SHEN GUO		3.0229554	0.09224009	11.01239	2.7441175	5.2559061	12.80	6.6	0.00000
2028	JANEQUEO		2.2967620	0.11312031	7.95310	2.0369515	3.4807594	15.20	2.8	0.00000
2029	BINOMI		2.3499725	0.12766948	5.59156	2.0499527	3.6024189	14.30	4.2	0.00000
2030	BELYAEV		2.2474866	0.09378689	2.58163	2.0367019	3.3693464	14.70	3.5	0.00000
2031	BAM		2.2341087	0.17274565	4.75240	1.8481761	3.3393075	14.40	4.0	0.00000
2032	ETHEL		3.0642686	0.13476439	1.52083	2.6513143	5.3640180	12.70	8.7	0.00000
2033	BASILEA		2.2255292	0.11131901	8.46426	1.9777853	3.3200903	14.80	3.3	0.00000
2034	BERNOULLI		2.2464142	0.17980328	8.56034	1.8425016	3.3669353	14.00	4.8	0.00000
2035	STEARNS		1.8841515	0.13129595	27.75335	1.6367700	2.5862689	13.71	5.6	0.00000
2036	SHERAGUL		2.2442851	0.18535568	3.97198	1.8282942	3.3621500	14.00	4.8	0.00000
2037	TRIPAXEPTAL		2.3009379	0.13187620	4.25328	1.9974989	3.4902565	14.80	3.3	0.00000
2038	BISTRO		2.4362650	0.08817323	14.76551	2.2214518	3.8026540	13.50	6.0	0.00000
2039	PAYNE-GAPOS		3.1653385	0.14636026	2.52757	2.7020586	5.6315794	14.00	7.9	0.00000
2040	CHALONGE		3.1042781	0.19774085	14.65914	2.4904356	5.4694157	12.80	8.3	0.00000
2041	LANCELOT		3.1536155	0.20014356	2.98214	2.5224397	5.6003232	13.64	9.3	0.00000
2042	SITARSKI		2.7517471	0.15187199	5.34084	2.3338339	4.5647058	14.03	4.7	0.00000
2043	ORTUTAY		3.1062469	0.10806768	3.09859	2.7705622	5.4746199	12.10	11.5	0.00000
2044	WIRT		2.3810458	0.34277734	24.05399	1.5648772	3.6741054	14.30	4.2	0.00000
2045	PEKING		2.3812399	0.05528615	6.90574	2.2495904	3.6745551	13.40	6.3	0.00000
2046	LENINGRAD		3.1528208	0.18298696	2.73874	2.5758958	5.5982065	12.10	18.9	0.00000
2047	SMETANA		1.8718922	0.00333687	25.27574	1.8656459	2.5610688	15.00	1.7	0.00000
2048	DWORNIK	E	1.9536577	0.04220963	23.75697	1.8711945	2.7306919	14.00	2.1	0.00000
2049	GRIETJE		1.9489677	0.08434398	24.42049	1.7845840	2.7208648	16.20	1.0	0.00000
2050	FRANCIS	S	2.3240707	0.23876017	26.59934	1.7691751	3.5430231	13.00	6.3	0.00000
2051	CHANG		2.8402829	0.07598200	1.34511	2.6244726	4.7867684	13.00	6.0	0.00000
2052	TAMRIKO		3.0081832	0.07741062	9.49102	2.7753179	5.2174273	11.00	15.1	0.00000
2053	NUKI		2.8026645	0.14140236	8.50437	2.4063611	4.6919856	12.70	8.7	0.00000
2054	GAWAIN		2.9646380	0.09809009	3.78222	2.6738362	5.1045504	13.63	5.7	0.00000
2055	DVORAK		2.3100216	0.31067753	21.52547	1.5923498	3.5109456	14.60	3.6	0.00000
2056	NANCY		2.2178576	0.13866436	3.93120	1.9103199	3.3029387	13.50	6.0	0.00000
2057	ROSEMARY		3.0765660	0.23629034	1.43946	2.3496034	5.3963408	14.00	4.8	0.00000
2058	ROKA		3.1108119	0.16021229	2.54338	2.6124215	5.4866929	12.00	19.8	0.00000
2059	BABOQUIVARI		2.6518974	0.52588367	10.99420	1.2573078	4.3185215	16.00	1.9	0.00000
2060	CHIRON		13.6419106	0.37976512	6.93575	8.4611883	50.3863373	6.24	109.9	5.91810
2061	ANZA	C	2.2647524	0.53742528	3.73962	1.0476172	3.4082472	18.00	1.2	0.00000
2062	ATEN	S	0.9664751	0.18251465	18.93460	0.7900792	0.9501365	18.36	0.4	0.00000
2063	BACCHUS		1.0776080	0.34942934	9.41988	0.7010601	1.1186422	18.70	0.6	0.00000

No.	Name	Type	A (A.U.)	E	I (deg.)	q (A.U.)	Period (years)	Mag.	Rad. (km)	Rot. (hours)
2064	THOMSEN		2.1788208	0.32921889	5.69413	1.4615119	3.2161205	15.00	3.0	0.00000
2065	SPICER		2.7000189	0.23345779	6.44945	2.0696783	4.4365993	13.30	6.6	0.00000
2066	PALALA		2.3946109	0.12528867	3.75897	2.0945933	3.7055478	14.10	4.6	0.00000
2067	AKSNES	EMP	3.9461753	0.18185809	3.06711	3.2285316	7.8390708	11.80	21.7	0.00000
2068	DANGREEN		2.7713084	0.09998122	12.89135	2.4942298	4.6134658	12.80	8.3	0.00000
2069	HUBBLE		3.1800208	0.17306750	9.16983	2.6296625	5.6708078	12.30	10.5	0.00000
2070	HUMASON		2.2503400	0.15465039	2.75638	1.9023240	3.3757651	14.70	3.5	0.00000
2071	NADEZHDA		2.2519767	0.15684082	3.63087	1.8987747	3.3794484	14.50	3.8	0.00000
2072	KOSMODEMYANSKAYA		2.4513154	0.16239880	4.75924	2.0532248	3.8379457	13.20	6.9	4.40000
2073	JANACEK		2.7157772	0.11024707	2.96379	2.4163706	4.4754963	13.80	5.3	0.00000
2074	SHOEMAKER	RA	1.7998412	0.08192485	30.07653	1.6523895	2.4146338	14.90	2.1	0.00000
2075	MARTINEZ		2.4039621	0.24770615	27.02831	1.8084860	3.7272751	15.00	3.0	0.00000
2076	LEVIN		2.2736788	0.15184461	4.99179	1.9284329	3.4284172	15.50	2.4	0.00000
2077	KIANGSU		2.3278999	0.29571137	28.07166	1.6395134	3.5517831	14.50	3.8	0.00000
2078	NANKING		2.3696988	0.37482744	20.13036	1.4814706	3.6478732	14.00	4.8	0.00000
2079	JACCHIA		2.5995827	0.08056062	13.30530	2.3901587	4.1913648	13.30	6.6	0.00000
2080	JIHLAVA		2.1765313	0.06137670	3.85072	2.0429430	3.2110519	14.70	3.5	0.00000
2081	SAZAVA	CFPD	2.4494185	0.16390754	3.92261	2.0479403	3.8334913	13.50	8.8	0.00000
2082	GALAHAD		2.9193721	0.16497426	3.06989	2.4377508	4.9880886	13.80	7.0	0.00000
2083	SMITHER	EMP	1.8719459	0.05109531	18.45170	1.7762982	2.5611789	14.13	2.4	0.00000
2084	OKAYAMA		2.3953273	0.10194380	4.84014	2.1511385	3.7072110	13.60	5.8	0.00000
2085	HENAN		2.6991842	0.08641843	3.83296	2.4659250	4.4345422	12.20	11.0	32.00000
2086	NEWELL		2.4003763	0.11338543	6.48714	2.1282086	3.7189386	13.10	7.3	0.00000
2087	KOCHERA		2.2061663	0.05769515	1.83052	2.0788810	3.2768559	14.50	3.8	0.00000
2088	SAHLIA		2.2070198	0.07977066	5.54070	2.0309646	3.2787580	14.30	4.2	10.37000
2089	CETACEA		2.5335965	0.15485515	15.41698	2.1412561	4.0327950	12.22	11.0	0.00000
2090	MIZUHO		3.0649767	0.14407232	11.83923	2.6233985	5.3658781	11.50	15.2	0.00000
2091	SAMPO		3.0137858	0.05690781	11.38175	2.8422780	5.2320104	12.00	12.1	0.00000
2092	SUMIANA		2.8482859	0.02761902	3.08535	2.7696190	4.8070140	12.70	8.7	0.00000
2093	GENICHESK		2.2683454	0.16913711	6.08660	1.8846840	3.4163611	14.30	4.2	0.00000
2094	MAGNITKA		2.2321975	0.09740834	5.02594	2.0147629	3.3350239	13.30	6.6	0.00000
2095	PARSIFAL		2.6425495	0.01019472	3.59392	2.6156094	4.2957077	13.90	5.0	0.00000
2096	VAINO		2.4455893	0.23147504	0.98737	1.8794963	3.8245051	14.30	4.2	0.00000
2097	1953PV		3.1203814	0.26429042	4.38684	2.2956946	5.5120292	12.80	8.3	0.00000
2098	ZYSKIN		2.4239764	0.12768339	6.51784	2.1144750	3.7739193	13.30	6.6	0.00000
2099	OPIK		2.3041761	0.36197934	26.91172	1.4701120	3.4976273	16.43	1.6	0.00000
2100	RA-SHALOM	U	0.8320881	0.43650323	15.76098	0.4688790	0.7590213	17.16	0.8	19.79000
2101	ADONIS		1.8745539	0.76401794	1.36187	0.4423610	2.5665333	19.50	1.0	0.00000
2102	TANTALUS		1.2900411	0.29847768	64.01132	0.9049926	1.4652284	17.50	1.0	0.00000
2103	1960FL		3.1478577	0.18544836	7.67710	2.5640926	5.5849929	12.50	9.6	0.00000
2104	TORONTO		3.1967187	0.10623148	18.37044	2.8571265	5.7155318	11.10	18.2	0.00000
2105	GUDY		2.3897221	0.14949253	29.28982	2.0324764	3.6942058	13.60	5.8	0.00000
2106	HUGO		2.7035213	0.09515955	8.03579	2.4462552	4.4452343	13.00	7.6	0.00000
2107	ILMARI		2.6267321	0.07816111	8.84377	2.4214239	4.2571964	13.00	7.6	0.00000
2108	OTTOSCHMIDT		2.4371428	0.00589320	10.77990	2.4227803	3.8047094	12.70	8.7	0.00000
2109	D'HOTEL		2.6899924	0.26139084	8.07991	1.9868530	4.4119091	12.90	8.0	32.00000
2110	MOORE-SITTERLY		2.1981061	0.17728379	1.13121	1.8084176	3.2589147	14.80	3.3	0.00000
2111	TSELINA		3.0190930	0.09082640	10.50584	2.7448795	5.2458363	11.30	16.6	0.00000
2112	ULYANOV		2.2542787	0.13763037	3.37052	1.9440215	3.3846316	13.80	5.3	0.00000
2113	EHRDNI		2.4738927	0.09663476	6.45347	2.2348287	3.8910899	13.40	6.3	0.00000
2114	WALLENQUIST		3.1899638	0.15406737	0.56027	2.6984944	5.6974249	12.50	9.6	0.00000
2115	IRAKLI		3.0075088	0.05795435	8.96996	2.8332105	5.2156730	12.50	9.6	0.00000
2116	MTSKHETA		2.5887992	0.05705169	9.07981	2.4411039	4.1653123	13.50	6.0	0.00000
2117	DANMARK		2.8704002	0.06880165	2.93611	2.6729119	4.8631053	13.00	7.6	0.00000
2118	FLAGSTAFF		2.5471895	0.21766417	6.32642	1.9927576	4.0652933	12.00	12.1	0.00000
2119	SCHWALL		2.2517314	0.15577263	3.83094	1.9009733	3.3788967	14.90	3.2	0.00000
2120	TYUMENIA		3.0644305	0.11969337	17.57655	2.6976385	5.3644433	11.80	13.2	0.00000
2121	SEVASTOPOL		2.1840167	0.17798449	4.37712	1.7952956	3.2276313	13.70	5.5	0.00000
2122	PYATILETKA		2.4011841	0.02917734	7.89406	2.3311238	3.7208157	13.30	6.6	0.00000
2123	VLTAVA		2.8601286	0.07633148	1.00927	2.6418109	4.8370252	13.00	7.6	0.00000
2124	NISSEN		3.0197401	0.09661097	10.72822	2.7280002	5.2475228	12.70	8.7	0.00000
2125	KARL-ØNTJES		2.7877922	0.10481677	1.68687	2.4955850	4.6546884	13.00	7.6	0.00000
2126	GERASIMOVIC		2.3903451	0.11960840	8.48274	2.1044395	3.6956503	13.60	5.8	0.00000
2127	TANYA		3.2074554	0.04870434	13.14936	3.0512383	5.7443504	12.00	12.1	0.00000
2128	WETHERILL		2.7355974	0.37993959	16.83805	1.6962357	4.5245800	15.20	2.8	0.00000
2129	COSICOSI		2.1812658	0.17363341	5.51801	1.8025252	3.2215352	15.20	2.8	0.00000
2130	EVDOKIYA		2.2533841	0.18808462	5.61452	1.8295571	3.3826168	15.00	3.0	0.00000
2131	MAYALL	RA	1.8870863	0.11080205	33.99046	1.6779933	2.5923142	13.50	3.7	0.00000
2132	ZHUKOV		2.7815511	0.08095289	5.87169	2.5563767	4.6390662	12.40	10.0	0.00000
2133	FRANCESWRIG		2.4101133	0.18433042	6.90073	1.9658562	3.7415900	14.50	3.8	0.00000
2134	DENNISPALM		2.6392758	0.25511420	31.28675	1.9659591	4.2877274	14.30	4.2	0.00000
2135	ARISTAEUS		1.6000447	0.50335413	23.03926	0.7946556	2.0239425	19.20	0.4	0.00000
2136	JUGTA		3.0275905	0.04168429	10.59388	2.9013875	5.2679992	12.80	8.3	0.00000
2137	PRISCILLA		3.1755896	0.07184306	11.73913	2.9474454	5.6589589	12.50	9.6	0.00000
2138	SWISSAIR		2.6856875	0.06542172	5.92474	2.5099852	4.4013228	12.80	8.3	0.00000
2139	MAKHARADZE		2.4602559	0.19021609	2.19075	1.9922756	3.8589611	13.50	6.0	0.00000
2140	KEMEROVO		2.9911637	0.05662068	6.97224	2.8218019	5.1732121	12.20	11.0	0.00000
2141	SIMFEROPOL		2.8047552	0.12755556	5.95805	2.4469931	4.6972365	12.50	9.6	0.00000
2142	LANDAU		3.1714492	0.10613174	0.66413	2.8348577	5.6478953	12.90	8.0	0.00000
2143	JIMARNOLD		2.2804716	0.23417272	8.36449	1.7464473	3.4437926	15.30	2.6	0.00000
2144	MARIETTA		2.8737471	0.06222915	2.82566	2.6949162	4.8716130	12.40	10.0	0.00000
2145	BLAAUW		3.2330897	0.07723478	15.04642	2.9833827	5.8133521	11.70	13.8	0.00000
2146	STENTOR		5.2075381	0.10315852	39.24384	4.6703362	11.8836184	11.50	24.9	0.00000

No.	Name	Type	A (A.U.)	E	I (deg.)	q (A.U.)	Period (years)	Mag.	Rad. (km)	Rot. (hours)
2147	KHARADZE		3.1799817	0.05943809	10.03404	2.9909694	5.6707029	13.00	10.1	0.00000
2148	EPEIOS		5.1887174	0.05844867	9.16717	4.8854437	11.8192520	12.00	19.8	0.00000
2149	SCHWAMBRANI		2.5509632	0.10470625	7.70007	2.2838614	4.0743308	13.50	6.0	0.00000
2150	1977TA69TM4		1.9132371	0.05713284	25.32580	1.8039284	2.6463859	15.00	1.7	0.00000
2151	HADWIGER		2.5614693	0.05623229	15.44640	2.4174318	4.0995264	12.00	12.1	0.00000
2152	HANNIBAL		3.1399486	0.20955259	13.93845	2.4819643	5.5639577	12.50	9.6	0.00000
2153	AKIYAMA		3.1087081	0.16799293	1.19229	2.5864670	5.4811277	13.00	7.6	0.00000
2154	UNDERHILL		2.6351075	0.12546179	7.75991	2.3045022	4.2775741	13.80	5.3	0.00000
2155	WODAN		2.8571079	0.07531906	2.54008	2.6419132	4.8293643	13.60	5.8	0.00000
2156	KATE		2.2427495	0.20116498	5.35133	1.7915868	3.3586996	14.00	4.8	5.62000
2157	ASHBROOK		2.7863448	0.11194702	8.62556	2.4744217	4.6510634	12.70	8.7	0.00000
2158	1933OS28SS		3.0792627	0.16886805	1.57943	2.5592735	5.4034371	12.60	9.1	0.00000
2159	KUKKAMAKI		2.4817514	0.04114643	3.27938	2.3796363	3.9096460	13.00	7.6	0.00000
2160	SPITZER		2.8995540	0.10153713	2.85935	2.6051416	4.9373832	13.70	5.5	0.00000
2161	GRISSOM		2.7476244	0.16132404	7.31214	2.3043666	4.5544510	13.40	6.3	0.00000
2162	ANHUI		2.2275367	0.12353658	3.05033	1.9523544	3.3245840	14.00	4.8	0.00000
2163	KORCZAK		3.1428874	0.18829754	2.51066	2.5510893	5.5717707	12.80	8.3	0.00000
2164	LYALYA		3.1802661	0.13776813	2.62621	2.7421267	5.6714640	13.10	7.3	0.00000
2165	YOUNG		3.1416643	0.15994164	0.95279	2.6391814	5.5685186	12.00	12.1	0.00000
2166	HANDAHL		2.3459680	0.21806538	5.12674	1.8343937	3.5932145	14.50	3.8	0.00000
2167	ERIN		2.5449643	0.18083853	6.04133	2.0847366	4.0599670	12.90	8.0	7.00000
2168	SWOPE		2.4518266	0.15562569	4.75232	2.0702593	3.8391457	14.50	3.8	0.00000
2169	TAIWAN		2.7875583	0.05219509	1.53193	2.6420612	4.6541023	13.30	6.6	0.00000
2170	BYELORUSSIA		2.4049520	0.18180257	2.08261	1.9677255	3.7295773	14.70	3.5	0.00000
2171	KIEV		2.2556572	0.16577211	7.51728	1.8817321	3.3877366	15.00	3.0	0.00000
2172	PLAVSK		2.8950665	0.14076991	3.32864	2.4875281	4.9259253	12.70	8.7	0.00000
2173	MARESJEV		3.1278086	0.12894607	14.47099	2.7244899	5.5317206	12.60	9.1	0.00000
2174	ASMODEUS		2.5361633	0.27348658	8.11865	1.8425566	4.0389252	14.50	3.8	0.00000
2175	ANDREA DORIA		2.2155230	0.20670076	3.69773	1.7575727	3.2977245	15.00	3.0	0.00000
2176	DONAR		2.9288623	0.05440582	3.04870	2.7695153	5.0124316	13.40	6.3	0.00000
2177	OLIVER		3.1943612	0.09437865	1.54190	2.8928816	5.7092099	12.90	8.0	0.00000
2178	KAZAKHSTANI		2.2074952	0.15459913	3.07685	1.8662184	3.2798176	15.00	3.0	0.00000
2179	PLATZEK		3.0080366	0.10083423	10.49226	2.7047236	5.2170463	13.00	7.6	0.00000
2180	MARJALEENA		3.0109532	0.08093111	9.23066	2.7672734	5.2246356	12.00	12.1	0.00000
2181	FOGELIN		2.5891447	0.12076224	13.04551	2.2764738	4.1661458	13.40	6.3	0.00000
2182	SEMIROT		3.1302571	0.13025853	2.29742	2.7225144	5.5382180	12.50	9.6	0.00000
2183	1959OB		2.9881198	0.37759718	18.26573	1.8598142	5.1653175	12.60	9.1	0.00000
2184	FUJIAN		3.1707306	0.11496581	5.21250	2.8062048	5.6459756	12.00	12.1	0.00000
2185	GUANGDONG		2.7087028	0.16030627	9.60493	2.2744806	4.4580202	13.00	7.6	0.00000
2186	KELDYSH		2.6830132	0.10174401	2.36285	2.4100327	4.3947506	13.50	6.0	0.00000
2187	LA SILLA		2.5366578	0.11733976	13.25148	2.2390070	4.0401063	14.00	4.8	0.00000
2188	ORLENOK		2.9003909	0.08940564	2.65854	2.6410797	4.9395204	13.00	7.6	0.00000
2189	ZARAGOZA		2.4024811	0.22501343	13.94741	1.8618906	3.7238309	14.00	4.8	0.00000
2190	COUBERTIN		2.4694123	0.09259445	0.82995	2.2407584	3.8805244	13.60	5.8	0.00000
2191	UPPSALA		3.0165455	0.08925572	9.04674	2.7473016	5.2391982	12.40	10.0	0.00000
2192	PYATIGORIYA		3.1344137	0.08638481	9.74865	2.8636477	5.5492520	12.50	9.6	0.00000
2193	JACKSON		3.1081831	0.06136669	11.71042	2.9174442	5.4797397	11.50	15.2	0.00000
2194	ARPOLA		2.3277810	0.04267909	8.52565	2.2284334	3.5515110	13.50	6.0	0.00000
2195	TENGSTROM		2.2218690	0.10544282	4.57641	1.9875889	3.3119037	13.50	6.0	0.00000
2196	ELLICOTT	C	3.4301698	0.06174609	10.31580	3.2183702	6.3529205	11.00	30.0	0.00000
2197	SHANGHAI		3.1715882	0.11706088	2.51763	2.8003192	5.6482668	11.50	15.2	0.00000
2198	CEPLECHA		2.5919490	0.19909719	3.64026	2.0758994	4.1729164	15.70	2.2	0.00000
2199	KLET		2.2406218	0.20146751	8.18457	1.7892094	3.3539214	14.50	3.8	0.00000
2200	PASADENA		2.4076915	0.14528032	4.59039	2.0579014	3.7359519	13.90	5.0	0.00000
2201	OLJATO		2.1736083	0.71196914	2.51535	0.6260663	3.2045863	16.70	1.4	24.00000
2202	PELE		2.2903321	0.51260096	8.78615	1.1163056	3.4661524	18.50	0.6	0.00000
2203	1935SQ1		3.1169078	0.17476234	1.65210	2.5721898	5.5028281	12.50	9.6	0.00000
2204	LYYLI		2.5933449	0.40282357	20.54106	1.5486845	4.1762881	13.00	7.6	0.00000
2205	GLINKA		3.0080273	0.11633706	10.48682	2.6580822	5.2170219	12.90	8.0	0.00000
2206	GABROVA		3.0154071	0.04959370	10.93646	2.8658619	5.2362328	12.80	8.3	0.00000
2207	ANTENOR	D	5.1439772	0.01583650	6.80846	5.0625148	11.6667147	10.00	61.0	0.00000
2208	PUSHKIN	D	3.4968636	0.04816763	5.41984	3.3284278	6.5391011	11.50	24.0	0.00000
2209	TIANJIN		2.8449099	0.06765076	2.61009	2.6524496	4.7984700	12.50	9.6	0.00000
2210	9597PL75TY2		2.4033313	0.22789726	2.93405	1.8556186	3.7258077	15.60	2.3	0.00000
2211	1951W02		3.1750288	0.09262986	17.37813	2.8809264	5.6574602	14.00	4.8	0.00000
2212	HEPHAISTOS	U	2.1633406	0.83499062	11.88277	0.3569715	3.1819060	15.20	2.7	0.00000
2213	MEEUS		2.1983643	0.22669910	5.33768	1.6999971	3.2594888	14.50	3.8	0.00000
2214	CAROL		3.1831646	0.25066155	14.01768	2.3852677	5.6792197	12.90	8.0	0.00000
2215	SICHUAN		2.7892299	0.26637554	10.75569	2.0462472	4.6582894	12.80	8.3	0.00000
2216	KERCH		3.0206606	0.09600582	10.42775	2.7306595	5.2499223	12.50	9.6	0.00000
2217	ELTIGEN		3.1689782	0.14849184	2.21627	2.6984107	5.6412959	12.20	11.0	0.00000
2218	WOTHO		3.0436728	0.16311625	14.96149	2.5472002	5.3100295	12.90	8.0	0.00000
2219	MANNUCCI		3.1475596	0.12213274	7.58675	2.7631395	5.5841999	12.00	12.1	0.00000
2220	HICKS		3.1460366	0.17456894	2.59007	2.5968363	5.5801473	13.20	6.9	0.00000
2221	CHILTON		2.5905788	0.13800763	13.81913	2.2330592	4.1696081	14.30	4.2	0.00000
2222	LERMONTOV		3.1078129	0.17945078	2.58322	2.5501134	5.4787598	12.50	9.6	0.00000
2223	SARPEDON	D	5.1507378	0.01501862	16.03228	5.0733809	11.6897221	9.80	48.0	0.00000
2224	TUCSON		2.8802052	0.04655484	2.67263	2.7461176	4.8880444	13.10	7.3	0.00000
2225	SERKOWSKI		2.8526058	0.03248981	3.26488	2.7599251	4.8179541	13.20	6.9	0.00000
2226	CUNITZA		2.8679731	0.07977804	2.54356	2.6391718	4.8569384	13.50	6.0	0.00000
2227	OTTO STRUVE		2.2361093	0.17377396	4.95255	1.8475316	3.3437941	15.00	3.0	0.00000
2228	SOYUZ-APOLL		3.1403146	0.18076596	1.98811	2.5726526	5.5649304	12.50	9.6	0.00000
2229	MEZZARCO		2.6953888	0.26390406	12.71428	1.9840648	4.4251924	14.00	4.8	0.00000

No.	Name	Type	A (A.U.)	E	I (deg.)	q (A.U.)	Period (years)	Mag.	Rad. (km)	Rot. (hours)
2230	YUNNAN		2.8566153	0.06444289	2.56403	2.6725266	4.8281150	13.20	6.9	0.00000
2231	DURRELL		2.7277210	0.25256360	8.27196	2.0387979	4.5050530	13.50	6.0	0.00000
2232	KALTAJ		2.6673384	0.14103375	3.69196	2.2911537	4.3562937	13.50	6.0	0.00000
2233	KUZNETSOV		2.2779639	0.08065697	3.41290	2.0942302	3.4381137	14.00	4.8	0.00000
2234	SCHMADEL		2.6997843	0.19890001	25.24272	2.1627972	4.4360209	13.40	6.3	0.00000
2235	VITTORE		3.2207892	0.19853437	18.77102	2.5813520	5.7802081	11.50	15.2	0.00000
2236	AUSTRASIA		2.3455970	0.21822134	10.11387	1.8337377	3.5923622	13.50	6.0	0.00000
2237	MELNIKOV		3.1577268	0.20616995	2.39723	2.5066984	5.6112785	12.50	9.6	0.00000
2238	STESHENKO		3.0712397	0.16704416	1.31800	2.5582068	5.3823328	13.00	7.6	0.00000
2239	PARACELSUS		3.2041190	0.09108353	10.86090	2.9122765	5.7353902	13.00	7.6	0.00000
2240	TSAI		3.1469176	0.15699852	0.84643	2.6528561	5.5824914	13.00	7.6	0.00000
2241	1979WM50NC	D	5.2480164	0.06592360	16.57442	4.9020481	12.0224438	9.50	66.0	0.00000
2242	1936TG72RE2		2.2080412	0.11718762	2.53665	1.9492860	3.2810340	14.50	3.8	0.00000
2243	LONNROT		2.2482967	0.19609541	6.84611	1.8074161	3.3711686	14.00	4.8	0.00000
2244	TESLA		2.8121428	0.17857926	7.82292	2.3099525	4.7158074	13.40	6.3	0.00000
2245	HEKATOSTOS		2.6366484	0.13407584	11.86297	2.2831373	4.2813263	12.50	9.6	0.00000
2246	BOWELL	D	3.9382703	0.09511575	6.49919	3.5636787	7.8155274	11.80	21.0	0.00000
2247	6512PL77AR1		2.4481456	0.10994486	5.94426	2.1789846	3.8305035	14.80	3.3	0.00000
2248	KANDA		3.0947924	0.12127033	1.63752	2.7194860	5.4443655	13.00	7.6	0.00000
2249	YAMAMOTO		3.1819749	0.09957515	4.09746	2.8651295	5.6760359	12.00	12.1	0.00000
2250	STALINGRAD		3.1691899	0.20151410	1.51025	2.5305536	5.6418614	12.50	9.6	0.00000
2251	TIKHOV		2.7111506	0.14885056	7.43541	2.3075943	4.4640646	12.80	8.3	0.00000
2252	CERGA		2.6177511	0.07307012	4.23727	2.4264717	4.2353816	13.00	7.6	0.00000
2253	ESPINETTE		2.2845573	0.27761739	3.87684	1.6503245	3.4530518	14.50	3.8	0.00000
2254	REQUIEM		2.3427572	0.14908692	5.04805	1.9934828	3.5858402	14.00	4.8	0.00000
2255	QINGHAI		3.0965281	0.15547647	14.20685	2.6150908	5.4489465	12.50	9.6	0.00000
2256	4519PL650H		3.1024580	0.16176496	0.46809	2.6005888	5.4646063	13.10	7.3	0.00000
2257	KAARINA		2.4875534	0.23856290	5.03311	1.8941154	3.9233642	14.30	4.2	0.00000
2258	VIIPURI		2.6930659	0.08084256	1.48055	2.4753516	4.4194732	13.00	7.6	0.00000
2259	SOFIEVKA		2.2930775	0.18619744	4.68372	1.8661124	3.4723868	14.00	4.8	0.00000
2260	NEOPTOLEMUS	D	5.1903691	0.04509369	17.77985	4.9563165	11.8248968	10.00	49.0	0.00000
2261	1977HC		2.3769596	0.23878172	22.71740	1.8093851	3.6646516	14.00	4.8	0.00000
2262	MITIDIKA		2.5847220	0.28245848	13.44862	1.8546454	4.1554756	13.50	6.0	0.00000
2263	SHAANXI		3.0167782	0.11142069	11.42291	2.6806467	5.2398043	12.50	9.6	0.00000
2264	SABRINA		3.1387813	0.16189650	0.14932	2.6306238	5.5608554	12.00	12.1	0.00000
2265	VERBAANDERT		2.6191082	0.21074006	19.84216	2.0671573	4.2386756	14.00	4.8	0.00000
2266	TCHAIKOVSKY	D	3.3836946	0.19728488	13.24066	2.7161429	6.2242460	12.00	19.0	0.00000
2267	AGASSIZ		2.2177718	0.13783628	1.95299	1.9120823	3.3027465	14.50	3.8	0.00000
2268	SZMYTOWNA		2.9379263	0.11647467	3.30773	2.5957322	5.0357175	13.00	7.6	0.00000
2269	EFREMIANA		3.1280024	0.08463463	15.39264	2.8632653	5.5322351	12.00	12.1	0.00000
2270	YAZHI		3.1541288	0.13061151	2.14066	2.7421632	5.6016903	12.20	11.0	0.00000
2271	KISO		2.7580311	0.06028908	3.38863	2.5917521	4.5803509	12.00	12.1	0.00000
2272	1972FA		1.8668934	0.08991234	24.33603	1.6990366	2.5508170	16.00	1.1	0.00000
2273	YARILO		2.4516551	0.16323511	0.39551	2.0514588	3.8387432	14.00	4.8	0.00000
2274	EHRSSON		2.4072154	0.23167080	2.24921	1.8495338	3.7348435	14.50	3.8	0.00000
2275	1979MH31XS		2.2957213	0.16992532	6.38578	1.9056201	3.4783938	15.50	2.4	0.00000
2276	WARCK		2.3748262	0.16972013	2.46485	1.9717705	3.6597192	14.00	4.8	0.00000
2277	MOREAU		2.6004148	0.12436028	11.57018	2.2770264	4.1933770	13.70	5.5	0.00000
2278	1953GE53GR1		2.4505746	0.15264931	4.22547	2.0764961	3.8362057	14.00	4.8	0.00000
2279	BARTO		2.4565833	0.16067179	2.98025	2.0618796	3.8503234	14.00	4.8	0.00000
2280	KUNIKOV		2.1788826	0.14097492	3.57072	1.8717147	3.2162569	14.80	3.3	0.00000
2281	1971UQ174SU		2.1884360	0.14444984	1.48437	1.8723168	3.2374332	14.70	3.5	0.00000
2282	ANDRES BELL		2.2029474	0.07909454	4.98291	2.0287063	3.2696867	15.00	3.0	0.00000
2283	BUNKE		2.2487991	0.08766927	6.73181	2.0516486	3.3722985	13.70	5.5	0.00000
2284	SAN JUAN		2.3255763	0.05097809	5.28569	2.2070229	3.5464666	14.00	4.8	0.00000
2285	RON HELIN		2.2203026	0.20753539	5.32929	1.7595112	3.3084018	15.00	3.0	0.00000
2286	FESENKOV		2.1926670	0.09339571	1.34720	1.9878812	3.2468257	14.50	3.8	0.00000
2287	KALMYKIA		2.2405424	0.16968805	5.28802	1.8603491	3.3537426	14.50	3.8	0.00000
2288	KAROLINUM		2.9089384	0.16067830	14.54870	2.4415350	4.9613719	12.00	12.1	0.00000
2289	6567P-L		2.6371722	0.14214034	2.15072	2.2623236	4.2826023	14.60	3.6	0.00000
2290	1932CD153FR		2.5906186	0.23705406	11.51647	1.9765019	4.1697035	13.50	6.0	0.00000
2291	KEVO		3.0464396	0.05775239	24.49467	2.8705003	5.3172717	11.50	15.2	0.00000
2292	SEILI		2.6173468	0.24147737	14.49350	1.9853168	4.2344003	13.00	7.6	0.00000
2293	GUERNICA		3.1339197	0.12768808	0.59232	2.7337556	5.5479407	12.00	12.1	0.00000
2294	1977PL144SA		2.5803468	0.11977913	6.31714	2.2712750	4.1449294	12.60	9.1	0.00000
2295	1977QD177TQ		2.9009213	0.09565308	2.51091	2.6234393	4.9408755	13.00	7.6	0.00000
2296	1975BA141FM		3.1845322	0.16558607	1.25449	2.6572180	5.6828794	13.00	7.6	0.00000
2297	DAGHESTAN		3.1569722	0.14183407	1.60686	2.7092059	5.6092672	12.50	9.6	0.00000
2298	CINDIJON		2.4066596	0.17203844	5.15213	1.9926217	3.7335505	16.00	1.9	0.00000
2299	HANKO		2.5858314	0.29636526	5.26285	1.8194807	4.1581511	14.50	3.8	0.00000
2300	STEBBINS		2.8384676	0.07999806	2.32505	2.6113956	4.7821798	14.00	4.8	0.00000
2301	WHITFORD		3.1510966	0.23277089	11.83724	2.4176128	5.5936146	12.50	9.6	0.00000
2302	1972TL235BF		2.6460626	0.19253948	12.10365	2.1365912	4.3042769	13.00	7.6	0.00000
2303	RETSINA		2.9985793	0.11386688	18.90590	2.6571403	5.1924615	13.50	6.0	0.00000
2304	SLAVIA		2.6124420	0.13490525	13.57660	2.2600100	4.2225032	13.50	6.0	0.00000
2305	KING		2.7853982	0.02856802	7.45935	2.7058251	4.6486940	12.50	9.6	0.00000
2306	1939PM67TK		2.7321966	0.06189353	4.22907	2.5630913	4.5161452	13.00	7.6	0.00000
2307	1957HJ77AH		3.0447047	0.06126814	7.71438	2.8581612	5.3127303	12.10	11.5	0.00000
2308	SCHILT		2.5457008	0.17656863	14.22896	2.0962100	4.0617299	13.50	6.0	0.00000
2309	MR. SPOCK		3.0088799	0.09552208	10.98166	2.7214656	5.2192402	12.50	9.6	0.00000
2310	OLSHANIYA		3.1542313	0.14559194	2.64775	2.6950006	5.6019635	12.50	9.6	0.00000
2311	EL LEONCITO	D	3.6302414	0.03845050	6.59537	3.4906569	6.9167686	11.50	24.0	0.00000

No.	Name	Type	A (A.U.)	E	I (deg.)	q (A.U.)	Period (years)	Mag.	Rad. (km)	Rot. (hours)
2312	DUBOSHIN		3.9790661	0.13729455	5.21046	3.4327619	7.9372807	11.00	31.0	0.00000
2313	1976TA49TE		2.4582701	0.18996079	1.82937	1.9912951	3.8542900	14.20	4.4	0.00000
2314	FIELD	X	2.2606368	0.02436314	5.72794	2.2055607	3.3989613	14.00	4.8	0.00000
2315	CZECHOSLOVA		3.0077059	0.11151314	10.73313	2.6723073	5.2161860	12.00	12.1	0.00000
2316	JO-ANN		2.4534762	0.16192420	1.81705	2.0561988	3.8430209	13.60	5.8	0.00000
2317	GALYA		2.5238862	0.16616984	4.17997	2.1044922	4.0096331	14.50	3.8	0.00000
2318	LUBARSKY		2.2520707	0.13186550	3.60014	1.9551002	3.3796599	15.00	3.0	0.00000
2319	7631PL75UC		2.9055088	0.08950123	2.97323	2.6454620	4.9526000	13.20	6.9	0.00000
2320	1979QJ50NF1		3.1673975	0.12743969	11.54957	2.7637453	5.6370754	11.90	12.6	0.00000
2321	LUZNICE		3.0032611	0.06812579	7.81297	2.7986617	5.2046275	12.90	8.0	0.00000
2322	KITT PEAK		2.2920825	0.04106692	2.40406	2.1979537	3.4701271	14.00	4.8	0.00000
2323	ZVEREV		3.1250517	0.16683102	4.64073	2.6036961	5.5244093	12.00	12.1	0.00000
2324	JANICE		3.0916443	0.17125858	0.40256	2.5621736	5.4360604	12.20	11.0	0.00000
2325	CHERNYKH		3.1519711	0.16081722	1.91611	2.6450799	5.5959439	12.50	9.6	0.00000
2326	TOLOLO		2.8622377	0.15591814	15.13465	2.4159629	4.8423762	12.00	12.1	0.00000
2327	GERSHBERG		2.3680463	0.12937899	4.03604	2.0616708	3.6440580	15.00	3.0	0.00000
2328	ROBESON		2.3412349	0.14640515	10.00733	1.9984660	3.5823455	14.00	4.8	0.00000
2329	ORTHOS		2.4045856	0.65857363	24.39599	0.8209889	3.7287247	16.30	1.7	0.00000
2330	ONTAKE		3.1769574	0.04520255	8.65583	3.0333507	5.6626153	12.00	12.1	0.00000
2331	PARVULESCO		2.4224987	0.22577000	3.71383	1.8755713	3.7704687	14.00	4.8	0.00000
2332	KALM		3.0688372	0.06841440	14.57605	2.8588846	5.3760185	11.50	15.2	0.00000
2333	PORTHAN		2.6465688	0.13540414	11.94501	2.2882123	4.3055120	13.00	7.6	0.00000
2334	CUFFEY		2.2683983	0.07396217	4.08975	2.1006224	3.4164805	14.50	3.8	0.00000
2335	JAMES	X	2.1234679	0.35952297	36.33140	1.3600324	3.0943437	14.30	4.2	0.00000
2336	XINJIANG		3.2076485	0.13999213	2.78922	2.7586029	5.7448692	12.50	9.6	0.00000
2337	1976UH155TV		2.5942810	0.16973163	14.36702	2.1539495	4.1785493	12.50	9.6	0.00000
2338	BOKHAN		2.8330112	0.06024523	3.20527	2.6623359	4.7683973	13.00	7.6	0.00000
2339	2509PL48TH1		2.5281794	0.19497088	4.85241	2.0352578	4.0198679	14.50	3.8	0.00000
2340	HATHOR	U	0.8439395	0.44979271	5.85577	0.4643417	0.7752950	21.50	0.1	0.00000
2341	AOLUTA		2.2118142	0.15164436	4.07643	1.8764050	3.2894475	14.00	4.8	0.00000
2342	LEBEDEV		3.2178607	0.13820551	0.33510	2.7731347	5.7723260	12.50	9.6	0.00000
2343	SIDING SPRI		2.3338983	0.25530034	1.76129	1.7380532	3.5655198	15.00	3.0	0.00000
2344	XIZANG		2.7559619	0.18581671	3.90016	2.2438581	4.5751967	13.00	7.6	0.00000
2345	FUCIK		3.0162871	0.07111308	9.14899	2.8017895	5.2385244	11.90	12.6	0.00000
2346	LILIO		2.3710737	0.15605806	5.92022	2.0010486	3.6510482	13.50	6.0	0.00000
2347	1936TK76EG		3.0993862	0.20245332	13.06437	2.4719052	5.4564924	12.50	9.6	0.00000
2348	MICHKOVITCH		2.3985245	0.16860829	4.67316	1.9941134	3.7146358	13.50	6.0	0.00000
2349	KURCHENKO		2.7705166	0.11654121	17.47640	2.4476373	4.6114888	12.50	9.6	0.00000
2350	VONLUDE		2.2414412	0.12739971	5.07783	1.9558822	3.3557611	14.50	3.8	0.00000
2351	O'HIGGINS		2.5297623	0.18793949	3.73173	2.0543201	4.0236440	14.50	3.8	0.00000
2352	KURCHATOV		3.1023664	0.12055687	14.77978	2.7283547	5.4643641	12.20	11.0	0.00000
2353	1975UD66VG		2.8047183	0.11264566	4.79093	2.4887788	4.6971436	13.00	7.6	0.00000
2354	LAVROV		2.7301693	0.10477533	3.26568	2.4441147	4.5111194	13.00	7.6	0.00000
2355	NEIMONGGOL		3.0232294	0.11019050	10.02063	2.6900983	5.2566209	12.50	9.6	0.00000
2356	HIRONS		3.2340736	0.03837387	15.61309	3.1099699	5.8160062	11.90	12.6	0.00000
2357	PHERECLOS	D	5.1762471	0.04439464	2.67564	4.9464493	11.7766705	10.00	48.0	0.00000
2358	BAHNER		3.0196517	0.10269470	9.71814	2.7095494	5.2472925	12.00	12.1	0.00000
2359	DEBEHOGNE		2.4263017	0.11322851	4.34042	2.1515751	3.7793508	13.50	6.0	0.00000
2360	VOLGO-DON		2.6724606	0.19539939	3.39960	2.1502633	4.3688478	13.50	6.0	0.00000
2361	GOGOL		3.1399267	0.13936286	1.62289	2.7023375	5.5638995	13.00	7.6	0.00000
2362	MARK TWAIN		2.1955695	0.19322146	3.95211	1.7713383	3.2532754	15.00	3.0	0.00000
2363	CEBRIONES		5.1301565	0.03656024	32.25758	4.9425969	11.6197271	10.00	49.8	0.00000
2364	SEILLIER		3.1817579	0.13052775	10.71351	2.7664504	5.6754556	12.00	12.1	0.00000
2365	INTERKOSMOS		2.5430703	0.11765014	5.32594	2.2438779	4.0554361	13.50	6.0	0.00000
2366	AARYN		2.2411747	0.12766720	1.08167	1.9550501	3.3551624	15.00	3.0	0.00000
2367	PRAHA		2.2076116	0.09887762	1.87642	1.9893281	3.2800767	14.50	3.8	0.00000
2368	BELTROVATA	DU	2.1044805	0.41323951	5.25843	1.2348260	3.0529337	16.80	2.4	5.90000
2369	CHEKHOV		2.7822261	0.04415242	2.63778	2.6593840	4.6407547	13.00	7.6	0.00000
2370	VAN ALTENA		2.7155259	0.18205914	8.26422	2.2211394	4.4748750	11.50	15.2	0.00000
2371	DIMITROV		2.4419503	0.01211072	1.77610	2.4123766	3.8159728	13.00	7.6	0.00000
2372	PROSKURIN		3.1043320	0.18573707	2.74946	2.5277424	5.4695582	13.00	7.6	0.00000
2373	1929PC71OJ		2.7938035	0.17289154	10.11021	2.3107784	4.6697512	14.50	3.8	0.00000
2374	VLADVYSOTSKIJ		3.0853643	0.21139985	15.12631	2.4331188	5.4195061	12.00	12.1	0.00000
2375	1975AA75AT		3.1701436	0.22012492	15.01954	2.4723160	5.6444082	12.00	12.1	0.00000
2376	MARTYNOV		3.2146585	0.09658705	3.85755	2.9041641	5.7637119	12.00	12.1	0.00000
2377	SHCHEGLOV		2.8786585	0.05641768	1.00624	2.7162514	4.8841076	13.00	7.6	0.00000
2378	1935CY		2.8878627	0.14331786	14.26625	2.4739802	4.9075508	12.00	12.1	0.00000
2379	HEISKANEN	CFPD	3.1771524	0.26915336	0.47078	2.3220112	5.6631374	12.00	29.0	0.00000
2380	HEILONGJIANG		2.1921561	0.05985662	1.92089	2.0609410	3.2456913	14.00	4.8	0.00000
2381	LANDI		2.6098962	0.16757876	13.62791	2.1725330	4.2163324	12.50	9.6	0.00000
2382	NONIE		2.7598207	0.32912201	31.03480	1.8515029	4.5848093	12.00	12.1	0.00000
2383	BRADLEY		2.2174039	0.10544537	3.57048	1.9835888	3.3019247	14.50	3.8	0.00000
2384	SCHULHOF		2.6090147	0.12206052	13.58906	2.2905569	4.2141967	13.50	6.0	0.00000
2385	MUSTEL		2.2424042	0.16090320	4.07883	1.8815942	3.3579240	14.50	3.8	0.00000
2386	NIKONOV		2.8165209	0.15503091	9.06431	2.3798733	4.7268248	13.00	7.6	0.00000
2387	1975FX		3.0213056	0.07835089	10.97416	2.7845836	5.2516041	13.00	7.6	0.00000
2388	GASE		2.4519992	0.17963751	2.21559	2.0115283	3.8395517	14.00	4.8	0.00000
2389	DIBAJ		2.4449883	0.23037872	7.79346	1.8817151	3.8230958	14.50	3.8	0.00000
2390	NEZARKA		2.6191916	0.14632301	10.34437	2.2359436	4.2388778	13.00	7.6	0.00000
2391	1957AA		2.4418359	0.13525307	3.00182	2.1115701	3.8157043	13.50	6.0	0.00000
2392	JONATHAN MU		2.3435621	0.15474954	3.36659	1.9808969	3.5876884	15.50	2.4	0.00000
2393	1955WB		3.2200384	0.19872975	10.12266	2.5801210	5.7781868	11.80	13.2	0.00000
2394	NADEEV		3.1946580	0.19482455	1.62688	2.5722604	5.7100062	12.60	9.1	0.00000

No.	Name	Type	A (A.U.)	E	I (deg.)	q (A.U.)	Period (years)	Mag.	Rad. (km)	Rot. (hours)
2395	AHO		3.0768032	0.05637099	0.30281	2.9033608	5.3969646	13.60	5.8	0.00000
2396	KOCHI		2.7929058	0.07385637	12.59757	2.5866320	4.6675010	12.50	9.6	0.00000
2397	LAPPAJARVI		3.0905561	0.17247491	10.30478	2.5575128	5.4331908	11.50	15.2	0.00000
2398	JILIN		2.3898375	0.23885147	3.74108	1.8190213	3.6944735	14.50	3.8	0.00000
2399	TERRADAS		2.2396173	0.16891642	5.12582	1.8613092	3.3516662	14.50	3.8	0.00000
2400	DEREVSKAYA		3.0035505	0.09707633	10.38177	2.7119770	5.2053800	13.00	7.6	0.00000
2401	AEHLITA		2.7707996	0.05946715	4.33100	2.6060281	4.6121955	13.50	6.0	0.00000
2402	SATPAEV		2.2224352	0.13076282	5.16796	1.9318233	3.3131695	13.50	6.0	0.00000
2403	SUMAVA		2.5477479	0.12758897	3.28739	2.2226832	4.0666299	13.50	6.0	0.00000
2404	ANTARCTICA		3.1209657	0.13827658	2.69271	2.6894093	5.5135779	13.00	7.6	0.00000
2405	WELCH		3.2230682	0.11980559	2.22334	2.8369267	5.7863441	12.50	9.6	0.00000
2406	ORELSKAYA		2.1924202	0.16344057	2.30785	1.8340899	3.2462780	15.00	3.0	0.00000
2407	1973DH		2.9204121	0.22157599	2.48512	2.2733190	4.9907546	12.00	12.1	0.00000
2408	ASTAPOVICH		2.6379042	0.24074817	17.69303	2.0028336	4.2843852	14.00	4.8	0.00000
2409	CHAPMAN		2.2662716	0.19104838	3.51286	1.8333042	3.4116774	14.00	4.8	0.00000
2410	MORRISON		2.2155228	0.06298122	2.39720	2.0759864	3.2977242	14.00	4.8	0.00000
2411	ZELLNER		2.2248726	0.08718028	1.61679	2.0309076	3.3186216	14.00	4.8	0.00000
2412	WIL		2.6809201	0.14675827	7.11607	2.2874730	4.3896089	13.00	7.6	0.00000
2413	6816P-L		3.0199475	0.11383148	10.64931	2.6761825	5.2480636	13.00	7.6	0.00000
2414	VIBEKE		3.2076175	0.11867728	16.70624	2.8269463	5.7447863	12.10	11.5	0.00000
2415	1978UJ		2.6597257	0.03676361	2.37460	2.5619445	4.3376575	13.10	7.3	0.00000
2416	SHARONOV		3.0107632	0.04853256	10.51414	2.8646431	5.2241411	12.20	11.0	0.00000
2417	MCVITTIE		3.1961653	0.20972860	3.10126	2.5258379	5.7140474	13.00	7.6	0.00000
2418	1971UV		3.1087289	0.17545690	1.32863	2.5632811	5.4811826	13.50	6.0	0.00000
2419	MOLDAVIA		2.2963071	0.09155537	6.39834	2.0860677	3.4797249	14.50	3.8	0.00000
2420	CIURLIONIS		2.5599797	0.13313699	14.65218	2.2191517	4.0959511	14.00	4.8	0.00000
2421	NININGER		3.2315052	0.05836407	10.15355	3.0429015	5.8090792	12.00	12.1	0.00000
2422	PEROVSKAYA		2.3291473	0.19658220	6.41260	1.8712785	3.5546386	15.00	3.0	0.00000
2423	IBARRURI		2.1884327	0.28276142	4.05769	1.5696284	3.2374253	15.00	3.0	0.00000
2424	TAUTENBURG		2.3488863	0.13613412	8.90709	2.0291226	3.5999212	14.00	4.8	0.00000
2425	1975FW		3.0020502	0.08990184	10.88855	2.7321603	5.2014794	13.00	7.6	0.00000
2426	SIMONOV		2.9095166	0.11537255	8.47383	2.5738380	4.9628510	12.50	9.6	0.00000
2427	KOBZAR		2.7415712	0.16315025	4.14779	2.2942832	4.5394092	14.00	4.8	0.00000
2428	KAMENYAR		3.1729887	0.07912800	9.31877	2.9219162	5.6520081	12.50	9.6	0.00000
2429	1977TZ		2.5714667	0.10085943	15.01457	2.3121099	4.1235504	14.00	4.8	0.00000
2430	BRUCE HELIN		2.3633161	0.21492574	23.41772	1.8553787	3.6331449	13.50	6.0	0.00000
2431	SKOVORODA		2.6477132	0.27837533	2.96408	1.9106551	4.3083048	13.50	6.0	0.00000
2432	SOOMANA		2.3523693	0.11313684	6.76497	2.0862298	3.6079316	14.00	4.8	0.00000
2433	SOOTIYO		2.6100738	0.21695565	10.40194	2.0438035	4.2167630	13.00	7.6	0.00000
2434	BATESON		3.0891609	0.15989459	15.64820	2.5952208	5.4295120	12.00	12.1	0.00000
2435	HOREMHEB		2.2030551	0.20506714	3.96491	1.7512809	3.2699270	16.00	1.9	0.00000
2436	HATSHEPSUT		3.1624959	0.12019201	4.09529	2.7823892	5.6239953	13.50	6.0	0.00000
2437	AMNESTIA		2.1889937	0.14781635	2.94115	1.8654248	3.2386706	14.70	3.5	0.00000
2438	OLESHKO		2.2432857	0.10920680	4.90553	1.9983037	3.3599041	14.30	4.2	0.00000
2439	ULUGBEK		3.1232557	0.16730019	0.28125	2.6007345	5.5196476	12.50	9.6	0.00000
2440	EDUCATIO		2.2160950	0.16237031	4.10255	1.8562669	3.2990017	15.00	3.0	0.00000
2441	HIBBS		2.4085684	0.19253229	3.74718	1.9448411	3.7379928	14.50	3.8	0.00000
2442	CORBETT		2.3881662	0.11674359	5.08904	2.1093633	3.6905990	14.00	4.8	0.00000
2443	TOMEILEEN		3.0068717	0.05346048	11.45164	2.8461230	5.2140160	11.50	15.2	0.00000
2444	LEDERLE		2.7295904	0.13195208	15.11268	2.3694153	4.5096850	13.50	6.0	0.00000
2445	BLAZHKO		2.2692668	0.14739589	6.07383	1.9347862	3.4184430	14.00	4.8	0.00000
2446	LUNACHARSKY		2.3538406	0.16192845	3.31852	1.9726869	3.6113167	13.50	6.0	0.00000
2447	KRONSTADT		2.5376098	0.26148555	8.78131	1.8740613	4.0423808	14.50	3.8	0.00000
2448	1975BU		2.7945900	0.11224648	17.73788	2.4809070	4.6717234	13.00	7.6	0.00000
2449	1978GC	E	1.9088807	0.16823813	24.98293	1.5877342	2.6373525	14.50	1.3	0.00000
2450	IOANNISIANI		3.1118677	0.11867257	2.52236	2.7425742	5.4894862	12.50	9.6	0.00000
2451	DOLLFUS		2.7230124	0.15306500	8.58476	2.3062146	4.4933934	13.00	7.6	0.00000
2452	LYOT		3.1527419	0.12868167	11.80412	2.7470417	5.5979967	13.00	7.6	0.00000
2453	A921SA		3.0216730	0.10762883	10.30254	2.6964538	5.2525616	12.50	9.6	0.00000
2454	OLAUS MAGNUS		2.2517197	0.20220445	4.72990	1.7964120	3.3788702	14.50	3.8	0.00000
2455	SOMVILLE		2.7301555	0.08704414	7.53673	2.4925115	4.5110855	13.00	7.6	0.00000
2456	PALAMEDES		5.1974492	0.07736199	13.86433	4.7953644	11.8491011	10.50	39.5	0.00000
2457	1975TU2		2.6398547	0.06700276	6.25879	2.4629772	4.2891378	14.00	4.8	0.00000
2458	1977RC7		3.1411176	0.12361843	2.06810	2.7528174	5.5670648	13.00	7.6	0.00000
2459	SPELLMANN		3.0200744	0.07317086	9.67514	2.7990930	5.2483945	13.00	7.6	0.00000
2460	MITLINCOLN		2.2566810	0.11067816	3.73988	2.0069156	3.3900433	13.50	6.0	0.00000
2461	1981EC1		3.1915505	0.15407921	2.51276	2.6997988	5.7016764	12.50	9.6	0.00000
2462	NEHALENNIA		2.4115767	0.13876681	2.99166	2.0769300	3.7449985	15.00	3.0	0.00000
2463	1934FF		2.6000636	0.15365702	13.41349	2.2005455	4.1925278	13.50	6.0	0.00000
2464	NORDENSKIOLD		3.1532309	0.22110508	0.86110	2.4560356	5.5992990	12.50	9.6	0.00000
2465	1949PK		2.7545862	0.07600502	3.85100	2.5452237	4.5717721	13.50	6.0	0.00000
2466	GOLSON		2.6373262	0.16408044	5.08692	2.2045927	4.2829771	13.50	6.0	0.00000
2467	KOLLONTAI		2.2136362	0.16078500	5.79943	1.8577166	3.2935126	14.00	4.8	0.00000
2468	1969TO1		2.3262753	0.15608171	5.71836	1.9631864	3.5480659	14.00	4.8	0.00000
2469	TADJIKISTAN		3.1172926	0.12399961	9.62333	2.7307496	5.5038471	13.00	7.6	0.00000
2470	AGEMATSU		2.8879840	0.00944196	3.11029	2.8607156	4.9078598	12.50	9.6	0.00000
2471	ULTRAJECTUM		3.0013931	0.08882629	10.29618	2.7347903	5.1997719	13.00	7.6	0.00000
2472	1973DG		2.2648726	0.09468163	5.11116	2.0504308	3.4085183	15.00	3.0	0.00000
2473	1977RX7		2.2414856	0.13533039	5.18956	1.9381444	3.3558607	14.00	4.8	0.00000
2474	RUBY		2.6841507	0.22037114	7.51849	2.0926414	4.3975458	13.00	7.6	0.00000
2475	SEMENOV		3.0357566	0.10848811	9.09588	2.7064133	5.2893271	12.30	10.5	0.00000
2476	ANDERSEN		3.0255356	0.11355727	10.83232	2.6819639	5.2626367	12.20	11.0	0.00000
2477	BIRYUKOV		2.5567186	0.15393667	6.07759	2.1631458	4.0881271	13.50	6.0	0.00000

No.	Name	Type	A (A.U.)	E	I (deg.)	q (A.U.)	Period (years)	Mag.	Rad. (km)	Rot. (hours)
2478	TOKAI		2.2251933	0.06777679	4.13889	2.0743766	3.3193388	13.80	5.3	0.00000
2479	SODANKYLA		2.3890040	0.19621503	2.91435	1.9202455	3.6925409	14.00	4.8	0.00000
2480	1976YS1		2.2257066	0.11974565	2.91962	1.9591877	3.3204875	15.00	3.0	0.00000
2481	1977UQ		2.5667720	0.26573017	2.26263	1.8847032	4.1122637	14.50	3.8	0.00000
2482	PERKIN		2.9265573	0.06522557	3.13391	2.7356710	5.0065155	12.50	9.6	0.00000
2483	GUINEVERE		3.9759238	0.27439576	4.49805	2.8849471	7.9278798	11.50	24.9	0.00000
2484	PARENAGO		2.3424976	0.25434071	1.19405	1.7467051	3.5852444	14.50	3.8	0.00000
2485	1932BH		3.1974015	0.23100866	2.79790	2.4587741	5.7173629	13.00	7.6	0.00000
2486	METSAHOVI		2.2682266	0.07979918	8.40855	2.0872240	3.4160931	13.50	6.0	0.00000
2487	JUHANI		2.3975201	0.18378632	2.81783	1.9568887	3.7123027	14.00	4.8	0.00000
2488	BRYAN		2.2639327	0.22453727	6.89074	1.7555953	3.4063969	15.00	3.0	0.00000
2489	SUVOROV		3.1120019	0.14978898	1.78810	2.6458583	5.4898410	13.00	7.6	0.00000
2490	BUSSOLINI		2.6087842	0.13214189	12.97873	2.2640543	4.2136378	13.00	7.6	0.00000
2491	1977CB		1.8776453	0.05411414	22.86777	1.7760381	2.5728848	13.50	3.5	0.00000
2492	KUTUZOV		3.1758962	0.15721317	0.83983	2.6766036	5.6597786	13.00	7.6	0.00000
2493	ELMER		2.7872627	0.17213240	8.72437	2.3074844	4.6533618	14.00	4.8	0.00000
2494	INGE		3.1564558	0.06935938	11.53303	2.9375260	5.6078906	12.00	12.1	0.00000
2495	NOVIOMAGUM		1.9175684	0.10260459	21.12166	1.7208172	2.6553779	16.50	0.9	0.00000
2496	FERNANDUS		2.1702499	0.03341201	0.91862	2.0977376	3.1971619	14.50	3.8	0.00000
2497	KULIKOVSKIJ		2.5388429	0.23274729	5.85963	1.9479342	4.0453281	14.10	4.6	0.00000
2498	TSESEVICH		2.9168966	0.08223368	1.23825	2.6770294	4.9817452	13.10	7.3	0.00000
2499	BRUNK		3.1006773	0.12479144	0.73004	2.7137392	5.4599018	13.10	7.3	0.00000
2500	1926GC		2.2404742	0.09962664	6.99026	2.0172634	3.3535895	14.00	4.8	0.00000
2501	LOHJA		2.4198983	0.19830748	3.32586	1.9400142	3.7643988	13.50	6.0	0.00000
2502	NUMMELA		2.9420204	0.22165887	17.78159	2.2898953	5.0462470	12.50	9.6	0.00000
2503	LIAONING		2.1928363	0.21260735	7.11080	1.7266232	3.2472022	15.50	2.4	0.00000
2504	GAVIOLA		2.7615976	0.08720855	4.08078	2.5207627	4.5892382	13.50	6.0	0.00000
2505	HEBEI		3.1565342	0.16653685	2.07972	2.6308548	5.6080999	12.00	12.1	0.00000
2506	PIROGOV		2.9038658	0.02004435	2.16392	2.8456597	4.9484000	13.00	7.6	0.00000
2507	1976WB1		2.7784326	0.08411064	10.32320	2.5447369	4.6312671	13.00	7.6	0.00000
2508	ALUPKA		2.3674331	0.12715831	6.08412	2.0663943	3.6426427	15.00	3.0	0.00000
2509	CHUKOTKA		2.4561429	0.19076747	2.83851	1.9875907	3.8492882	15.00	3.0	0.00000
2510	SHANDONG		2.2531121	0.19691610	5.27245	1.8094380	3.3820045	13.50	6.0	0.00000
2511	PATTERSON		2.2989600	0.10337879	8.04822	2.0612962	3.4857569	14.00	4.8	0.00000
2512	TAVASTIA		2.2438500	0.12023812	6.38166	1.9740536	3.3611717	14.00	4.8	0.00000
2513	BAETSLE		2.2868221	0.18040203	3.15934	1.8742747	3.4581878	14.00	4.8	0.00000
2514	TAIYUAN		2.6502652	0.10020892	2.35038	2.3846850	4.3145351	13.50	6.0	0.00000
2515	GANSU		3.1680934	0.20977952	4.06925	2.5034924	5.6389332	14.00	4.8	0.00000
2516	ROMAN		2.2799597	0.16448987	1.09477	1.9049294	3.4426332	15.00	3.0	0.00000
2517	1968SB		3.1697762	0.18730377	2.63088	2.5760653	5.6434269	13.00	7.6	0.00000
2518	RUTLLANT		2.3081408	0.17332292	5.92809	1.9080871	3.5066583	14.50	3.8	0.00000
2519	1975VD2		3.1438577	0.17424954	2.42414	2.5960419	5.5743513	12.00	12.1	0.00000
2520	NOVOROSSIJSK		3.1069026	0.09078917	6.24884	2.8248293	5.4763532	13.50	6.0	0.00000
2521	1979DK		2.7950242	0.09003786	7.72116	2.5433662	4.6728125	13.00	7.6	0.00000
2522	1980PP		3.0237684	0.04935670	8.74712	2.8745253	5.2580271	13.00	7.6	0.00000
2523	1980PV		3.0204043	0.03707671	8.88319	2.9084177	5.2492547	13.00	7.6	0.00000
2524	BUDOVICIUM		3.1215856	0.15141612	0.28403	2.6489272	5.5152206	12.50	9.6	0.00000
2525	O'STEEN		3.1330254	0.19733867	2.77873	2.5147583	5.5455661	12.00	12.1	0.00000
2526	ALISARY		3.1269639	0.18705726	3.28586	2.5420425	5.5294800	13.50	6.0	0.00000
2527	GREGORY		2.4658594	0.18441142	2.61424	2.0111268	3.8721528	14.50	3.8	0.00000
2528	MOHLER		3.1381299	0.18261702	0.50775	2.5650539	5.5591240	12.80	8.3	0.00000
2529	ROCKWELL KENT		2.5322859	0.09702183	4.39482	2.2865989	4.0296664	14.30	4.2	0.00000
2530	SHIPKA		3.0218618	0.12072072	10.06431	2.6570604	5.2530541	13.00	7.6	0.00000
2531	CAMBRIDGE		3.0166488	0.04958584	11.02712	2.8670657	5.2394671	12.00	12.1	0.00000
2532	SUTTON		2.3727083	0.17135072	4.34548	1.9661431	3.6548247	13.90	5.0	0.00000
2533	A905VA		3.0982845	0.17271572	1.55998	2.5631621	5.4535832	12.50	9.6	0.00000
2534	HOUZEAU		3.1340413	0.18401849	0.80617	2.5573199	5.5482635	12.50	9.6	0.00000
2535	HAMEENLINNA		2.2396386	0.08087199	3.42983	2.0585146	3.3517137	13.50	6.0	0.00000
2536	KOZYREV		2.3064477	0.22748525	4.86438	1.7817649	3.5028007	14.50	3.8	0.00000
2537	GILMORE		2.6563473	0.17287517	12.96704	2.1971307	4.3293953	13.50	6.0	0.00000
2538	VANDERLINDEN		2.2390790	0.13028961	4.76688	1.9473501	3.3504572	14.50	3.8	0.00000
2539	NINGXIA		2.2626545	0.16834636	3.98013	1.8817447	3.4035127	15.50	2.4	0.00000
2540	1971TH2		2.1965065	0.05183807	1.26493	2.0826437	3.2553580	14.50	3.8	0.00000
2541	1973DE		2.9367442	0.07873784	3.19907	2.7055113	5.0326786	13.50	6.0	0.00000
2542	CALPURNIA		3.1232257	0.08640562	4.62936	2.8533616	5.5195680	12.50	9.6	0.00000
2543	1980LJ		3.0846288	0.29220286	15.05363	2.1832914	5.4175682	12.50	9.6	0.00000
2544	GUBAREV		2.3732781	0.24077247	22.53207	1.8018581	3.6561413	14.00	4.8	0.00000
2545	VERBIEST		2.2294984	0.12532228	5.96778	1.9500926	3.3289766	14.00	4.8	0.00000
2546	1950FC		2.6018009	0.18800795	10.38051	2.1126416	4.1967301	13.50	6.0	0.00000
2547	HUBEI		2.3859541	0.12891707	6.18615	2.0783639	3.6854722	14.50	3.8	0.00000
2548	1975DA		2.6320870	0.10102302	18.16954	2.3661854	4.2702208	13.50	6.0	0.00000
2549	BAKER		3.1816061	0.18520875	0.06050	2.5923448	5.6750488	14.00	4.8	0.00000
2550	1976UP20		3.1817245	0.17884190	10.41833	2.6126988	5.6753659	12.50	9.6	0.00000
2551	DECABRINA		3.1370494	0.19007871	0.64019	2.5407631	5.5562534	13.00	7.6	0.00000
2552	REMEK		2.1471870	0.18809269	0.90090	1.7433167	3.1463337	15.50	2.4	0.00000
2553	VILJEV		3.0882013	0.05991848	5.25141	2.9031610	5.4269824	12.50	9.6	0.00000
2554	SKIFF		2.2640896	0.14388378	4.85383	1.9383237	3.4067509	14.50	3.8	0.00000
2555	THOMAS		2.8684824	0.08224298	0.89942	2.6325698	4.8582320	13.00	7.6	0.00000
2556	LOUISE		2.1627839	0.03662706	2.79095	2.0835674	3.1806777	15.00	3.0	0.00000
2557	PUTNAM		2.3504574	0.15520792	6.06633	1.9856479	3.6035337	13.50	6.0	0.00000
2558	VIV		2.2163620	0.15553825	5.15158	1.8716329	3.2995982	15.00	3.0	0.00000
2559	1981UH		2.7883987	0.15197004	8.88375	2.3646457	4.6562076	13.50	6.0	0.00000
2560	1932CW		2.7501955	0.03437001	5.93744	2.6556714	4.5608454	13.00	7.6	0.00000

No.	Name	Type	A (A.U.)	E	I (deg.)	q (A.U.)	Period (years)	Mag.	Rad. (km)	Rot. (hours)
2561	MARGOLIN		2.4324296	0.13630550	2.49040	2.1008761	3.7936773	15.00	3.0	0.00000
2562	1973FF1		3.0072677	0.04791763	10.26530	2.8631666	5.2150455	12.50	9.6	0.00000
2563	BOYARCHUK		3.2014685	0.12939657	2.03957	2.7872095	5.7282748	12.50	9.6	0.00000
2564	1977QX		2.2368059	0.10969973	1.96372	1.9914289	3.3453567	14.50	3.8	0.00000
2565	1977TB1		2.3574393	0.23410572	2.04134	1.8055493	3.6196015	16.00	1.9	0.00000
2566	KIRGHIZIA		2.4495921	0.07893293	5.08636	2.2562387	3.8338988	14.00	4.8	0.00000
2567	ELBA		2.7379348	0.13930634	8.90516	2.3565233	4.5303802	12.50	9.6	0.00000
2568	MAKSUTOV		2.2055705	0.18755746	8.03393	1.7918993	3.2755287	15.00	3.0	0.00000
2569	MADELINE		2.6266773	0.16079099	11.48731	2.2043312	4.2570629	12.50	9.6	0.00000
2570	PORPHYRO		2.7659235	0.10448550	16.24576	2.4769247	4.6000257	13.00	7.6	0.00000
2571	GEISEI		2.2285821	0.19413140	2.87310	1.7959445	3.3269248	15.50	2.4	0.00000
2572	1950DL		2.3908091	0.14663696	5.14976	2.0402281	3.6967268	14.00	4.8	0.00000
2573	HANNU OLAVI		3.0218801	0.10002290	12.91740	2.7196231	5.2531023	12.50	9.6	0.00000
2574	LADOGA		2.8497665	0.07049339	2.12034	2.6488769	4.8107629	13.00	7.6	0.00000
2575	BULGARIA		2.2403436	0.12209404	4.67660	1.9668109	3.3532965	14.50	3.8	0.00000
2576	YESENIN		3.0810969	0.13738349	12.19746	2.6578050	5.4082656	12.00	12.1	0.00000
2577	LITVA		1.9041348	0.13826866	22.90872	1.6408526	2.6275229	14.00	2.8	0.00000
2578	1975VW3		3.0024140	0.09097338	10.55950	2.7292743	5.2024250	12.50	9.6	0.00000
2579	SPARTACUS		2.2104445	0.07412577	5.77349	2.0465937	3.2863925	14.50	3.8	0.00000
2580	1977QP4		2.1824787	0.19529833	1.61580	1.7562442	3.2242227	14.50	3.8	0.00000
2581	1980VX		2.2359562	0.09876636	2.49244	2.0151188	3.3434505	15.00	3.0	0.00000
2582	HARIMAYA-BA		3.2078228	0.06313047	18.14198	3.0053115	5.7453375	12.00	12.1	0.00000
2583	1975XA3		2.2522452	0.20969567	6.87896	1.7799591	3.3800528	14.10	4.6	0.00000
2584	TURKMENIA		2.2276502	0.06513964	1.43823	2.0825417	3.3248377	14.20	4.4	0.00000
2585	IRPEDINA		2.4254568	0.23392323	5.99819	1.8580860	3.7773767	13.80	5.3	0.00000
2586	MATSON		2.3874054	0.08814729	4.36612	2.1769621	3.6888351	14.00	4.8	0.00000
2587	GARDNER		3.1771598	0.15065344	2.63240	2.6985097	5.6631570	12.40	10.0	0.00000
2588	FLAVIA		2.4585903	0.20865139	2.26740	1.9456019	3.8550429	14.30	5.5	0.00000
2589	DANIEL		2.8804963	0.08064234	2.61786	2.6482062	4.8887854	12.70	11.6	0.00000
2590	MOURAO		2.3429365	0.11727835	6.12583	2.0681608	3.5862520	13.80	7.0	0.00000
2591	1949PS		2.9374137	0.04119501	1.54950	2.8164070	5.0343995	12.50	12.7	0.00000
2592	HUNAN		3.1176655	0.12344164	1.33401	2.7328157	5.5048347	12.00	16.0	0.00000
2593	BURYATIA		2.1693840	0.07958206	0.21529	1.9967397	3.1952484	15.00	4.0	0.00000
2594	1978TB		5.1543341	0.08673322	5.50041	4.7072821	11.7019672	12.50	12.7	0.00000
2595	GUDIACHVILI		2.7859733	0.14322335	9.87125	2.3869569	4.6501336	13.50	8.0	0.00000
2596	VAINU BAPPU		3.0310335	0.06670334	10.24870	2.8288536	5.2769880	13.50	8.0	0.00000
2597	ARTHUR		3.0056298	0.15148865	1.09291	2.5503111	5.2107859	13.00	10.1	0.00000
2598	MERLIN		2.7831354	0.21505933	7.78427	2.1845961	4.6430297	13.50	8.0	0.00000
2599	VESELI		2.5361068	0.16246872	15.31810	2.1240690	4.0387907	12.50	12.7	0.00000
2600	LUMME		3.0105796	0.09233094	11.72594	2.7326100	5.2236633	12.50	12.7	0.00000
2601	BOLOGNA		3.1287053	0.05413355	9.58799	2.9593372	5.5340996	12.00	16.0	0.00000
2602	MOORE		2.3848376	0.10529724	5.54842	2.1337209	3.6828856	14.00	6.4	0.00000
2603	TAYLOR		2.7806799	0.04253437	3.05432	2.6624055	4.6368866	13.00	10.1	0.00000
2604	1972LD1		2.3893888	0.23043714	14.83086	1.8387848	3.6934330	14.00	6.4	0.00000
2605	1974QA		3.0890887	0.08262280	9.42353	2.8338594	5.4293218	14.00	6.4	0.00000
2606	ODESSA		2.7613137	0.26363423	12.37684	2.0333369	4.5885301	12.50	12.7	0.00000
2607	YAKUTIA		2.3770473	0.22751357	2.09754	1.8362368	3.6648545	15.00	4.0	0.00000
2608	SENECA		2.4789059	0.58642077	15.63654	1.0252240	3.9029236	18.80	0.7	18.50000
2609	KIRIL-METODI		2.2209949	0.08845637	5.71552	2.0245337	3.3099494	14.00	6.4	0.00000
2610	TUVA		2.1593721	0.09882331	0.67137	1.9459759	3.1731548	15.00	4.0	0.00000
2611	BOYCE		3.0402815	0.05352646	3.33231	2.8775461	5.3011575	12.50	12.7	0.00000
2612	KATHRYN		2.9000628	0.16269761	20.21236	2.4282296	4.9386826	12.00	16.0	0.00000
2613	PLZEN		3.0393696	0.04629069	13.00715	2.8986752	5.2987723	12.00	16.0	0.00000
2614	TORRENCE		2.3388872	0.16754118	6.91954	1.9470272	3.5769584	14.50	5.1	0.00000
2615	1951RJ		3.1623864	0.16703486	4.27529	2.6341574	5.6237030	13.50	8.0	0.00000
2616	LESYA		2.1627452	0.07668477	1.44533	1.9968957	3.1805928	13.60	7.6	0.00000
2617	JIANGXI		3.1605115	0.23528824	12.88350	2.4169436	5.6187029	11.50	20.1	0.00000
2618	COONABARABRAN		3.0230899	0.11356179	9.22745	2.6797824	5.2562571	13.40	8.4	0.00000
2619	SKALNATE PLESO		3.0056822	0.04647483	1.11391	2.8659937	5.2109222	13.80	7.0	0.00000
2620	1980TN		2.8584046	0.07062457	3.09109	2.6565311	4.8326526	13.50	8.0	0.00000
2621	GOTO		3.0918894	0.16362870	13.05618	2.5859675	5.4367070	11.80	17.5	0.00000
2622	BOLZANO		3.0060062	0.09619389	11.00755	2.7168467	5.2117648	12.80	11.1	0.00000
2623	A919SA		2.2545266	0.23458304	4.06131	1.7256528	3.3851898	14.50	5.1	0.00000
2624	1962RE		3.9702983	0.11769837	2.77642	3.5030005	7.9110603	12.00	16.0	0.00000
2625	JACK LONDON		2.1956744	0.14115404	4.46073	1.8857461	3.2535081	14.00	6.4	0.00000
2626	BELNIKA		2.8501735	0.02550832	1.49148	2.7774704	4.8117933	13.00	10.1	0.00000
2627	CHURYUMOV		3.1147912	0.16920120	2.50077	2.5877647	5.4972234	13.00	10.1	0.00000
2628	KOPAL		2.9085770	0.14913899	1.32367	2.4747946	4.9604473	14.00	6.4	0.00000
2629	1980RB1		1.7403622	0.22906430	23.43702	1.3417073	2.2959342	16.00	2.5	0.00000
2630	1980TF3		3.0742934	0.12197151	1.93433	2.6993172	5.3903623	13.00	10.1	0.00000
2631	ZHEJIANG		2.7986748	0.15911464	9.59113	2.3533647	4.6819706	13.00	10.1	0.00000
2632	GUIZHOU		3.0362418	0.10907064	10.45784	2.7050769	5.2905951	12.50	12.7	0.00000
2633	BISHOP		2.2251256	0.13855927	3.12754	1.9168139	3.3191876	14.00	6.4	0.00000
2634	JAMES BRADLEY		3.4468956	0.07143922	6.44483	3.2006519	6.3994431	11.50	20.1	0.00000
2635	HUGGINS		2.2314892	0.07809224	4.16764	2.0572271	3.3334365	14.00	6.4	0.00000
2636	LASSELL		3.0036232	0.07820619	10.47566	2.7687213	5.2055688	12.50	12.7	0.00000
2637	BOBROVNIKOFF		2.2546828	0.23533088	4.93388	1.7240862	3.3855414	14.50	5.1	0.00000
2638	GADOLIN		2.5550952	0.08211331	14.38215	2.3452878	4.0842338	13.50	8.0	0.00000
2639	PLANMAN		2.4482806	0.18668269	9.63041	1.9912289	3.8308201	14.00	6.4	0.00000
2640	HALLSTROM		2.3975999	0.08736485	6.65508	2.1881340	3.7124882	13.50	8.0	0.00000
2641	1949GJ		2.3780718	0.13222460	9.01308	2.0636322	3.6672244	13.50	8.0	0.00000
2642	1961RA		2.4264274	0.18487453	14.48254	1.9778427	3.7796445	14.00	6.4	0.00000
2643	1973SD		2.3780179	0.27313569	22.97886	1.7284964	3.6670997	16.50	2.0	0.00000

No.	Name	Type	A (A.U.)	E	I (deg.)	q (A.U.)	Period (years)	Mag.	Rad. (km)	Rot. (hours)
2644	VICTOR JARA		2.1706605	0.16566581	2.68069	1.8110563	3.1980691	15.00	4.0	0.00000
2645	DAPHNE PLANE		2.3906882	0.10706675	13.79251	2.1347251	3.6964467	13.50	8.0	0.00000
2646	ABETTI		3.0147676	0.09091685	9.67305	2.7406745	5.2345667	12.50	12.7	0.00000
2647	1980SP		2.2433133	0.13767377	3.93635	1.9344679	3.3599660	14.00	6.4	0.00000
2648	OWA		2.2496095	0.17476989	4.79553	1.8564454	3.3741212	14.00	6.4	0.00000
2649	OONGAQ		2.6268642	0.14275694	12.22455	2.2518611	4.2575173	13.00	10.1	0.00000
2650	1931EG		2.6365921	0.19593842	13.94453	2.1199825	4.2811894	12.50	12.7	0.00000
2651	KAREN		2.9887991	0.32151118	17.77348	2.0278668	5.1670790	13.50	8.0	0.00000
2652	1953GM		2.6362684	0.08273634	6.99675	2.4181533	4.2804008	13.00	10.1	0.00000
2653	1964VP		2.4437420	0.08096678	4.72568	2.2458801	3.8201733	13.50	8.0	0.00000
2654	RISTENPART		3.0422375	0.10196044	7.45508	2.7320497	5.3062744	13.50	8.0	0.00000
2655	GUANGXI		3.2026932	0.14732857	17.08059	2.7308452	5.7315626	12.50	12.7	0.00000
2656	EVENKIA		2.2551429	0.08028171	3.20025	2.0740962	3.3865781	15.00	4.0	0.00000
2657	BASHKIRIA		3.1856658	0.14044797	2.25138	2.7382455	5.6859145	13.00	10.1	0.00000
2658	GINGERICH		3.0540905	0.29736990	9.30605	2.1458960	5.3373151	13.00	10.1	0.00000
2659	MILLIS		3.1185577	0.11833876	1.31645	2.7495115	5.5071979	12.50	12.7	0.00000
2660	WASSERMAN		2.6185753	0.16838579	12.35540	2.1776445	4.2373819	13.50	8.0	0.00000
2661	1982FC1		3.0266519	0.09038266	9.94736	2.7530949	5.2655497	13.00	10.1	0.00000
2662	KANDINSKY		2.4375508	0.16138557	2.91896	2.0441654	3.8056645	15.00	4.0	0.00000
2663	6561PL		2.2331042	0.13970853	6.22016	1.9211206	3.3370562	14.50	5.1	0.00000
2664	EVERHART		2.3801043	0.18263334	3.26112	1.9454178	3.6719263	15.00	4.0	0.00000
2665	1938DW1		2.2476544	0.08401898	4.78901	2.0588088	3.3697240	14.50	5.1	0.00000
2666	1951TA		3.1934483	0.20972779	13.41064	2.5236936	5.7067633	13.00	10.1	0.00000
2667	1967UO		3.2292078	0.18469596	2.23280	2.6327863	5.8028855	13.50	8.0	0.00000
2668	TARTARIA		2.3163676	0.07821223	3.15431	2.1351993	3.5254230	14.50	5.1	0.00000
2669	SHOSTAKOVICH		2.7804837	0.21789876	7.78626	2.1746199	4.6363959	13.50	8.0	0.00000
2670	CHUVASHIA		3.1702621	0.07412141	9.84436	2.9352779	5.6447248	12.00	16.0	0.00000
2671	ABKHAZIA		2.6083901	0.11907190	1.46826	2.2978041	4.2126837	15.00	4.0	0.00000
2672	PISEK		2.6117022	0.15237194	14.13926	2.2137520	4.2207098	14.50	5.1	0.00000
2673	1980KN		3.2065392	0.14220461	2.30472	2.7505546	5.7418895	13.50	8.0	0.00000
2674	PANDARUS		5.1796765	0.06776190	1.85877	4.8286915	11.7883749	10.00	40.1	0.00000
2675	TOLKIEN		2.2125018	0.10193501	2.75240	1.9869704	3.2909815	13.80	7.0	0.00000
2676	AARHUS		2.4028699	0.12615561	4.55262	2.0997343	3.7247353	13.90	6.7	0.00000
2677	1935FF		2.9926353	0.05204958	10.08284	2.8368700	5.1770301	12.50	12.7	0.00000
2678	AAVASAKSA		2.2593043	0.08708665	3.44558	2.0625491	3.3959560	13.50	8.0	0.00000
2679	KITTISVAARA		2.6204557	0.10319940	10.07912	2.3500264	4.2419472	13.00	10.1	0.00000
2680	1975NF		2.4022508	0.21478230	2.43257	1.8862898	3.7232957	14.50	5.1	0.00000
2681	OSTROVSKIJ		2.7473204	0.19208725	3.99502	2.2195952	4.5536952	13.50	8.0	0.00000
2682	SOROMUNDI		2.2702911	0.17052516	5.49829	1.8831493	3.4207578	15.00	4.0	0.00000
2683	BRIAN		2.9168999	0.05714839	1.48840	2.7502038	4.9817543	13.00	10.1	0.00000
2684	DOUGLAS		3.0496800	0.04529100	9.92083	2.9115567	5.3257575	13.00	10.1	0.00000
2685	MASURSKY		2.5673378	0.11457799	12.10736	2.2731774	4.1136231	13.50	8.0	0.00000
2686	LINDA SUSAN		3.0027955	0.05503989	9.31495	2.8375218	5.2034168	13.00	10.1	0.00000
2687	1982HG		2.5201631	0.12042099	10.08687	2.2166824	4.0007639	13.00	10.1	0.00000
2688	HALLEY		3.1558950	0.15756387	3.47472	2.6586399	5.6063962	13.00	10.1	0.00000
2689	1935CF		2.2328465	0.11707619	5.50261	1.9714334	3.3364785	14.50	5.1	0.00000
2690	RISTIINA		3.0345583	0.11678337	11.42707	2.6801724	5.2861958	12.00	16.0	0.00000
2691	1974KB		2.2438927	0.11350960	3.59067	1.9891893	3.3612678	14.50	5.1	0.00000
2692	CHKALOV		2.5802069	0.18466361	9.30257	2.1037364	4.1445923	13.50	8.0	0.00000
2693	YAN'AN		2.2389700	0.18095390	7.30299	1.8338196	3.3502128	14.50	5.1	0.00000
2694	PINO TORINESE		2.3085673	0.10381520	1.58793	2.0689027	3.5076301	14.50	5.1	0.00000
2695	CHRISTABEL		2.7096341	0.07641718	14.89295	2.5025716	4.4603195	13.00	10.1	0.00000
2696	MAGION		2.4509692	0.11237908	25.32482	2.1755316	3.8371327	13.00	10.1	0.00000
2697	1969TC3		3.5589044	0.08873041	3.58660	3.2431214	6.7138944	11.00	25.3	0.00000
2698	AZERBAJDZHAN		2.6628275	0.05395248	6.88319	2.5191612	4.3452477	13.00	10.1	0.00000
2699	KALININ		2.6397150	0.16694894	16.13788	2.1990173	4.2887979	13.20	9.2	0.00000
2700	BAIKONUR		2.9057984	0.04194142	2.39864	2.7839251	4.9533410	13.30	8.8	0.00000
2701	CHERSON		3.1686289	0.14742129	6.26334	2.7015057	5.6403632	13.50	8.0	0.00000
2702	1978SZ2		3.4310327	0.07366666	1.58690	3.1782799	6.3553176	12.70	11.6	0.00000
2703	RODARI		2.1934078	0.05639784	6.03238	2.0697043	3.2484715	14.80	4.4	0.00000
2704	JULIAN LOEWE		2.3839309	0.09838385	4.50926	2.1493907	3.6807854	14.20	5.8	0.00000
2705	WU		2.1896544	0.15984727	4.52265	1.8396441	3.2401369	14.90	4.2	0.00000
2706	1980VW		3.0170350	0.04206853	10.85276	2.8901126	5.2404733	13.10	9.6	0.00000
2707	UEFERJI		3.1741767	0.14324234	2.68241	2.7195003	5.6551833	12.80	11.1	0.00000
2708	BURNS		3.0951519	0.16968958	2.76988	2.5699370	5.4453144	13.10	9.6	0.00000
2709	SAGAN		2.1955781	0.06903197	2.72990	2.0440130	3.2532942	14.00	6.4	0.00000
2710	VEVERKA		2.4239068	0.13145117	3.11832	2.1052814	3.7737565	14.80	4.4	0.00000
2711	ALEKSANDROV		3.0093360	0.09403389	10.25233	2.7263563	5.2204266	12.70	11.6	0.00000
2712	1937YD		2.1620274	0.03713867	0.82008	2.0817325	3.1790090	15.00	4.0	0.00000
2713	1938EA		2.8552411	0.02453691	1.35475	2.7851822	4.8246317	12.50	12.7	0.00000
2714	MATTI		2.2439404	0.20627403	6.09374	1.7810737	3.3613749	14.50	5.1	0.00000
2715	MIELIKKI		2.7360067	0.15060177	6.73690	2.3239594	4.5255957	13.00	10.1	0.00000
2716	TUULIKKI		2.3690021	0.10685929	5.95161	2.1158524	3.6462648	14.50	5.1	0.00000
2717	TELLERVO		2.2148905	0.21819346	3.28025	1.7316158	3.2963123	13.50	8.0	0.00000
2718	1951OM		3.1079881	0.16595837	1.50434	2.5921915	5.4792237	13.00	10.1	0.00000
2719	1965SU		2.1881921	0.12332462	0.62555	1.9183341	3.2368917	14.50	5.1	0.00000
2720	PYOTR PERVYJ		2.3300254	0.20382893	3.29231	1.8550988	3.5566490	15.50	3.2	0.00000
2721	VSEKHSVYATSKIJ		3.2164314	0.19360796	2.23219	2.5937047	5.7684808	13.00	10.1	0.00000
2722	ABALAKIN		3.1930480	0.15827017	1.67356	2.6876838	5.7056899	13.50	8.0	0.00000
2723	GORSHKOV		3.1349893	0.18332691	2.06339	2.5602615	5.5507808	13.50	8.0	0.00000
2724	ORLOV		2.9252353	0.12070745	3.96521	2.5721376	5.0031233	12.50	12.7	0.00000
2725	DAVID BENDER		3.0297711	0.15517946	15.61148	2.5596130	5.2736917	12.00	16.0	0.00000
2726	KOTELNIKOV		2.8602712	0.07294809	1.56137	2.6516199	4.8373866	13.50	8.0	0.00000

No.	Name	Type	A (A.U.)	E	I (deg.)	q (A.U.)	Period (years)	Mag.	Rad. (km)	Rot. (hours)
2727	PATON		2.6094193	0.10139803	3.51577	2.3448293	4.2151771	13.50	8.0	0.00000
2728	YATSKIV		2.4566848	0.16712514	2.59661	2.0461111	3.8505626	13.50	8.0	0.00000
2729	1979UA2		2.8891323	0.06779561	3.17508	2.6932616	4.9107871	13.00	10.1	0.00000
2730	BARKS		2.7199447	0.13134150	6.44162	2.3627031	4.4858022	13.00	10.1	0.00000
2731	CUCULA		3.1685803	0.20886883	13.25011	2.5067625	5.6402330	13.00	10.1	0.00000
2732	WITT		2.7596743	0.02543303	6.49168	2.6894875	4.5844450	13.00	10.1	0.00000
2733	HAMINA		2.3471901	0.13687405	10.40971	2.0259206	3.5960226	14.00	6.4	0.00000
2734	HASEK		3.1623149	0.02743684	16.53490	3.0755508	5.6235127	12.50	12.7	0.00000
2735	ELLEN		1.8571002	0.05477866	23.05519	1.7553709	2.5307722	14.50	5.1	0.00000
2736	OPS		2.2907474	0.08531491	7.45714	2.0953126	3.4670956	14.50	5.1	0.00000
2737	KOTKA		2.7503493	0.19394812	8.81096	2.2169242	4.5612283	12.50	12.7	0.00000
2738	1940EC		2.7203445	0.11287833	1.11502	2.4132767	4.4867916	13.20	9.2	0.00000
2739	1952UZ1		2.4562237	0.13102597	1.16900	2.1343946	3.8494785	14.40	5.3	0.00000
2740	1974SY4		3.0019188	0.06791668	9.35394	2.7980385	5.2011385	12.50	12.7	0.00000
2741	VALDIVIA		2.6058724	0.18492866	10.28001	2.1239717	4.2065854	12.80	11.1	0.00000
2742	GIBSON		2.9087853	0.06694827	3.15920	2.7140472	4.9609804	13.20	9.2	0.00000
2743	1965WR		2.6541963	0.17617248	12.26402	2.1866000	4.3241377	13.50	8.0	0.00000
2744	BIRGITTA		2.3019123	0.33175236	6.76244	1.5382475	3.4924741	16.50	2.0	0.00000
2745	1976SR10		2.2882798	0.19066104	22.39767	1.8519940	3.4614947	14.50	5.1	0.00000
2746	HISSAO		2.2483296	0.08396067	3.97580	2.0595584	3.3712425	14.50	5.1	0.00000
2747	1980DW		3.1071241	0.11454323	5.81532	2.7512240	5.4769387	12.50	12.7	0.00000
2748	PATRICK GENE		2.8053362	0.13453074	4.23035	2.4279323	4.6986961	14.00	6.4	0.00000
2749	1937TD		3.1722763	0.17397235	0.32602	2.6203878	5.6501050	12.50	12.7	0.00000
2750	LOVIISA		2.2121494	0.07470102	5.17593	2.0468996	3.2901950	14.50	5.1	0.00000
2751	1962RP		2.4076211	0.17149010	1.48062	1.9947380	3.7357881	14.00	6.4	0.00000
2752	1965SP		3.0260706	0.11199444	10.12311	2.6871676	5.2640328	12.00	16.0	0.00000
2753	1966DH		2.7882307	0.03997220	6.88431	2.6767788	4.6557860	13.00	10.1	0.00000
2754	1966PD		2.2279446	0.23237447	5.71192	1.7102270	3.3254969	14.00	6.4	0.00000
2755	AVICENNA		2.8475366	0.25720522	4.54213	2.1151352	4.8051167	13.50	8.0	0.00000
2756	DZHANGAR		2.5516925	0.11500299	5.76004	2.2582400	4.0760779	14.50	5.1	0.00000
2757	1977VN		3.1618321	0.20328575	0.68344	2.5190766	5.6222243	12.00	16.0	0.00000
2758	CORDELIA		2.5526488	0.27354738	2.81595	1.8543785	4.0783701	15.00	4.0	0.00000
2759	IDOMENEUS		5.1628714	0.06659458	21.98173	4.8190522	11.7310514	11.00	25.3	0.00000
2760	KACHA		3.9637275	0.12515616	13.44672	3.4676425	7.8914294	11.00	25.3	0.00000
2761	EDDINGTON		3.0766971	0.18633994	3.17859	2.5033855	5.3966856	14.00	6.4	0.00000
2762	FOWLER		2.3311799	0.15207744	4.70591	1.9766600	3.5592926	15.00	4.0	0.00000
2763	JEANS		2.4039545	0.21685065	3.54396	1.8826553	3.7272570	13.50	8.0	0.00000
2764	MOELLER		2.2476945	0.08278738	1.98946	2.0616138	3.3698144	14.90	4.2	0.00000
2765	1981EY		3.1488054	0.05315742	14.02168	2.9814229	5.5875149	12.90	10.6	0.00000
2766	1982FE1		2.5492272	0.17813598	6.51424	2.0951180	4.0701723	13.40	8.4	0.00000
2767	1967UM		3.0212677	0.08244797	10.87615	2.7721703	5.2515054	12.50	12.7	0.00000
2768	GORKY		2.2339265	0.17100696	6.27980	1.8519096	3.3388996	14.00	6.4	0.00000
2769	MENDELEEV		3.1385670	0.13113558	2.52456	2.7269890	5.5602856	12.50	12.7	0.00000
2770	1977SM1		2.1700242	0.06360385	2.86118	2.0320022	3.1966629	14.50	5.1	0.00000
2771	1978SP7		2.6775546	0.22648197	13.94076	2.0711367	4.3813457	13.50	8.0	0.00000
2772	DUGAN		2.3147244	0.20342144	9.78468	1.8438598	3.5216722	14.50	5.1	0.00000
2773	1981JZ2		2.3279371	0.14225535	3.67033	1.9967756	3.5518682	14.50	5.1	0.00000
2774	TENOJOKI		3.1794972	0.14459696	8.52533	2.7197516	5.6694074	12.00	16.0	0.00000
2775	1953TX2		2.4204998	0.18592329	3.73814	1.9704725	3.7658029	15.00	4.0	0.00000
2776	BAIKAL		2.3670182	0.17518263	4.77274	1.9523578	3.6416855	14.00	6.4	0.00000
2777	SHUKSHIN		2.3723097	0.08915477	4.91176	2.1608069	3.6539037	14.00	6.4	0.00000
2778	1979XP		2.2810247	0.12127386	4.61679	2.0043962	3.4450459	14.50	5.1	0.00000
2779	MARY		2.2119279	0.06298298	3.89300	2.0726140	3.2897010	14.50	5.1	0.00000
2780	MONNIG		2.1934099	0.11659967	5.46147	1.9376591	3.2484765	14.00	6.4	0.00000
2781	1982QH		3.1442876	0.18801814	2.31300	2.5531044	5.5754948	13.00	10.1	0.00000
2782	2605PL		2.6808717	0.22068729	3.78133	2.0892375	4.3894897	14.50	5.1	0.00000
2783	CHERNYSHEVSKIJ		2.5597227	0.16674364	0.77240	2.1329052	4.0953345	14.40	5.3	0.00000
2784	1975GA		2.2418706	0.17439762	6.69825	1.8508937	3.3567255	14.30	5.5	0.00000
2785	SEDOV		2.8730059	0.04339712	1.43991	2.7483256	4.8697286	13.00	10.1	0.00000
2786	GRINEVIA		2.6056504	0.17626622	13.27932	2.1463623	4.2060480	13.30	8.8	0.00000
2787	TOVARISHCH		3.0191946	0.06329073	10.33419	2.8281076	5.2461009	12.60	12.1	0.00000
2788	1981EL		2.5583887	0.10307364	2.63306	2.2946863	4.0921335	14.20	5.8	0.00000
2789	1956XA		2.2275903	0.16401051	3.81364	1.8622421	3.3247042	15.00	4.0	0.00000
2790	1965UU1		2.6527879	0.18057597	14.63799	2.1737580	4.3206968	12.50	12.7	0.00000
2791	PARADISE		2.3992732	0.17076099	31.03698	1.9895709	3.7163751	13.00	10.1	0.00000
2792	1977EY1		2.2766092	0.12776365	9.37142	1.9857364	3.4350476	14.50	5.1	0.00000
2793	VALDAJ		3.1632257	0.04392374	22.12615	3.0242848	5.6259422	12.00	16.0	0.00000
2794	KULIK		2.4447265	0.21836780	7.47615	1.9108769	3.8224814	14.00	6.4	0.00000
2795	LEPAGE		2.2952964	0.02842613	6.04059	2.2300501	3.4774282	14.00	6.4	0.00000
2796	KRON		2.6458018	0.11168420	14.03166	2.3503075	4.3036404	13.50	8.0	0.00000
2797	TEUCER		5.1904216	0.09522819	22.28507	4.6961470	11.8250761	9.50	50.5	0.00000
2798	2009PL		2.4178860	0.05969394	5.31709	2.2735527	3.7597044	14.00	6.4	0.00000
2799	JUSTUS		2.3884971	0.12790027	5.30318	2.0830076	3.6913655	15.50	3.2	0.00000
2800	4585PL		3.1518867	0.13895494	3.06867	2.7139163	5.5957189	13.50	8.0	0.00000
2801	1935SU1		2.7993057	0.17455998	9.56577	2.3106589	4.6835537	13.00	10.1	0.00000
2802	WEISELL		3.1236765	0.11283577	9.60195	2.7712140	5.5207629	12.00	16.0	0.00000
2803	VILHO		3.1551442	0.16314396	1.33888	2.6404014	5.6043959	13.00	10.1	0.00000
2804	YRJO		3.0119307	0.07798471	11.22611	2.7770462	5.2271800	12.50	12.7	0.00000
2805	KALLE		2.6931951	0.14489567	6.88260	2.3029628	4.4197912	13.50	8.0	0.00000
2806	1953GG		2.3785169	0.04738072	2.34256	2.2658212	3.6682539	13.50	8.0	0.00000
2807	KARL MARX		2.7929041	0.18134886	7.90085	2.2864141	4.6674972	13.50	8.0	0.00000
2808	1976HS		3.0071585	0.08411945	8.95467	2.7541981	5.2147622	12.00	16.0	0.00000
2809	1978QW2		2.4282351	0.17861912	2.46743	1.9945059	3.7838690	14.50	5.1	0.00000

No.	Name	Type	A (A.U.)	E	I (deg.)	q (A.U.)	Period (years)	Mag.	Rad. (km)	Rot. (hours)
2810	LEV TOLSTOJ		2.6079564	0.15216209	12.72198	2.2111244	4.2116327	14.00	6.4	0.00000
2811	1980JA		2.8645389	0.03786314	1.03732	2.7560785	4.8482170	13.00	10.1	0.00000
2812	SCALTRITI		2.2240956	0.09178075	6.81668	2.0199664	3.3168831	14.50	5.1	0.00000
2813	ZAPPALA		3.1515262	0.13933609	14.76065	2.7124050	5.5947590	12.50	12.7	0.00000
2814	1982FA3		2.8695066	0.07013107	2.45188	2.6682651	4.8608346	13.00	10.1	0.00000
2815	SOMA		2.2326789	0.16867626	5.70643	1.8560790	3.3361025	14.50	5.1	0.00000
2816	PIEN		2.7253809	0.18981911	7.71646	2.2080514	4.4992571	13.00	10.1	0.00000
2817	PEREC		2.3575194	0.17874978	2.26931	1.9361132	3.6197858	15.00	4.0	0.00000
2818	2580PL		2.3764460	0.15013321	2.96117	2.0196626	3.6634641	15.50	3.2	0.00000
2819	11933UR		2.7598200	0.20207056	2.43973	2.2021415	4.5848079	13.50	8.0	0.00000
2820	IISALMI		2.2290399	0.16171078	2.94417	1.8685800	3.3279495	14.00	6.4	0.00000
2821	1978SQ		2.4385931	0.19808754	6.75807	1.9555383	3.8081062	15.00	4.0	0.00000
2822	SACAJAWEA		2.5821176	0.12583414	14.72846	2.2571990	4.1491966	13.50	8.0	0.00000
2823	VAN DER LAAN		2.4108746	0.09168339	3.82645	2.1898375	3.7433629	14.50	5.1	0.00000
2824	FRANKE		2.3257403	0.20688181	3.36724	1.8445868	3.5468419	14.50	5.1	0.00000
2825	BEATTY		2.2458692	0.17337073	3.51880	1.8565012	3.3657100	14.00	6.4	0.00000
2826	AHTI		3.2219892	0.03466736	15.51164	3.1102912	5.7834382	12.00	16.0	0.00000
2827	VELLAMO		2.3092673	0.03085006	8.62622	2.2380261	3.5092256	13.50	8.0	0.00000
2828	IKU-TURSO		2.2414558	0.09092181	3.30730	2.0376587	3.3557937	14.00	6.4	0.00000
2829	1948PK		3.0873387	0.19007754	14.32758	2.5005050	5.4247084	12.00	16.0	0.00000
2830	GREENWICH		2.3763762	0.20846631	25.35098	1.8809818	3.6633027	13.50	8.0	0.00000
2831	1930SZ		2.2250040	0.19771732	4.22208	1.7850821	3.3189151	13.90	6.7	0.00000
2832	1975EC1		2.4756014	0.08627354	4.16938	2.2620225	3.8951223	13.70	7.3	0.00000
2833	RADISHCHEV		2.8767750	0.06598338	1.33500	2.6869557	4.8793149	13.20	9.2	0.00000
2834	CHRISTY CAROL		2.5431099	0.15539820	6.45205	2.1479151	4.0555305	13.20	9.2	0.00000
2835	RYOMA		2.7471225	0.08233903	1.34000	2.5209272	4.5532036	13.10	9.6	0.00000
2836	SOBOLEV		3.0022268	0.09022450	9.67271	2.7313523	5.2019391	12.30	13.9	0.00000
2837	GRIBOEDOV		2.9056942	0.05884720	2.88778	2.7347023	4.9530749	13.00	10.1	0.00000
2838	1971UM1		2.3411710	0.18854825	2.13171	1.8997474	3.5821991	15.00	4.0	0.00000
2839	1929TP		2.2165582	0.14943585	4.81058	1.8853250	3.3000364	13.50	8.0	0.00000
2840	KALLAVESI		2.3977754	0.09309575	8.52637	2.1745529	3.7128959	13.50	8.0	0.00000
2841	PUIJO		2.2520163	0.08518567	4.91587	2.0601768	3.3795376	14.00	6.4	0.00000
2842	19500D		2.6182032	0.09898243	11.71552	2.3590472	4.2364788	13.00	10.1	0.00000
2843	1975XQ		2.2980435	0.12976734	5.46086	1.9998326	3.4836731	14.00	6.4	0.00000
2844	HESS		2.2206633	0.17090517	2.95356	1.8411403	3.3092079	14.50	5.1	0.00000
2845	FRANKLINKEN		2.2607543	0.16247237	6.00981	1.8934442	3.3992260	14.50	5.1	0.00000
2846	1942CJ		3.2253952	0.06225316	11.39402	3.0246041	5.7926116	11.80	17.5	0.00000
2847	1959CC1		2.1693630	0.11600837	2.45031	1.9176987	3.1952021	13.80	7.0	0.00000
2848	1959VF		3.1884029	0.20389725	0.92119	2.5382962	5.6932440	12.50	12.7	0.00000
2849	SHKLOVSKIJ		2.5661004	0.01080425	6.79253	2.5383756	4.1106496	13.70	7.3	0.00000
2850	1978TM7		2.4492149	0.04827732	7.85423	2.3309734	3.8330133	13.20	9.2	0.00000
2851	1978UQ2		2.4787958	0.12436730	8.55939	2.1705146	3.9026637	13.50	8.0	0.00000
2852	1981QU2		2.7863500	0.08707219	1.70262	2.5437365	4.6510768	13.40	8.4	0.00000
2853	1963RG		2.3451619	0.14395083	4.15422	2.0075738	3.5913625	14.20	5.8	0.00000
2854	1964JE		2.2049768	0.12244680	5.76110	1.9349844	3.2742064	14.20	5.8	0.00000
2855	1931TB2		2.4553616	0.16608724	8.13389	2.0475574	3.8474517	14.00	6.4	0.00000
2856	1933GB		3.0271120	0.00518838	9.92009	3.0114062	5.2667503	12.50	12.7	0.00000
2857	1942DA		2.4000885	0.09512983	5.73763	2.1717687	3.7182698	14.00	6.4	0.00000
2858	1975XB		2.2664623	0.19412814	6.69386	1.8264782	3.4121082	15.50	3.2	0.00000
2859	1978RW1		2.2384880	0.11885238	3.55637	1.9724383	3.3491309	14.00	6.4	0.00000
2860	PASACENTENNIUM		2.3325133	0.21463187	22.70047	1.8318816	3.5623467	14.00	6.4	0.00000
2861	LAMBRECHT		2.4720924	0.07161286	4.04970	2.2950587	3.8868434	13.50	8.0	0.00000
2862	1977JP		2.2010128	0.11546001	3.48117	1.9468838	3.2653809	14.00	6.4	0.00000
2863	BEN MAYER		3.1637113	0.19402388	1.97017	2.5498757	5.6272378	13.50	8.0	0.00000
2864	SODERBLOM		2.7465765	0.14892022	3.14250	2.3375556	4.5518460	14.00	6.4	0.00000
2865	1935OK		2.5602100	0.07364512	14.27954	2.3716629	4.0965037	12.50	12.7	0.00000
2866	1961TA		2.9101925	0.20590387	8.24605	2.3109727	4.9645810	13.00	10.1	0.00000
2867	1969VC		2.3637536	0.14580899	9.93326	2.0190971	3.6341538	15.00	4.0	0.00000
2868	1972UA		2.8168583	0.17558575	7.55889	2.3222580	4.7276735	14.50	5.1	0.00000
2869	1980RM2		2.6351030	0.17596003	12.85789	2.1714301	4.2775626	13.00	10.1	0.00000
2870	HAUPT		2.3918610	0.21111646	4.15816	1.8868997	3.6991665	14.00	6.4	0.00000
2871	SCHOBER		2.2585740	0.13892066	5.77380	1.9448113	3.3943098	14.00	6.4	0.00000
2872	GENTELEC		2.7404268	0.11910168	2.86704	2.4140372	4.5365667	13.50	8.0	0.00000
2873	BINZEL		2.2512932	0.15809289	5.90344	1.8953797	3.3779099	14.00	6.4	0.00000
2874	JIM YOUNG		2.2444863	0.13361572	4.89363	1.9445877	3.3626020	15.00	4.0	0.00000
2875	LAGERKVIST		2.7956431	0.10160448	9.04289	2.5115931	4.6743646	13.00	10.1	0.00000
2876	AESCHYLUS		2.6009457	0.11870376	14.86506	2.2922037	4.1946611	14.50	5.1	0.00000
2877	LIKHACHEV		3.1031778	0.19781901	2.34273	2.4893103	5.4665079	13.50	8.0	0.00000
2878	PANACEA		3.0430262	0.08945426	10.26300	2.7708144	5.3083372	12.50	12.7	0.00000
2879	SHIMIZU		2.7666483	0.14437778	10.74534	2.3672056	4.6018338	12.50	12.7	0.00000
2880	NIHONDAIRA		2.2037666	0.16734341	5.74119	1.8349808	3.2715111	14.00	6.4	0.00000
2881	1983AA1		2.2475951	0.15514702	4.61606	1.8988873	3.3695903	15.00	4.0	0.00000
2882	TEDESCO		3.1674175	0.18097337	0.28641	2.5941994	5.6371293	12.90	10.6	0.00000
2883	BARABASHOV		2.2455146	0.08280101	1.41004	2.0595837	3.3649130	14.50	5.1	0.00000
2884	REDDISH		3.1187627	0.16853409	1.94586	2.5931449	5.5077410	13.00	10.1	0.00000
2885	1939TC		2.2378919	0.19460469	2.87066	1.8023876	3.3477931	15.00	4.0	0.00000
2886	1965YG		2.3662579	0.15592007	1.30983	1.9973109	3.6399310	14.50	5.1	0.00000
2887	1977QD5		2.2595079	0.14908594	4.37606	1.9226470	3.3964155	14.00	6.4	0.00000
2888	HODGSON		2.2573593	0.13099629	7.62672	1.9616536	3.3915720	14.50	5.1	0.00000
2889	1981WT1		3.0219197	0.12188884	9.47903	2.6535814	5.2532053	12.70	11.6	0.00000
2890	VILYUJSK		2.2604504	0.16005655	6.62677	1.8986505	3.3985407	14.00	6.4	0.00000
2891	MCGETCHIN		3.3713858	0.10865337	9.29617	3.0050735	6.1903143	12.50	12.7	0.00000
2892	FILIPENKO		3.1535704	0.22352597	16.95965	2.4486656	5.6002035	11.50	20.1	0.00000

No.	Name	Type	A (A.U.)	E	I (deg.)	q (A.U.)	Period (years)	Mag.	Rad. (km)	Rot. (hours)
2893	1975QD		5.2272892	0.07783307	14.61554	4.8204336	11.9512911	10.00	40.1	0.00000
2894	KAKHOVKA		3.1103532	0.14596488	2.58854	2.6563511	5.4854794	13.50	8.0	0.00000
2895	MEMNON		5.2072482	0.04989909	27.25318	4.9474115	11.8826265	10.50	31.9	0.00000
2896	1931RN		2.2207148	0.18658973	5.99833	1.8063523	3.3093233	14.00	6.4	0.00000
2897	OLE ROMER		2.2472954	0.10130837	5.83917	2.0196254	3.3689163	14.00	6.4	0.00000
2898	1938DN		2.5554152	0.02126459	14.30897	2.5010753	4.0850010	13.00	10.1	0.00000
2899	1964TR2		2.2626138	0.15527301	3.22428	1.9112910	3.4034207	14.00	6.4	0.00000
2900	LUBOS PEREK		3.0200596	0.10658828	10.16412	2.6981566	5.2483559	13.00	10.1	0.00000
2901	1973DP		2.8627536	0.04991801	3.17641	2.7198505	4.8436856	13.00	10.1	0.00000
2902	1980FN3		2.2036238	0.19843850	4.37363	1.7663400	3.2711930	15.50	3.2	0.00000
2903	1981UV9		2.5614686	0.05781019	14.35543	2.4133897	4.0995255	13.00	10.1	0.00000
2904	MILLMAN		2.6027145	0.13844021	15.40209	2.2423942	4.1989417	13.00	10.1	0.00000
2905	PLASKETTT		2.8043888	0.09628495	8.89538	2.5343683	4.6963158	13.00	10.1	0.00000
2906	CALTECH		3.1581388	0.12017212	30.72733	2.7786186	5.6123767	11.00	25.3	0.00000
2907	NEKRASOV		3.0151272	0.09276005	10.21188	2.7354436	5.2355032	12.50	12.7	0.00000
2908	SHIMOYAMA		2.9795301	0.15432115	13.34231	2.5197258	5.1430612	13.00	10.1	0.00000
2909	HOSHI-NO-IE		3.0223932	0.11165847	11.44763	2.6849174	5.2544403	12.00	16.0	0.00000
2910	YOSHKAR-OLA		2.2020104	0.15616481	2.94229	1.8581338	3.2676010	14.50	5.1	0.00000
2911	1938GJ		2.7980478	0.09215120	9.62835	2.5402043	4.6803970	13.80	7.0	0.00000
2912	1942DM		2.2886691	0.07161031	7.27980	2.1247768	3.4623783	13.50	8.0	0.00000
2913	1931TK		2.7059641	0.19794266	16.06735	2.1703384	4.4512606	14.00	6.4	0.00000
2914	1965SB		2.2612338	0.12917364	2.95948	1.9691421	3.4003079	15.50	3.2	0.00000
2915	MOSKVINA		2.5614934	0.18547793	13.19862	2.0863929	4.0995846	14.50	5.1	0.00000
2916	VORONVELIYA		2.2344270	0.09917208	3.59409	2.0128341	3.3400211	14.50	5.1	0.00000
2917	SAWYER HOGG		2.7949109	0.10953033	12.82130	2.4887834	4.6725283	13.00	10.1	0.00000
2918	SALAZAR		3.1659856	0.16037656	2.09524	2.6582358	5.6333070	13.50	8.0	0.00000
2919	DALI		3.1353366	0.14739314	1.40534	2.6732097	5.5517039	12.50	12.7	0.00000
2920	AUTOMEDON		5.1949339	0.03183387	21.02526	5.0295587	11.8404989	10.00	40.1	0.00000
2921	SOPHOCLES		3.2406330	0.16301347	1.44868	2.7123661	5.8337088	14.50	5.1	0.00000
2922	1976GY1		2.3708065	0.14631009	2.98263	2.0239336	3.6504312	14.50	5.1	0.00000
2923	SCHUYLER		2.4557652	0.13085806	2.87918	2.1344085	3.8484006	14.00	6.4	0.00000
2924	MITAKE-MURA		2.8877041	0.04849232	3.14112	2.7476728	4.9071469	13.00	10.1	0.00000
2925	1978VC5		2.3856938	0.19125970	2.22066	1.9294066	3.6848691	15.00	4.0	0.00000
2926	1980KG		2.2748377	0.11844568	3.48126	2.0053930	3.4310389	14.50	5.1	0.00000
2927	ALAMOSA		2.5321710	0.16701733	16.96726	2.1092546	4.0293922	13.50	8.0	0.00000
2928	1976GN8		3.0073991	0.06797762	9.53657	2.8029633	5.2153878	12.50	12.7	0.00000
2929	HARRIS		3.1197312	0.06697481	14.90235	2.9107876	5.5103064	12.50	12.7	0.00000
2930	EURIPIDES		2.7801671	0.02426123	4.06646	2.7127168	4.6356039	14.00	6.4	0.00000
2931	MAYAKOVSKY		2.8759069	0.05628276	2.22163	2.7140429	4.8771062	13.00	10.1	0.00000
2932	KEMPCHINSKY		3.6272721	0.10413295	2.24450	3.2495537	6.9082847	13.00	10.1	0.00000
2933	AMBER		2.6090586	0.04769422	7.22323	2.4846215	4.2143025	13.00	10.1	0.00000
2934	ARISTOPHANES		3.1683378	0.04462420	8.83045	3.0269532	5.6395860	12.50	12.7	0.00000
2935	1976UU		2.5987937	0.12550171	13.01419	2.2726407	4.1894565	14.00	6.4	0.00000
2936	1979SF		2.6792824	0.07405116	8.45593	2.4808786	4.3855872	13.50	8.0	0.00000
2937	GIBBS		2.3206472	0.30226380	21.77599	1.6191996	3.5351977	14.00	6.4	0.00000
2938	HOPI		3.1392832	0.33783853	41.46791	2.0787125	5.5621891	12.50	12.7	0.00000
2939	COCONINO		2.4411395	0.16112585	3.97665	2.0478086	3.8140719	14.00	6.4	0.00000
2940	BACON		2.7841954	0.23395449	6.43632	2.1328206	4.6456833	15.00	4.0	0.00000
2941	1930YV		2.1512842	0.09002963	3.24109	1.9576049	3.1553438	15.00	4.0	0.00000
2942	1932BG		2.2386744	0.15297835	6.81976	1.8962058	3.3495498	14.00	6.4	0.00000
2943	1933QU		2.4472115	0.15452354	12.97034	2.0690596	3.8283114	14.50	5.1	0.00000
2944	1935QF		2.6488237	0.16537704	10.66832	2.2107692	4.3110156	14.00	6.4	0.00000
2945	1935ST1		2.6694603	0.13743863	2.62843	2.3025734	4.3614931	13.00	10.1	0.00000
2946	1941UV		2.4530094	0.17473935	0.58405	2.0243721	3.8419242	14.00	6.4	0.00000
2947	1955QP1		2.3072565	0.12530898	3.13407	2.0181365	3.5046430	14.50	5.1	0.00000
2948	AMOSOV		2.8628380	0.10812974	12.32036	2.5532801	4.8439002	13.50	8.0	0.00000
2949	KAVERZNEV		2.1944730	0.13973838	4.86406	1.8878210	3.2508383	14.50	5.1	0.00000
2950	1974VQ2		2.7546008	0.26333308	9.59293	2.0292232	4.5718079	13.50	8.0	0.00000
2951	1977RB8		3.1212335	0.13055371	14.75249	2.7137449	5.5142875	11.50	20.1	0.00000
2952	LILLIPUTIA		2.3139422	0.17122866	3.31942	1.9177289	3.5198872	15.50	3.2	0.00000
2953	1979SV11		2.8294270	0.02303735	1.07205	2.7642446	4.7593508	13.00	10.1	0.00000
2954	1982BT1		2.2873104	0.19387853	3.93927	1.8438500	3.4592955	15.00	4.0	0.00000
2955	1982BX1		2.1802204	0.11519006	3.59550	1.9290807	3.2192194	14.50	5.1	0.00000
2956	1982HN1		2.7646110	0.09216619	2.86456	2.5098071	4.5967517	13.50	8.0	0.00000
2957	1934CB1		3.0216632	0.08845026	8.70661	2.7543962	5.2525363	11.50	20.1	0.00000
2958	1981DG		2.8738806	0.01493108	1.02371	2.8309705	4.8719530	13.00	10.1	0.00000
2959	SCHOLL		3.9397194	0.27701604	5.23471	2.8483541	7.8198414	12.50	12.7	0.00000
2960	OHTAKI		2.2205808	0.11287212	4.50808	1.9699390	3.3090234	15.10	3.8	0.00000
2961	KATSURAHAMA		2.2679877	0.13704908	4.54800	1.9571621	3.4155533	14.20	5.8	0.00000
2962	1940YF		2.5682127	0.03797431	15.67515	2.4706867	4.1157265	12.50	12.7	0.00000
2963	1964VM1		2.8703401	0.07433365	2.70989	2.6569772	4.8629527	13.50	8.0	0.00000
2964	1974OA1		2.5924294	0.20082918	13.58192	2.0717938	4.1740761	13.50	8.0	0.00000
2965	1975BX		2.3911691	0.21997942	24.26002	1.8651611	3.6975617	14.50	5.1	0.00000
2966	1977EB2		2.4487412	0.14011428	2.54432	2.1056376	3.8319013	15.00	4.0	0.00000
2967	1977SS1		3.2057436	0.12458635	18.01855	2.8063517	5.7397528	12.50	12.7	0.00000
2968	1978QJ		2.3661673	0.30993715	9.17375	1.6328042	3.6397214	15.50	3.2	0.00000
2969	1978RU1		2.8450098	0.02713360	1.87801	2.7678144	4.7987227	14.00	6.4	0.00000
2970	1978UC		2.6383495	0.15281884	12.10961	2.2351601	4.2854705	13.50	8.0	0.00000
2971	1980YL		2.2461462	0.11777050	6.99829	1.9816163	3.3663323	14.50	5.1	0.00000
2972	1939TB		2.1508029	0.16962638	1.09374	1.7859699	3.1542847	14.50	5.1	0.00000
2973	1951AJ		2.4701240	0.15086938	1.55399	2.0974579	3.8822024	13.50	8.0	0.00000
2974	1955QK		2.3119965	0.14054301	6.41899	1.9870615	3.5154486	15.00	4.0	0.00000
2975	1970AF1		2.2477789	0.09557457	6.89649	2.0329483	3.3700035	13.50	8.0	0.00000

No.	Name	Type	A (A.U.)	E	I (deg.)	q (A.U.)	Period (years)	Mag.	Rad. (km)	Rot. (hours)
2976	LAUTARO		3.3555343	0.13033888	9.66899	2.9181776	6.1467071	12.00	16.0	0.00000
2977	CHIVILIKIN		2.7861788	0.16884844	9.58575	2.3157370	4.6506481	13.50	8.0	0.00000
2978	1978SR		3.1077700	0.17430913	1.24297	2.5660572	5.4786468	13.00	10.1	0.00000
2979	MURMANSK		3.1337471	0.14356567	11.35187	2.6838486	5.5474825	12.50	12.7	0.00000
2980	1981EU7		2.5681221	0.17935497	7.29913	2.1075165	4.1155086	14.00	6.4	0.00000
2981	CHAGALL		3.1477935	0.17122474	0.84436	2.6088133	5.5848222	13.00	10.1	0.00000
2982	MURIEL		2.9979615	0.06671780	10.25402	2.7979441	5.1908569	13.00	10.1	0.00000
2983	1981RW2		2.8494112	0.06140949	4.25743	2.6744304	4.8098626	13.00	10.1	0.00000
2984	CHAUCER		2.4705002	0.13374828	3.05355	2.1400752	3.8830893	13.50	8.0	0.00000
2985	SHAKESPEARE		2.8477509	0.04227193	2.66272	2.7273710	4.8056593	13.00	10.1	0.00000
2986	MRINALINI		3.1874862	0.14070296	2.54794	2.7389975	5.6907887	13.00	10.1	0.00000
2987	SARABHAI		2.8859782	0.06569829	1.01449	2.6963744	4.9027481	13.50	8.0	0.00000
2988	1943EM		2.6070638	0.12535886	14.68952	2.2802453	4.2094703	13.00	10.1	0.00000
2989	1976UF1		2.2384810	0.17412455	3.63157	1.8487066	3.3491156	14.50	5.1	0.00000
2990	1981EN27		2.4395275	0.12358088	2.78789	2.1380486	3.8102951	14.50	5.1	0.00000
2991	1982HV		2.3372924	0.22081599	5.15659	1.8211809	3.5733006	15.00	4.0	0.00000
2992	VONDEL		2.7447622	0.18911304	7.04984	2.2256918	4.5473361	14.50	5.1	0.00000
2993	1970PA		2.5855634	0.19696026	12.28557	2.0763102	4.1575050	13.50	8.0	0.00000
2994	1975PA		2.4167421	0.23020484	2.48594	1.8603964	3.7570369	15.50	3.2	0.00000
2995	1978QK		2.6152544	0.13834625	14.87266	2.2534437	4.2293239	14.00	6.4	0.00000
2996	BOWMAN		2.7825112	0.02981470	3.67508	2.6995516	4.6414685	13.00	10.1	0.00000
2997	1974 MJ		2.5546563	0.19990519	7.21840	2.0439672	4.0831814	15.00	4.0	0.00000
2998	1975 TR3		2.4223332	0.19518520	3.07042	1.9495296	3.7700822	15.00	4.0	0.00000
2999	DANTE		2.2720139	0.10510148	6.76854	2.0332220	3.4246521	14.50	5.1	0.00000
3000	LEONARDO		2.3513670	0.18148468	2.74867	1.9246299	3.6056256	14.00	6.4	0.00000
3001	MICHELANGELO		2.3562932	0.07023950	18.36905	2.1907883	3.6169624	13.50	8.0	0.00000
3002	1982 FB3		2.2396684	0.12998088	6.14668	1.9485544	3.3517807	14.00	6.4	0.00000
3003	1983 YH		3.0206282	0.12199197	11.59062	2.6521358	5.2498379	12.50	12.7	0.00000
3004	1976 DD		2.5909154	0.26523912	30.27231	1.9037033	4.1704206	16.00	2.5	0.00000
3005	1979 QK2		2.3685706	0.18411128	2.35793	1.9324899	3.6452682	15.00	4.0	0.00000
3006	LIVADIA		2.4323170	0.18728648	3.04395	1.9767768	3.7934139	15.00	4.0	0.00000
3007	REAVES		2.3685730	0.13342905	8.34417	2.0525365	3.6452737	14.00	6.4	0.00000
3008	NOJIRI		3.1603320	0.14529732	0.80014	2.7011442	5.6182241	13.00	10.1	0.00000
3009	COVENTRY		2.1964748	0.20516992	4.55440	1.7458243	3.2552876	15.50	3.2	0.00000
3010	1978 SB5		3.2055364	0.18871112	2.03862	2.6006160	5.7391958	13.50	8.0	0.00000
3011	1978 WM14		3.2144279	0.19108281	6.18645	2.6002059	5.7630920	13.00	10.1	0.00000
3012	MINSK		3.2232163	0.05949377	18.30148	3.0314550	5.7867432	12.00	16.0	0.00000
3013	1979 SD7		2.3598261	0.14055470	3.66037	2.0281415	3.6250999	14.50	5.1	0.00000
3014	1979 TM		2.3633583	0.22810563	0.98597	1.8242629	3.6332419	14.00	6.4	0.00000
3015	1980 VN		3.4063983	0.16196981	17.39152	2.8546646	6.2869954	12.50	12.7	0.00000
3016	1981 EK		2.8343637	0.04413104	2.89153	2.7092803	4.7718120	13.50	8.0	0.00000
3017	1981 UL		2.6063523	0.13021711	11.85271	2.2669606	4.2077479	13.50	8.0	0.00000
3018	GODIVA		2.3679523	0.18710972	4.73314	1.9248855	3.6438413	14.00	6.4	0.00000
3019	1940 AC		2.8652668	0.05424020	3.22043	2.7098539	4.8500652	12.50	12.7	0.00000
3020	1949 PR		2.7635829	0.06231929	6.27294	2.5913584	4.5941877	13.00	10.1	0.00000
3021	1967 CB		3.1659715	0.26239124	16.58660	2.3352482	5.6332688	13.00	10.1	0.00000
3022	1980 SH		1.9309167	0.10338494	23.52734	1.7312890	2.6831522	15.00	4.0	0.00000
3023	HEARD		2.2154024	0.08574774	3.99291	2.0254366	3.2974553	14.50	5.1	0.00000
3024	1981 UW9		3.4279878	0.11010746	14.79742	3.0505407	6.3468595	12.00	16.0	0.00000
3025	HIGSON		3.2004242	0.08436359	20.95602	2.9304249	5.7254725	13.00	10.1	0.00000
3026	1977 TA1		3.0288160	0.02316714	9.64809	2.9586470	5.2711983	13.00	10.1	0.00000
3027	1978 PQ2		2.6725659	0.21916232	1.96509	2.0868402	4.3691068	14.50	5.1	0.00000
3028	1978 TA2		3.0186288	0.02964456	9.51067	2.9291430	5.2446270	12.00	16.0	0.00000
3029	1981 EA8		2.2392797	0.11237843	3.42069	1.9876330	3.3509080	14.50	5.1	0.00000
3030	VEHRENBERG		2.2699802	0.24545293	3.48483	1.7128069	3.4200552	15.50	3.2	0.00000
3031	1984 CX		2.2354054	0.09879763	4.33842	2.0145526	3.3422153	14.00	6.4	0.00000
3032	1984 CA1		2.8945880	0.08496417	3.22582	2.6486516	4.9247041	13.00	10.1	0.00000
3033	HOLBAEK		2.2349634	0.09595180	4.74136	2.0205147	3.3412244	14.00	6.4	0.00000
3034	A917 SE		2.3246944	0.20999168	4.92305	1.8365279	3.5444498	13.50	8.0	0.00000
3035	A924 EJ		2.6341431	0.13085642	2.58799	2.2894485	4.2752256	13.50	8.0	0.00000
3036	KRAT		3.2190402	0.09079467	22.78556	2.9267683	5.7754998	11.00	25.3	0.00000
3037	1944 BA		2.6768837	0.18707177	18.98308	2.1761141	4.3796988	13.00	10.1	0.00000
3038	1978 QB3		2.4390535	0.20345029	4.69905	1.9428273	3.8091846	14.50	5.1	0.00000
3039	YANGEL		2.5599258	0.14083017	15.27485	2.1994109	4.0958219	13.50	8.0	0.00000
3040	KOZAI		1.8407664	0.20061946	46.64610	1.4714729	2.4974570	17.00	1.6	0.00000
3041	WEBB		2.5870540	0.14609268	14.61428	2.2091043	4.1611009	13.50	8.0	0.00000
3042	1981 EF10		2.2766259	0.21051691	4.99924	1.7973576	3.4350851	14.50	5.1	0.00000
3043	SAN DIEGO		1.9266938	0.10616839	21.78832	1.7221398	2.6743548	14.50	5.1	0.00000
3044	1983 RE3		2.8526320	0.15656991	13.55300	2.4059956	4.8180203	13.00	10.1	0.00000
3045	ALOIS		3.1295648	0.11349020	3.34903	2.7743897	5.5363803	12.50	12.7	0.00000
3046	MOLIERE		3.1264238	0.15912822	18.36946	2.6289215	5.5280476	13.50	8.0	0.00000
3047	GOETHE		2.6442258	0.02669716	1.61183	2.5736327	4.2997956	14.50	5.1	0.00000
3048	1964 TH1		2.4006371	0.14534806	1.93705	2.0517092	3.7195446	14.00	6.4	0.00000
3049	1968 FH		3.1290231	0.13073730	2.49812	2.7199430	5.5349431	12.50	12.7	0.00000
3050	CARRERA		2.2244098	0.18882869	1.30626	1.8043773	3.3175857	15.00	4.0	0.00000
3051	1974 YP		2.5935400	0.25879839	13.34498	1.9223360	4.1767588	14.00	6.4	0.00000
3052	1976 YJ3		2.3769083	0.18065241	3.89643	1.9475142	3.6645334	14.50	5.1	0.00000
3053	DRESDEN		2.3801544	0.20521308	4.61388	1.8917156	3.6720426	14.50	5.1	0.00000
3054	STRUGATSKIA		3.1082661	0.19782265	2.06690	2.4933808	5.4799590	12.50	12.7	0.00000
3055	1978 TR3		2.5623500	0.10794918	15.00940	2.2857463	4.1016412	13.50	8.0	0.00000
3056	INAG		2.4191453	0.11541164	5.63621	2.1399479	3.7626424	14.00	6.4	0.00000
3057	MALAREN		2.2609489	0.07422811	7.27905	2.0931227	3.3996646	14.50	5.1	0.00000
3058	DELMARY		2.2496915	0.15697712	3.55051	1.8965414	3.3743060	15.50	3.2	0.00000

No.	Name	Type	A (A.U.)	E	I (deg.)	q (A.U.)	Period (years)	Mag.	Rad. (km)	Rot. (hours)
3059	1981 EF23		2.2690296	0.12929368	2.36221	1.9756583	3.4179068	15.00	4.0	0.00000
3060	1982 RD1		2.2775526	0.17719591	7.25870	1.8739794	3.4371824	14.50	5.1	0.00000
3061	1982 UB1		3.0899050	0.19344302	3.25765	2.4921844	5.4314737	13.00	10.1	0.00000
3062	1982 XC		3.0159028	0.11829021	11.33255	2.6591508	5.2375236	12.00	16.0	0.00000
3063	MAKHAON		5.1540565	0.06038073	12.19692	4.8428507	11.7010212	10.00	40.1	0.00000
3064	1984 BB1		2.4571514	0.11568092	2.93690	2.1729057	3.8516591	14.50	5.1	0.00000
3065	1984 CV		2.7217541	0.06345386	4.30116	2.5490482	4.4902792	13.50	8.0	0.00000
3066	1984 EO		2.5259781	0.13418882	15.55064	2.1870201	4.0146189	12.50	12.7	0.00000
3067	1982 TE2		2.2453909	0.13721770	4.52337	1.9372834	3.3646345	14.50	5.1	0.00000
3068	KHANINA		2.2295654	0.10288179	6.44614	2.0001836	3.3291266	14.50	5.1	0.00000
3069	1982 UG2		2.3517694	0.24121833	1.68059	1.7844795	3.6065512	14.50	5.1	0.00000
3070	1949 GK		2.3048749	0.19822481	2.34663	1.8479913	3.4992180	15.00	4.0	0.00000
3071	1973 FT1		3.2015202	0.09209437	2.20602	2.9066782	5.7284131	12.50	12.7	0.00000
3072	VILNIUS		2.2385814	0.17938673	5.64077	1.8370095	3.3493409	15.00	4.0	0.00000
3073	KURSK		2.2426755	0.13592224	5.03764	1.9378461	3.3585334	14.50	5.1	0.00000
3074	1979 YE9		2.3384359	0.11100164	2.41864	2.0788658	3.5759237	14.00	6.4	0.00000
3075	1981 EY15		2.2733357	0.13117325	9.96909	1.9751347	3.4276409	15.00	4.0	0.00000
3076	1982 RB1		2.2375510	0.18908142	7.71097	1.8144717	3.3470285	15.00	4.0	0.00000
3077	1982 SK		2.2404168	0.05470421	1.47099	2.1178565	3.3534608	14.00	6.4	0.00000
3078	1984 FG		3.1701207	0.08600025	7.06347	2.8974895	5.6443472	12.50	12.7	0.00000
3079	SCHILLER		2.6885641	0.21732482	3.91406	2.1042724	4.4083958	14.50	5.1	0.00000
3080	MOISSEIEV		2.6092389	0.19616683	13.84740	2.0973928	4.2147393	13.00	10.1	0.00000
3081	1971 UP		2.4105339	0.18310997	5.29079	1.9691410	3.7425692	14.50	5.1	0.00000
3082	DZHALIL		2.5759356	0.07682053	10.36344	2.3780508	4.1343045	13.50	8.0	0.00000
3083	1974 MH		2.2837903	0.15260652	6.48192	1.9352690	3.4513130	15.50	3.2	0.00000
3084	1977 QB1		2.4378436	0.22539508	4.14223	1.8883656	3.8063505	14.50	5.1	0.00000
3085	1980 DA		2.3880446	0.09976237	3.82421	2.1498075	3.6903167	14.00	6.4	0.00000
3086	KALBAUGH		1.9358118	0.02689117	19.00173	1.8837554	2.6933615	15.00	4.0	0.00000
3087	BEATRICE TINSLEY		3.0718098	0.11971300	19.84353	2.7040741	5.3838315	14.00	6.4	0.00000
3088	1981 UX9		3.0237660	0.04635291	10.24574	2.8836057	5.2580204	13.00	10.1	0.00000
3089	1981 XK2		2.9259162	0.18807904	16.73427	2.3756125	5.0048699	12.00	16.0	0.00000
3090	TJOSSEM		3.1645873	0.09097577	9.60998	2.8766866	5.6295753	13.50	8.0	0.00000
3091	6081 P-L		2.3483689	0.16763993	2.01077	1.9546884	3.5987315	15.00	4.0	0.00000
3092	6550 P-L		3.5452175	0.10752614	10.89209	3.1640139	6.6752005	12.50	12.7	0.00000
3093	1971 MG		2.6764579	0.20555831	12.78696	2.1262898	4.3786540	12.50	12.7	0.00000
3094	1979 FE2		2.6474569	0.07208999	14.56982	2.4566019	4.3076792	13.00	10.1	0.00000
3095	1980 RT2		3.5042174	0.06805655	2.99059	3.2657323	6.5597391	12.50	12.7	0.00000
3096	1981 QC1		2.6660602	0.19603598	12.15817	2.1434166	4.3531632	14.00	6.4	0.00000
3097	2011 P-L		2.9274497	0.09010326	7.45305	2.6636767	5.0088053	13.00	10.1	0.00000
3098	4579 P-L		2.3027723	0.21097508	1.34697	1.8169448	3.4944315	16.00	2.5	0.00000
3099	1940 GF		2.8865874	0.20098916	15.42487	2.3064146	4.9043002	12.50	12.7	0.00000
3100	1977 EQ1		2.2586696	0.08725871	2.82152	2.0615809	3.3945253	15.50	3.2	0.00000
3101	GOLDBERGER		1.9789912	0.04613908	28.56219	1.8876823	2.7839777	15.00	4.0	0.00000
3102	1981 QA		2.1520267	0.44878551	8.41371	1.1862283	3.1569774	17.00	1.6	0.00000
3103	1982 BB		1.4068080	0.35481307	20.94345	0.9076541	1.6686000	16.00	2.5	0.00000
3104	1982 BB1		2.9633281	0.08586161	24.18889	2.7088921	5.1011682	12.50	12.7	0.00000
3105	A907 PB		2.2615991	0.19342761	6.47679	1.8241433	3.4011314	14.20	5.8	0.00000
3106	1981 EE		3.1619077	0.22290619	14.85308	2.4570987	5.6224260	12.00	16.0	0.00000
3107	WEAVER		2.2021859	0.20736051	1.60060	1.7455395	3.2679918	14.50	5.1	0.00000
3108	1972 QM		2.2288449	0.16775444	3.28747	1.8549463	3.3275130	15.00	4.0	0.00000
3109	1974 DC		2.4517815	0.08553897	7.18443	2.2420588	3.8390403	12.50	12.7	0.00000
3110	1975 SC		2.5612979	0.12427639	2.24313	2.2429888	4.0991149	14.50	5.1	0.00000
3111	1977 DX8		2.2235878	0.16067754	2.01315	1.8663073	3.3157473	15.00	4.0	0.00000
3112	1977 QC5		2.3788834	0.19574074	3.95698	1.9132389	3.6691012	14.50	5.1	0.00000
3113	1978 RO		2.4272366	0.07516639	4.96620	2.2447901	3.7815354	14.50	5.1	0.00000
3114	1980 FB12		2.4195609	0.19741400	2.23307	1.9419059	3.7636123	14.50	5.1	0.00000
3115	1981 PL		2.5790007	0.14201330	10.17164	2.2127483	4.1416860	12.50	12.7	0.00000
3116	1983 CF		2.2284310	0.19999765	5.46788	1.7827500	3.3265862	13.50	8.0	0.00000
3117	NIEPCE		2.8447425	0.05882988	3.24204	2.6773868	4.7980466	13.00	10.1	0.00000
3118	1974 OD		3.0359883	0.05985504	13.25931	2.8542693	5.2899327	12.00	16.0	0.00000
3119	DOBRONRAVIN		3.0543082	0.20728141	4.74011	2.4212070	5.3378859	13.50	8.0	0.00000
3120	1979 RZ		3.0280011	0.09495654	12.92139	2.7404726	5.2690711	13.00	10.1	0.00000
3121	1981 EV		2.2276597	0.08597609	6.36496	2.0361342	3.3248594	14.50	5.1	0.00000
3122	1981 ET3		1.7685165	0.42259106	22.18710	1.0211573	2.3518722	15.50	3.2	0.00000
3123	1981 QF2		2.4618292	0.13288464	1.99385	2.1346898	3.8626633	14.00	6.4	0.00000
3124	KANSAS		2.7442460	0.07938673	5.91483	2.5263894	4.5460539	14.00	6.4	0.00000
3125	1982 BJ1		2.6033897	0.19941698	12.73235	2.0842297	4.2005754	13.50	8.0	0.00000
3126	1969 TP1		3.0063188	0.10214190	9.68165	2.6992476	5.2125773	12.50	12.7	0.00000
3127	1973 ST4		2.5957677	0.20297670	4.84300	2.0688872	4.1821418	13.50	8.0	0.00000
3128	1979 FJ2		3.1169777	0.15966175	2.93831	2.6193156	5.5030131	12.50	12.7	0.00000
3129	BONESTELL		2.6981480	0.21507762	6.92352	2.1178367	4.4319887	13.50	8.0	0.00000
3130	1981 YO		2.4665167	0.19893284	4.21065	1.9758456	3.8737011	14.00	6.4	0.00000
3131	1982 BM1		2.9229980	0.04328552	2.41301	2.7964745	4.9973845	13.00	10.1	0.00000
3132	1940 WL		3.1535692	0.11474226	4.47082	2.7917216	5.6002002	12.50	12.7	0.00000
3133	SENDAI		2.1810088	0.16017227	6.56391	1.8316717	3.2209659	14.00	6.4	0.00000
3134	KOSTINSKY		3.9750500	0.22286037	7.62683	3.0891688	7.9252667	12.00	16.0	0.00000
3135	1981 EC9		2.4191055	0.14146806	5.99543	2.0768793	3.7625494	15.00	4.0	0.00000
3136	1981 WD4		3.1757922	0.12691548	4.55130	2.7727349	5.6595006	12.90	10.6	0.00000
3137	1982 SM1		2.4016633	0.19114317	2.46963	1.9426017	3.7219296	14.60	4.8	0.00000
3138	1980 KL		2.2257111	0.07522521	4.61413	2.0582814	3.3204978	14.50	5.1	0.00000
3139	1980 VL1		3.1914973	0.03589910	20.50475	3.0769255	5.7015343	12.00	16.0	0.00000
3140	1983 AO		3.0166893	0.11343875	11.26627	2.6744797	5.2395725	12.00	16.0	0.00000
3141	1984 RH		3.4093697	0.07010074	10.89879	3.1703703	6.2952237	11.50	20.1	0.00000

No.	Name	Type	A (A.U.)	E	I (deg.)	q (A.U.)	Period (years)	Mag.	Rad. (km)	Rot. (hours)
3142	KILOPI		2.5550025	0.08730461	14.14549	2.3319390	4.0840120	13.50	8.0	0.00000
3143	1980 UA		2.8466246	0.08140594	3.09761	2.6148922	4.8028083	14.00	6.4	0.00000
3144	1931 TY1		2.2255144	0.20949078	5.50438	1.7592897	3.3200576	14.50	5.1	0.00000
3145	1955 RY		2.1927433	0.23569913	5.03213	1.6759157	3.2469957	15.50	3.2	0.00000
3146	1972 KG		2.4349947	0.19648419	8.39354	1.9565567	3.7996800	14.50	5.1	0.00000
3147	1976 YU3		2.6229441	0.19482213	3.55155	2.1119366	4.2479906	14.50	5.1	0.00000
3148	1979 SA12		3.1154768	0.17790797	0.74539	2.5612085	5.4990387	13.00	10.1	0.00000
3149	1981 SH		2.2480769	0.15679298	7.13206	1.8955941	3.3706737	15.00	4.0	0.00000
3150	1983 CB		3.1970880	0.11896125	22.07658	2.8167584	5.7165222	12.00	16.0	0.00000
3151	TALBOT		2.7639835	0.13741773	19.49105	2.3841631	4.5951867	13.00	10.1	0.00000
3152	1983 LF		2.6261814	0.08978524	11.33088	2.3903892	4.2558575	12.50	12.7	0.00000
3153	1984 SH3		2.4234393	0.12875468	7.70758	2.1114101	3.7726645	14.00	6.4	0.00000
3154	1984 SO3		3.1068921	0.16348585	2.47767	2.5989592	5.4763255	13.50	8.0	0.00000
3155	1984 SP3		2.3426600	0.10063385	7.19962	2.1069090	3.5856168	13.00	10.1	0.00000
3156	1953 EE		2.8560767	0.19709264	15.81805	2.2931652	4.8267503	12.50	12.7	0.00000
3157	1973 SX3		3.1554248	0.13879782	7.55497	2.7174587	5.6051440	12.50	12.7	0.00000
3158	1976 SU2		2.5498312	0.10271632	14.58132	2.2879219	4.0716190	14.00	6.4	0.00000
3159	1976 US2		2.5692222	0.10962675	14.63508	2.2875669	4.1181536	14.00	6.4	0.00000
3160	1980 LE		2.3779402	0.15470827	5.06967	2.0100532	3.6669197	14.50	5.1	0.00000
3161	1980 TB5		2.5702596	0.17503428	14.90219	2.1203761	4.1206474	13.00	10.1	0.00000
3162	1980 YH		3.1637781	0.14339758	17.92313	2.7100999	5.6274157	12.50	12.7	0.00000
3163	1981 QM		2.3961077	0.33118474	3.08249	1.6025532	3.7090225	15.00	4.0	0.00000
3164	6562 P-L		3.1404409	0.17278475	2.34697	2.5978208	5.5652661	13.50	8.0	0.00000
3165	MIKAWA		2.2449658	0.17873944	3.92346	1.8437020	3.3636796	14.00	6.4	0.00000
3166	1940 FG		2.2375052	0.11778132	5.24002	1.9739687	3.3469255	14.00	6.4	0.00000
3167	1955 RS		2.5410268	0.10465126	15.62886	2.2751052	4.0505490	12.50	12.7	0.00000
3168	1980 XM		2.9919972	0.09818900	10.54388	2.6982162	5.1753745	12.50	12.7	0.00000
3169	1981 LA		1.8916614	0.06695589	24.90969	1.7650034	2.6017473	13.50	8.0	0.00000
3170	1979 SS11		2.9292440	0.08810150	2.02604	2.6711731	5.0134115	13.00	10.1	0.00000
3171	1979 WO		3.1933069	0.12824406	11.37760	2.7837842	5.7063837	12.00	16.0	0.00000
3172	1981 WW		2.4269795	0.22262056	3.64948	1.8866839	3.7809348	14.50	5.1	0.00000
3173	1981 WY		2.2044036	0.20992269	7.79937	1.7416493	3.2729299	14.00	6.4	0.00000
3174	1984 UV		3.1532044	0.16769975	2.37379	2.6244128	5.5992284	13.00	10.1	0.00000
3175	1979 YP		2.3643043	0.21324219	0.64153	1.8601350	3.6354239	14.50	5.1	0.00000
3176	1980 VR1		2.8757260	0.03014803	18.10974	2.7890284	4.8766460	12.00	16.0	0.00000
3177	1934 AK		2.6335111	0.14982906	15.98597	2.2389345	4.2736874	13.00	10.1	0.00000
3178	1984 WA		2.7135239	0.37805602	6.82109	1.6876599	4.4699273	13.50	8.0	0.00000
3179	1962 FA		3.0961242	0.15968281	1.73769	2.6017263	5.4478803	13.50	8.0	0.00000
3180	1962 RO		2.2301364	0.14754014	5.27443	1.9011018	3.3304057	15.50	3.2	0.00000
3181	AHNERT		2.2287717	0.06519854	3.95869	2.0834589	3.3273489	14.00	6.4	0.00000
3182	1984 WC		2.6138918	0.14414020	12.58752	2.2371249	4.2260189	13.50	8.0	0.00000
3183	1949 PP		3.1901605	0.12934542	2.17778	2.7775280	5.6979523	13.00	10.1	0.00000
3184	1949 QC		2.6673234	0.26214263	8.22697	1.9681042	4.3562570	13.50	8.0	0.00000
3185	1953 VY1		2.3654172	0.19342951	3.96372	1.9078757	3.6379912	15.00	4.0	0.00000
3186	1973 SD3		3.1178863	0.16497029	0.77996	2.6035275	5.5054193	13.50	8.0	0.00000
3187	1977 TO3		2.2834201	0.05826511	2.75552	2.1503763	3.4504738	14.00	6.4	0.00000
3188	1978 OM		2.2890809	0.13344450	4.69481	1.9836155	3.4633124	14.50	5.1	0.00000
3189	1978 RF6		3.1112113	0.18428710	8.16658	2.5378551	5.4877491	14.00	6.4	0.00000
3190	1978 SR6		2.9999132	0.11114729	9.96060	2.6664810	5.1959271	14.00	6.4	0.00000
3191	1979 SX9		2.8750019	0.01361866	2.73775	2.8358481	4.8748040	13.50	8.0	0.00000
3192	1982 BY1		2.3760390	0.17104462	2.87831	1.9696302	3.6625228	14.50	5.1	0.00000
3193	1982 DJ		2.2953520	0.10641228	5.73055	2.0510986	3.4775548	14.50	5.1	0.00000
3194	DORSEY		3.0140829	0.09664773	10.96846	2.7227786	5.2327833	13.00	10.1	0.00000
3195	1978 PT2		2.9124627	0.06052949	0.86491	2.7361729	4.9703913	13.50	8.0	0.00000
3196	1978 RY		3.0296953	0.01976750	8.97005	2.9698057	5.2734933	14.00	6.4	0.00000
3197	1981 AD		2.6654077	0.18244655	16.40863	2.1791131	4.3515644	14.50	5.1	0.00000
3198	1981 YH1		2.1799629	0.23856737	17.98653	1.6598948	3.2186491	14.50	5.1	0.00000
3199	NEFERTITI		1.5747294	0.28376126	32.97521	1.1278822	1.9761000	16.30	2.2	0.00000
3200	PHAETHON		1.2713993	0.89015728	22.02741	0.1396539	1.4335833	16.00	2.5	0.00000
3201	6560 P-L		2.2577412	0.08842545	2.99177	2.0580993	3.3924325	15.00	4.0	0.00000
3202	A908 AA		3.9431431	0.10246359	11.04791	3.5391147	7.8300371	11.50	20.1	0.00000
3203	1938 SL		2.3232393	0.26294681	6.66052	1.7123511	3.5411227	14.50	5.1	0.00000
3204	1978 RH		3.1823938	0.26458085	2.06342	2.3403933	5.6771564	13.00	10.1	0.00000
3205	1979 MO6		2.6828525	0.19871494	12.30453	2.1497295	4.3943553	14.50	5.1	0.00000
3206	1980 VN1		2.5534449	0.23520280	8.67373	1.9528674	4.0802774	15.00	4.0	0.00000
3207	1981 EY25		2.9088707	0.05982992	2.20971	2.7348335	4.9611993	12.00	16.0	0.00000
3208	1981 JM		3.1121118	0.12028612	2.33695	2.7377682	5.4901323	13.00	10.1	0.00000
3209	1982 BL1		2.1921170	0.05323294	5.23456	2.0754242	3.2456043	14.50	5.1	0.00000
3210	1983 WH1		3.1087081	0.05921341	13.63333	2.9246311	5.4811277	12.50	12.7	0.00000
3211	1931 CE		2.7339485	0.25127068	10.48697	2.0469873	4.5204897	14.00	6.4	0.00000
3212	1938 DH2		2.2556894	0.15256758	7.81688	1.9115443	3.3878093	15.00	4.0	0.00000
3213	1977 NQ		3.2127166	0.14201826	0.96675	2.7564521	5.7584901	13.00	10.1	0.00000
3214	1978 TZ6		3.0139713	0.05674669	11.49773	2.8429384	5.2324934	12.00	16.0	0.00000
3215	LAPKO		3.1232731	0.10796699	7.22290	2.7860627	5.5196934	13.00	10.1	0.00000
3216	1980 RB		2.3958392	0.30519375	4.92317	1.6646440	3.7083993	15.00	4.0	0.00000
3217	1980 RK		2.3872714	0.26268601	6.14759	1.7601686	3.6885245	15.50	3.2	0.00000
3218	6611 P-L		2.5209119	0.21736908	2.71155	1.9729437	4.0025477	15.00	4.0	0.00000
3219	KOMAKI		3.0345974	0.13169540	6.80072	2.6349549	5.2862978	12.50	12.7	0.00000
3220	1951 WF		2.2255206	0.17371330	6.61623	1.8389181	3.3200715	14.50	5.1	0.00000
3221	1981 XF2		2.2044356	0.15330227	3.65393	1.8664907	3.2730012	14.50	5.1	0.00000
3222	1983 NJ		3.0887289	0.06549618	15.95764	2.8864288	5.4283729	12.50	12.7	0.00000
3223	1942 RN		2.6077731	0.14394185	10.02519	2.2324054	4.2111888	11.40	21.1	0.00000
3224	1977 RL6		2.7886965	0.16335449	4.29599	2.3331504	4.6569538	12.40	13.3	0.00000

No.	Name	Type	A (A.U.)	E	I (deg.)	q (A.U.)	Period (years)	Mag.	Rad. (km)	Rot. (hours)
3225	HOAG		1.8799102	0.05285457	25.06328	1.7805483	2.5775414	14.00	6.4	0.00000
3226	6565 P-L		2.8741541	0.07251528	3.06372	2.6657338	4.8726482	14.40	5.3	0.00000
3227	1928 DF		2.4442704	0.13994268	3.91910	2.1022127	3.8214121	13.50	8.0	0.00000
3228	1935 CL		2.4606354	0.13702396	1.92280	2.1234694	3.8598542	14.00	6.4	0.00000
3229	A916 PC		2.3135831	0.15301511	9.46442	1.9595701	3.5190682	13.50	8.0	0.00000
3230	1972 LE		3.1330314	0.32569489	15.61929	2.1126189	5.5455818	13.00	10.1	0.00000
3231	1972 RU2		2.4461541	0.12649086	6.39565	2.1367381	3.8258307	14.00	6.4	0.00000
3232	1974 SL		3.0208180	0.08203830	9.84757	2.7729952	5.2503333	13.00	10.1	0.00000
3233	1977 RA6		2.2261436	0.10381356	3.60362	1.9950397	3.3214657	14.00	6.4	0.00000
3234	1978 QO2		3.1122308	0.17970227	0.96644	2.5529559	5.4904470	13.50	8.0	0.00000
3235	1981 EL1		2.6854248	0.24282116	13.48401	2.0333469	4.4006772	14.50	5.1	0.00000
3236	1982 BH1		2.2012882	0.14513974	1.10745	1.8817939	3.2659941	15.00	4.0	0.00000
3237	VICTORPLATT		3.0151808	0.06458159	9.10980	2.8204556	5.2356434	12.50	12.7	0.00000
3238	1975 VB9		2.6652739	0.18547295	11.73350	2.1709375	4.3512373	14.50	5.1	0.00000
3239	1978 UJ2		2.1848884	0.22105491	3.02137	1.7019081	3.2295640	15.50	3.2	0.00000
3240	1978 VG6		5.2592525	0.12678443	2.32473	4.5924611	12.0610762	11.50	20.1	0.00000
3241	1978 WH14		3.0418940	0.16266938	1.64587	2.5470710	5.3053751	13.50	8.0	0.00000
3242	1979 SG9		2.6792741	0.16095857	12.38798	2.2480221	4.3855667	14.00	6.4	0.00000
3243	1980 DC		3.0390458	0.09740915	9.35185	2.7430151	5.2979259	12.50	12.7	0.00000
3244	4008 P-L		2.2434044	0.16411836	3.67929	1.8752204	3.3601706	15.50	3.2	0.00000
3245	JENSCH		3.1315644	0.15318984	0.33320	2.6518407	5.5416875	14.00	6.4	0.00000
3246	1976 GQ3		3.1956365	0.03461533	21.72647	3.0850186	5.7126298	13.00	10.1	0.00000
3247	1981 YE		2.3773220	0.12895048	3.93193	2.0707650	3.6654897	14.00	6.4	0.00000
3248	1982 FK		3.1901410	0.16905206	10.87744	2.6508410	5.6978998	12.00	16.0	0.00000
3249	1977 DT4		2.3466651	0.24696892	3.36920	1.7671118	3.5948162	14.80	4.4	0.00000
3250	1979 EB		3.0132987	0.10899272	9.55762	2.6848712	5.2307420	12.30	13.9	0.00000
3251	6536 P-L		3.1104810	0.16093110	0.72065	2.6099079	5.4858170	13.40	8.4	0.00000
3252	1981 EM4		2.6629076	0.11381208	12.71923	2.3598366	4.3454442	11.50	20.1	0.00000
3253	1982 HQ1		2.2482715	0.19833648	7.42710	1.8023571	3.3711114	15.00	4.0	0.00000
3254	BUS		3.9424629	0.17791848	4.46602	3.2410259	7.8280110	12.00	16.0	0.00000
3255	1980 RA		2.3721633	0.36286965	21.36917	1.5113772	3.6535654	15.00	4.0	0.00000
3256	DAGUERRE		2.7784343	0.09604807	7.83218	2.5115709	4.6312709	13.50	8.0	0.00000
3257	1982 GG		2.2506258	0.17019543	5.55679	1.8675796	3.3764083	14.50	5.1	0.00000
3258	1983 RJ		2.2056811	0.19629313	7.55201	1.7727211	3.2757754	14.50	5.1	0.00000
3259	1984 SZ4		3.1588953	0.13369089	15.52545	2.7365797	5.6143932	11.50	20.1	0.00000
3260	1974 SO2		2.2344539	0.09337802	5.21730	2.0258050	3.3400817	13.50	8.0	0.00000
3261	1979 SF9		2.9052527	0.07491337	2.74795	2.6876104	4.9519453	12.50	12.7	0.00000
3262	1983 WB		3.0069835	0.06673977	9.46345	2.8062980	5.2143068	12.00	16.0	0.00000
3263	1932 CN		2.4145651	0.06916675	7.74891	2.2475574	3.7519617	14.00	6.4	0.00000
3264	1934 AF		3.1452949	0.14775904	0.94096	2.6805491	5.5781736	13.50	8.0	0.00000
3265	1953 VN2		2.4101028	0.14198364	6.93721	2.0679078	3.7415657	14.00	6.4	0.00000
3266	1978 PA		1.9084493	0.11030985	26.37952	1.6979285	2.6364586	15.00	4.0	0.00000
3267	1981 AA		2.3285000	0.29683125	24.00062	1.6373284	3.5531566	14.00	6.4	0.00000
3268	1981 DD		2.3471942	0.12740992	6.35378	2.0481384	3.5960317	14.50	5.1	0.00000
3269	1981 EX16		2.7849774	0.15914412	17.18938	2.3417647	4.6476402	13.50	8.0	0.00000
3270	DUDLEY		2.1485317	0.33066267	27.63398	1.4380924	3.1492898	16.00	2.5	0.00000
3271	1982 RB		2.1021974	0.39490500	24.99721	1.2720292	3.0479672	18.00	1.0	0.00000
3272	1938 DB1		2.2435782	0.09244048	3.92782	2.0361807	3.3605611	14.00	6.4	0.00000
3273	1975 TS2		3.3922064	0.03992703	14.17302	3.2567658	6.2477469	12.50	12.7	0.00000
3274	1981 QO2		3.1561720	0.10211126	1.26822	2.8338914	5.6071348	13.00	10.1	0.00000
3275	1982 HE1		2.3333385	0.17984471	8.58041	1.9137000	3.5642376	14.50	5.1	0.00000
3276	1982 RZ1		3.1130877	0.17613745	2.68141	2.5647564	5.4927144	13.00	10.1	0.00000
3277	1984 AF1		3.1399074	0.27241251	8.57071	2.2845573	5.5638480	12.50	12.7	0.00000
3278	1984 BT		3.2110710	0.03854037	9.66321	3.0873151	5.7540665	12.50	12.7	0.00000
3279	9103 P-L		2.2024941	0.17400423	3.16143	1.8192509	3.2686782	14.50	5.1	0.00000
3280	1933 SJ		2.5827520	0.17377907	2.21790	2.1339235	4.1507254	13.50	8.0	0.00000
3281	1938 DZ		2.3508463	0.09748395	5.99072	2.1216767	3.6044283	13.50	8.0	0.00000
3282	1949 DA		2.1904020	0.04185595	3.16842	2.0987208	3.2417965	14.00	6.4	0.00000
3283	1979 QA10		2.3960981	0.10078035	6.88753	2.1546185	3.7090006	14.00	6.4	0.00000
3284	1953 NB		2.7618527	0.38854241	6.66698	1.6887559	4.5898743	14.00	6.4	0.00000
3285	RUTH WOLFE		2.5276985	0.21410356	20.51231	1.9865092	4.0187211	13.50	8.0	0.00000
3286	1980 BV		2.6386714	0.10344932	13.41081	2.3657026	4.2862549	14.00	6.4	0.00000
3287	1981 DK1		2.3654244	0.29938588	12.08611	1.6572497	3.6380076	15.00	4.0	0.00000
3288	SELEUCUS		2.0325189	0.45735368	5.93246	1.1029388	2.8976896	16.50	2.0	75.00000
3289	1934 RP		2.3268945	0.20737749	1.75979	1.8443489	3.5494823	15.40	3.3	0.00000
3290	1973 SZ1		3.9793172	0.12218674	2.76731	3.4930973	7.9380317	12.70	11.6	0.00000
3291	1982 VX3		3.1523669	0.09210833	1.99989	2.8620076	5.5969977	13.40	8.4	0.00000
3292	2631 P-L		3.1596465	0.17503448	1.57894	2.6065993	5.6163960	13.50	8.0	0.00000
3293	4650 P-L		2.4003668	0.13861267	2.13818	2.0676455	3.7189167	14.90	4.2	0.00000
3294	6563 P-L		2.6973963	0.07053705	6.98027	2.5071299	4.4301367	13.80	7.0	0.00000
3295	1950 DH		2.6962101	0.25411910	8.81375	2.0110517	4.4272151	14.00	6.4	0.00000
3296	1975 SF		2.6554863	0.19470941	13.92417	2.1384380	4.3272910	13.50	8.0	0.00000
3297	1978 WN14		3.1441236	0.16481031	2.37520	2.6259396	5.5750580	13.50	8.0	0.00000
3298	1979 OB15		2.3528616	0.19254149	2.57204	1.8998381	3.6090641	14.50	5.1	0.00000
3299	HALL		2.2797587	0.07889855	5.46941	2.0998890	3.4421780	14.50	5.1	0.00000
3300	1928 NA		3.1542001	0.21054246	18.78637	2.4901071	5.6018806	11.50	20.1	0.00000
3301	1978 CT		2.2340784	0.15282851	5.06566	1.8926474	3.3392396	14.50	5.1	0.00000
3302	1977 RS6		2.4534883	0.09580784	3.38399	2.2184248	3.8430495	14.10	6.1	0.00000
3303	1967 UN		2.8990004	0.07474758	2.76672	2.6823070	4.9359689	12.50	12.1	0.00000
3304	1981 EQ21		3.0607271	0.27421042	2.20398	2.2214439	5.3547220	14.00	6.1	0.00000
3305	1985 KB		2.6053102	0.15102024	13.44931	2.2118556	4.2052245	13.50	7.6	0.00000
3306	1979 SM11		2.2472098	0.14533156	4.52811	1.9206192	3.3687241	13.80	6.6	0.00000
3307	1981 DE1		2.2595909	0.09574083	6.37213	2.0432558	3.3966024	14.70	4.4	0.00000

No.	Name	Type	A (A.U.)	E	I (deg.)	q (A.U.)	Period (years)	Mag.	Rad. (km)	Rot. (hours)
3308	1981 EP		3.1521013	0.18035051	23.59079	2.5836182	5.5962901	12.50	12.1	0.00000
3309	1982 BH		1.8173424	0.05349827	21.13632	1.7201177	2.4499383	15.10	3.7	0.00000
3310	1931 TS2		3.0101738	0.05743832	11.08604	2.8372746	5.2226071	12.00	15.2	0.00000
3311	1976 QM1		2.7881951	0.03967588	0.92870	2.6775711	4.6556973	13.50	7.6	0.00000
3312	1984 SN		3.0062971	0.11701264	9.63125	2.6545224	5.2125216	13.00	9.6	0.00000
3313	1980 DG		2.6570399	0.13011698	11.36756	2.3113139	4.3310890	13.50	7.6	0.00000
3314	1981 FH		2.2179797	0.04455821	7.40890	2.1191504	3.3032110	14.50	4.8	0.00000
3315	1984 CZ		2.6416328	0.08600687	10.05287	2.4144342	4.2934723	13.50	7.6	0.00000
3316	1984 CN1		3.1286662	0.08913938	8.34779	2.8497789	5.5339961	12.50	12.1	0.00000
3317	PARIS		5.1943383	0.12631685	27.89591	4.5382061	11.8384638	9.50	48.2	0.00000
3318	1985 HB		3.0087824	0.04570288	11.57400	2.8712723	5.2189860	12.50	12.1	0.00000
3319	1977 EJ5		3.1604586	0.16030654	3.78556	2.6538162	5.6185617	13.20	8.8	0.00000
3320	1982 VZ4		2.4583817	0.04809184	4.07043	2.3401535	3.8545523	14.50	4.8	0.00000
3321	1975 TZ2		2.5455637	0.20201495	7.34660	2.0313220	4.0614023	14.50	4.8	0.00000
3322	1975 XY1		2.3930776	0.21360689	23.50306	1.8818997	3.7019894	13.50	7.6	0.00000
3323	1979 SY9		2.5616241	0.18596008	0.72024	2.0852642	4.0998983	15.00	3.8	0.00000
3324	1983 CW1		2.6979599	0.02582685	10.77496	2.6282799	4.4315252	13.50	7.6	0.00000
3325	1984 JZ		3.1856148	0.01357005	22.26948	3.1423860	5.6857781	12.50	12.1	0.00000
3326	1985 FL		2.3683131	0.17354278	3.37933	1.9573094	3.6446736	14.00	6.1	0.00000
3327	1985 PW		3.1659708	0.10766763	1.56299	2.8250980	5.6332669	13.00	9.6	0.00000
3328	1985 QD1		3.0146005	0.10973375	11.45566	2.6837969	5.2341318	12.50	12.1	0.00000
3329	1985 RT1		2.9964206	0.08174987	10.41050	2.7514637	5.1868558	13.00	9.6	0.00000
3330	1985 RU1		3.1332715	0.21836382	10.30592	2.4490783	5.5462189	12.00	15.2	0.00000
3331	1979 QS		2.4185665	0.09058040	3.56074	2.1994917	3.7612920	14.50	4.8	0.00000
3332	1980 NT1		2.5438607	0.08525970	14.89008	2.3269720	4.0573268	12.80	10.5	0.00000
3333	SCHABER		3.1216209	0.23049980	11.96883	2.4020879	5.5153141	12.70	11.0	0.00000
3334	1981 YR		2.8478661	0.02730450	3.26315	2.7701066	4.8059511	13.00	9.6	0.00000
3335	1966 AA		2.6091895	0.12770627	13.32533	2.2759798	4.2146206	13.50	7.6	0.00000
3336	1971 UX		2.3242631	0.18663076	0.85413	1.8904841	3.5434630	16.00	2.4	0.00000
3337	1971 UG1		2.8423243	0.07640189	1.98282	2.6251655	4.7919297	13.50	7.6	0.00000
3338	RICHTER		2.1460669	0.17018872	0.73593	1.7808305	3.1438720	15.50	3.0	0.00000
3339	1978 LB		3.1784754	0.13072349	17.80006	2.7629738	5.6666746	12.50	12.1	0.00000
3340	1979 TK		2.2333977	0.19460072	5.60337	1.7987770	3.3377142	15.50	3.0	0.00000
3341	1980 OD		3.0222471	0.23604099	10.47522	2.3088729	5.2540593	13.50	7.6	0.00000
3342	1982 BD3		3.1379950	0.06911983	6.19195	2.9210973	5.5587654	13.00	9.6	0.00000
3343	1982 HS		2.3492198	0.31112075	25.06916	1.6183288	3.6006880	14.50	4.8	0.00000
3344	MODENA		2.4159274	0.11809969	9.43987	2.1306071	3.7551374	14.00	6.1	0.00000
3345	1982 YC1		2.4751899	0.18544047	15.81122	2.0161896	3.8941510	13.00	9.6	0.00000
3346	1951 SD		3.1789110	0.05203630	21.58583	3.0134921	5.6678395	12.00	15.2	0.00000
3347	1975 VN1		3.1260712	0.10091986	4.77603	2.8105886	5.5271125	13.00	9.6	0.00000
3348	1978 EA3		3.1729953	0.16074401	10.39942	2.6629553	5.6520262	13.00	9.6	0.00000
3349	1979 FH2		2.7376654	0.03315597	4.23935	2.6468954	4.5297112	14.00	6.1	0.00000
3350	SCOBEE		2.3109493	0.20451510	3.40789	1.8383254	3.5130608	15.50	3.0	0.00000
3351	SMITH		3.0385706	0.27077559	13.20467	2.2157998	5.2966833	14.00	6.1	0.00000
3352	MCAULIFFE		1.8787979	0.36945680	4.77705	1.1846632	2.5752542	17.50	1.2	0.00000
3353	JARVIS		1.8630937	0.08467534	21.80774	1.7053356	2.5430331	14.50	4.8	0.00000
3354	MCNAIR		2.3239183	0.09680940	6.41051	2.0989411	3.5426748	14.00	6.1	0.00000
3355	ONIZUKA		2.1860380	0.06583677	4.06726	2.0421162	3.2321129	14.50	4.8	0.00000
3356	RESNIK		2.1923704	0.11397591	4.09669	1.9424930	3.2461674	14.50	4.8	0.00000
3357	1984 FT		3.0282009	0.05335973	11.25694	2.8666167	5.2695923	12.50	12.1	0.00000
3358	1978 RX		3.1994710	0.19076677	2.09552	2.5891182	5.7229147	12.80	10.5	0.00000
3359	1978 RA6		2.2568059	0.12139497	5.74100	1.9828410	3.3903246	15.30	3.3	0.00000
3360	1981 VA		2.4584634	0.74445021	22.01967	0.6282598	3.8547449	18.80	0.7	0.00000
3361	1982 HR		1.2094791	0.32276151	2.68724	0.8191058	1.3301406	19.80	0.4	0.00000
3362	KHUFU		0.9897225	0.46855494	9.92280	0.5259831	0.9846234	18.80	0.7	0.00000
3363	1960 EE		2.7755792	0.10014991	3.33147	2.4976051	4.6241341	12.80	10.5	0.00000
3364	1984 GF		2.1984701	0.10458526	5.55059	1.9685425	3.2597241	14.30	5.3	0.00000
3365	1985 CG2		2.7114279	0.17376661	7.79154	2.2402723	4.4647493	13.30	8.4	0.00000
3366	1985 SD1		3.0026572	0.08523679	9.95591	2.7467203	5.2030578	12.30	13.3	0.00000
3367	ALEX		2.7836823	0.06737732	5.31169	2.5961254	4.6443992	14.30	5.3	0.00000
3368	1985 QT		3.3813388	0.09742800	19.12054	3.0519018	6.2177472	11.80	16.7	0.00000
3369	1985 UZ		3.0465038	0.13038491	7.95281	2.6492856	5.3174396	13.30	8.4	0.00000
3370	1934 CU		2.2154069	0.10880671	7.11400	1.9743558	3.2974656	15.00	3.8	0.00000
3371	1955 RZ		2.7378113	0.01261043	9.67475	2.7032864	4.5300741	13.00	9.6	0.00000
3372	1976 SP4		2.6944261	0.13911100	3.28381	2.3196018	4.4228215	12.90	10.1	0.00000
3373	1978 QQ2		2.2458804	0.13012530	3.20357	1.9536345	3.3657348	13.80	6.6	0.00000
3374	1980 KO		2.9496264	0.01242431	3.02592	2.9129794	5.0658288	13.70	7.0	0.00000
3375	1981 JY1		2.1715386	0.02600962	1.08026	2.1150577	3.2000098	14.60	4.6	0.00000
3376	1982 UJ8		2.3481696	0.06762975	6.34130	2.1893635	3.5982735	13.20	8.8	0.00000
3377	4122 P-L		2.9131126	0.05843630	1.29277	2.7428811	4.9720545	13.30	8.4	0.00000
3378	A922 WB		2.3159575	0.09126011	8.07924	2.1046031	3.5244870	14.30	5.3	0.00000
3379	1931 TJ1		2.3545978	0.12992744	2.85303	2.0486710	3.6130593	13.30	8.4	0.00000
3380	1940 EF		2.8414693	0.02409444	3.24094	2.7730055	4.7897673	12.80	10.5	0.00000
3381	1941 UG		2.4537380	0.20421457	4.18989	1.9526488	3.8436360	14.30	5.3	0.00000
3382	1948 RD		2.2420247	0.18323046	6.00237	1.8312174	3.3570714	14.30	5.3	0.00000
3383	1951 AB		2.5660765	0.04536282	14.63770	2.4496720	4.1105919	13.30	8.4	0.00000
3384	1974 SB1		2.3848717	0.20981526	2.76061	1.8844893	3.6829648	14.30	5.3	0.00000
3385	1979 SK11		2.2204449	0.04165402	6.80622	2.1279545	3.3087199	13.80	6.6	0.00000
3386	1980 FA		2.8361542	0.08885150	2.16155	2.5841577	4.7763348	13.30	8.4	0.00000
3387	1981 WE		2.5990741	0.18968868	12.82414	2.1060593	4.1901350	13.80	6.6	0.00000
3388	1981 YR1		2.3627577	0.20155385	25.00687	1.8865349	3.6318576	14.80	4.2	0.00000
3389	1984 DU		2.7713177	0.13972403	7.07279	2.3840981	4.6134887	13.30	8.4	0.00000
3390	1984 ES1		2.2520187	0.11624436	3.38843	1.9902343	3.3795431	14.30	5.3	0.00000

No.	Name	Type	A (A.U.)	E	I (deg.)	q (A.U.)	Period (years)	Mag.	Rad. (km)	Rot. (hours)
3391	1977 DD3		5.2498493	0.08437435	14.90935	4.8068967	12.0287428	0.00	0.0	0.00000
3392	1979 YB		2.1396391	0.27998075	26.36067	1.5405813	3.1297581	0.00	365.3	0.00000
3393	1984 WY1		2.5850680	0.06538319	9.64090	2.4160480	4.1563101	0.00	763.2	0.00000
3394	1986 DB		2.3170824	0.19809000	7.08303	1.8580915	3.5270548	0.00	528.0	0.00000
3395	1985 UN		2.7912054	0.05543210	4.03760	2.6364830	4.6632390	0.00	0.0	0.00000
3396	A915 TE		3.3613575	0.20813726	8.38713	2.6617336	6.1627145	0.00	0.0	0.00000
3397	1964 XA		2.3491313	0.29781482	21.99736	1.6495252	3.6004844	0.00	528.0	0.00000
3398	1978 PC		2.2873991	0.23650607	24.17900	1.7464153	3.4594965	0.00	528.0	0.00000
3399	1979 SZ9		3.0987287	0.17644274	0.16425	2.5519805	5.4547558	0.00	836.9	0.00000
3400	1981 GX		1.9350888	0.09912682	20.22366	1.7432694	2.6918528	0.00	333.2	0.00000
3401	1981 PA		2.3691306	0.35783160	21.78650	1.5213808	3.6465611	0.00	528.0	0.00000
3402	1981 PB		2.1324244	0.27839538	4.84793	1.5387673	3.1139419	0.00	210.2	0.00000
3403	1981 SW		2.4118414	0.19357835	4.58344	1.9449611	3.7456150	0.00	664.7	0.00000
3404	1934 CY		2.6685126	0.12730756	10.04851	2.3287907	4.3591704	0.00	664.7	0.00000
3405	1964 UQ		2.6089947	0.11672750	13.17711	2.3044534	4.2141480	0.00	836.9	0.00000
3406	1969 DA		2.7936885	0.13378952	8.33413	2.4199224	4.6694636	0.00	0.0	0.00000
3407	1973 DT		2.6867533	0.15764809	13.10289	2.2631917	4.4039431	0.00	664.7	0.00000
3408	1977 QG4		2.3722072	0.22754517	2.86103	1.8324229	3.6536667	0.00	528.0	0.00000
3409	1977 RE6		2.8546312	0.08391337	1.39243	2.6150894	4.8230858	0.00	0.0	0.00000
3410	1978 SZ7		2.2599628	0.09772800	4.73456	2.0391011	3.3974414	0.00	528.0	0.00000
3411	1980 LK		2.2430117	0.11824423	5.38785	1.9777886	3.3592887	0.00	333.2	0.00000
3412	1983 AU2		2.2248397	0.10350888	2.97077	1.9945492	3.3185480	0.00	419.4	0.00000
3413	1983 CB3		2.2519829	0.12770580	5.79253	1.9643917	3.3794627	0.00	664.7	0.00000
3414	1983 DJ		2.1903782	0.10114237	5.29628	1.9688382	3.2417438	0.00	528.0	0.00000
3415	1928 SL		3.9680333	0.24891876	1.35481	2.9803154	7.9042916	11.30	21.0	0.00000
3416	1931 VP		1.9178053	0.20711431	22.05884	1.5206004	2.6558697	15.80	2.6	0.00000
3417	1937 GG		2.4235091	0.22589339	7.92524	1.8760545	3.7728279	15.30	3.3	0.00000
3418	1973 QZ1		3.1589181	0.18388407	1.89757	2.5780435	5.6144543	12.80	10.5	0.00000
3419	1981 JZ		3.2047234	0.07249836	17.63044	2.9723861	5.7370129	13.30	8.4	0.00000
3420	1984 EB		3.1103966	0.07164622	14.31144	2.8875484	5.4855938	12.80	10.5	0.00000
3421	1979 WK1		2.2337782	0.09257394	2.46245	2.0269885	3.3385670	14.80	4.2	0.00000
3422	1978 OJ		2.6920080	0.14882250	13.92792	2.2913766	4.4168687	13.30	8.4	0.00000
3423	1981 CK		3.0494289	0.11083382	0.42459	2.7114491	5.3251004	13.30	8.4	0.00000
3424	1982 CD		2.5481596	0.07180035	6.75215	2.3652008	4.0676155	13.80	6.6	0.00000
3425	1929 BD		3.0011933	0.08568957	9.23234	2.7440224	5.1992531	11.80	16.7	0.00000
3426	1932 CQ		2.6186759	0.09689175	13.12995	2.3649478	4.2376261	13.80	6.6	0.00000
3427	1938 AD		2.2811067	0.13428164	2.60040	1.9747961	3.4452319	14.30	5.3	0.00000
3428	1952 JH		2.6652050	0.16356352	8.85432	2.2292745	4.3510685	13.30	8.4	0.00000
3429	1974 SU1		2.3389194	0.18967596	1.32909	1.8952826	3.5770323	14.80	4.2	0.00000
3430	1980 TF4		2.7583640	0.09753098	4.43202	2.4893379	4.5811801	13.30	8.4	0.00000
3431	1984 QC		3.0952089	0.03993321	12.28434	2.9716074	5.4454651	11.30	21.0	0.00000
3432	1986 EE		3.1715095	0.25341034	13.38268	2.3678162	5.6480560	12.30	13.3	0.00000
3433	1963 TJ1		2.3938479	0.18661021	4.51721	1.9471314	3.7037768	13.30	8.4	0.00000
3434	1981 VO		2.6377208	0.23084261	3.45207	2.0288224	4.2839384	13.60	7.3	0.00000
3435	1981 XC2		2.3239610	0.04661330	7.72144	2.2156334	3.5427725	13.80	6.6	0.00000
3436	1976 SS3		2.8636253	0.05854836	1.73725	2.6959646	4.8458977	12.80	10.5	0.00000
3437	1982 UZ5		2.2710598	0.07471609	3.94046	2.1013751	3.4224951	14.30	5.3	0.00000
3438	1974 SD5		3.0492895	0.19928347	15.27665	2.4416165	5.3247347	13.80	6.6	0.00000
3439	1983 RL2		2.7443588	0.13561347	4.74386	2.3721867	4.5463338	15.30	3.3	0.00000
3440	1950 DD		2.8019779	0.05768392	7.54050	2.6403487	4.6902614	12.80	10.5	0.00000
3441	1969 TS1		3.1008251	0.18668365	2.76432	2.5219517	5.4602923	13.30	8.4	0.00000
3442	1978 TO7		3.1586361	0.12713061	12.24058	2.7570767	5.6137023	12.30	13.3	0.00000
3443	1979 SB1		2.3908160	0.30935213	12.66951	1.6512119	3.6967425	14.30	5.3	0.00000
3444	1980 RJ2		2.5555079	0.26643759	6.44574	1.8746244	4.0852232	13.30	8.4	0.00000
3445	1983 FC		2.6873245	0.12518309	11.30624	2.3509169	4.4053473	13.30	8.4	0.00000

The plates

Plates V and VI were obtained by Charles Kowal, at Palomar Observatory. All other plates are courtesy of the National Aeronautics and Space Administration.

Plate I — The Lost City meteorite. This meteorite is an H-type chondrite. The meteor was seen while it fell through the atmosphere, its orbit was calculated, and the meteorite was then found near Lost City, Oklahoma. Studies of this meteorite show that it was exposed to the cosmic rays of space for about five million years, before striking the Earth. Its orbit resembled the orbits of typical Apollo-type asteroids.

Plate II — A brecciated meteorite. Note how the meteorite is composed of many small inclusions cemented together. The meteorite evidently came from the regolith of some object. This particular meteorite, which was found in Antarctica, may actually have come from the Moon.

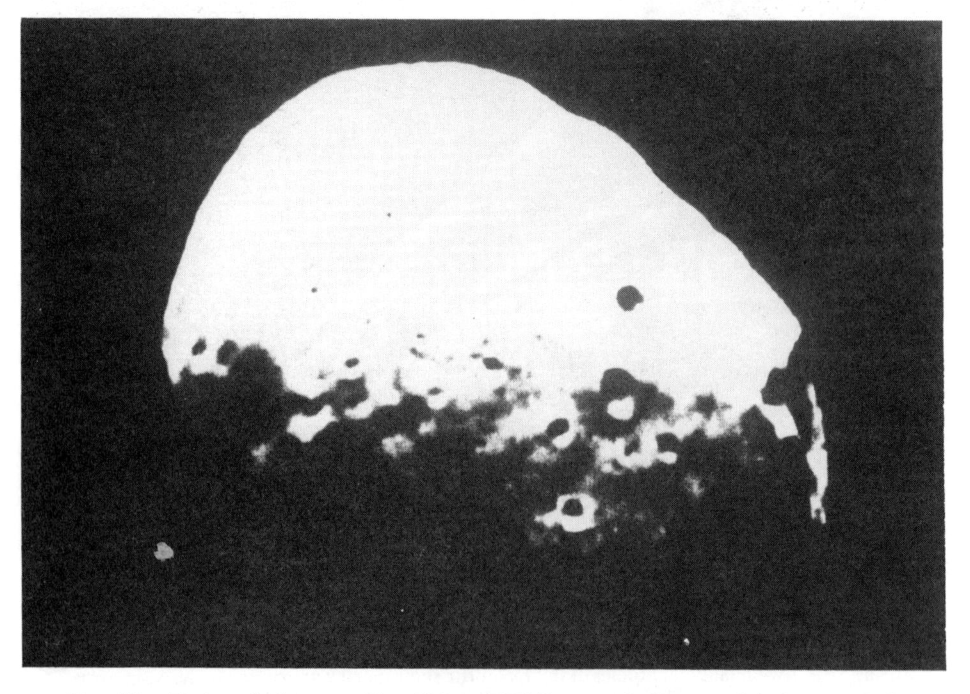

Plate III — Phobos, the inner satellite of Mars, is 27 kilometers in diameter. It is dust-covered, heavily cratered and fractured. We expect that typical asteroids will look much like this.

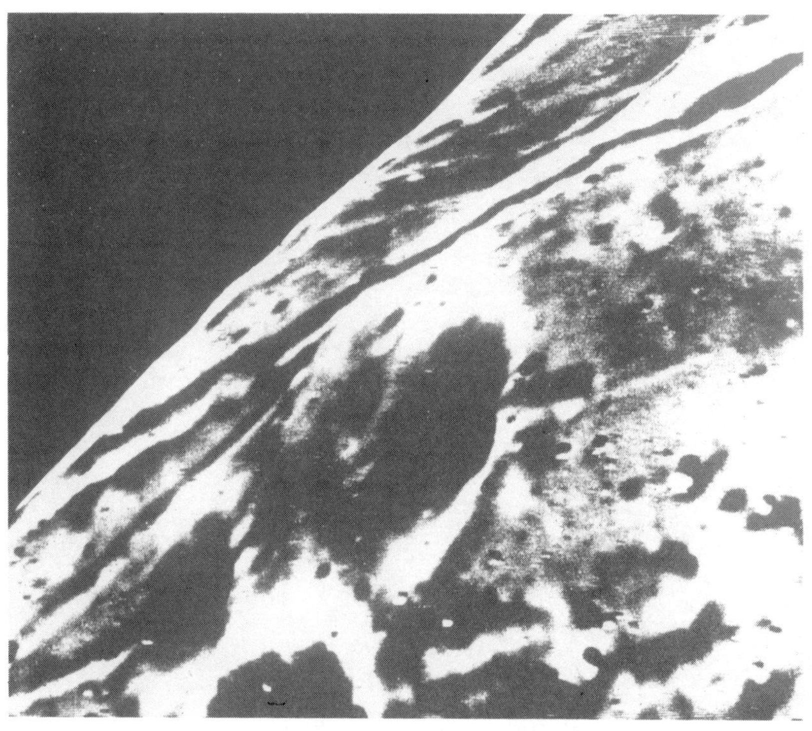

Plate IV — A close-up of Phobos. Note the long groove in Phobos's surface, which is thought to be an indication of fracturing of the surface of this satellite. Evidently, Phobos has undergone violent collisions, just like the asteroids.

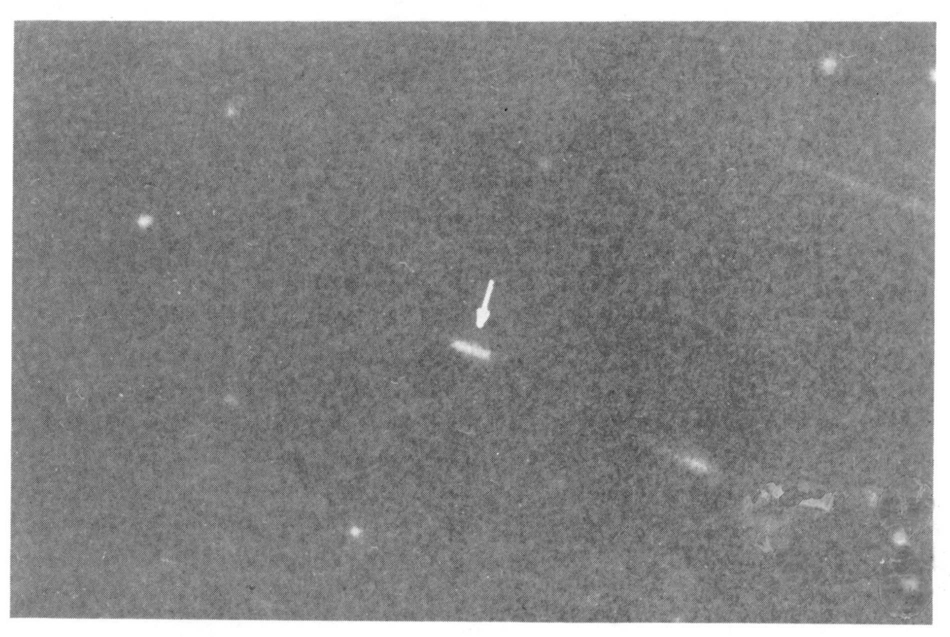

Plate V — (2060)Chiron. This object was discovered because of its slow motion across the sky. notice the short trail it made during this 75-minute exposure.

Plate VI — (2063)Bacchus. Unlike Chiron, Bacchus moves very quickly, and made a long trail during this exposure.

Plate VII — A mining camp on the Moon. The crew's living quarters, at top center, are covered with lunar soil to provide insulation and protection from cosmic rays and solar flares.

Plate VIII — An asteroid brought into Earth orbit. The asteroid's mass is being used to provide raw materials for construction of the Solar Power Satellite. Another asteroid is being used as reaction mass to drive the spacescraft in the background. The Solar Power Satellite is 5 kilometers wide, 15 kilometers long, and could deliver 10 gigawatts of electrical power to the Earth via microwaves.

Plate IX — A Solar Power Satellite under construction. Such enormous structures can be built in space because of the absence of gravity.

Plate X — A solar sail. This sail is 700 meters square, and would be suitable for sending unmanned research spacecraft to the asteroids. Larger sails would be needed for asteroid retrieval.

Plate XI — Exterior of the space colony. This colony is composed of two large cylinders. Each cylinder is 32 kilometers long, and is ringed by small agricultural stations. Three hinged mirrors on each cylinder direct sunlight into the interior of the twin colonies.

Plate XII — Interior of a space colony. Note the park-like environment, including even a flowing stream. Artificial gravity is produced by rotating the entire structure at a rate of two revolutions per minute. The Earth and Moon are visible through the enormous windows.

Plate XIII — The Hubble Space Telescope. This telescope will enable us to study objects in the solar system with detail matched only by interplanetary spacescraft.

Glossary

absolute magnitude — The brightness (or 'magnitude') that an object would have if we could see it from some standard distance. The absolute magnitude tells us how bright an object really is, with the effect of its distance removed. For asteroids, we must specify the distance of the asteroid from the Sun and from the Earth, and the viewing geometry, in order to calculate its absolute magnitude.

albedo — The reflectivity of an object. A bright, shiny object has a higher albedo than a dark, dull object.

astronomical unit — The average distance of the Earth from the Sun. This is 150 000 000 kilometers, or 93 000 000 miles. This unit is frequently used to describe distances within the solar system.

breccia — A rock formed from the compacted fragments of other rocks.

chondrules — Small grains of minerals imbedded within many meteorites.

commensurability — When the period of revolution of one object is a simple multiple of the period of another object, the two periods are said to be 'commensurate'. Since Jupiter has a period of revolution of about 12 years, an object having a period of 6 years would be in a '1:2 commensurability'.

Delta-V (ΔV) — The amount of velocity that must be imparted to a spacecraft in order to reach a particular object. Objects which are relatively easy to reach are said to require a low ΔV.

differentiation — The separation of different types of materials into different portions of a planet or asteroid. Generally, heavy elements sink to the center of a molten body, while lighter elements float to the top.

eccentricity — The degree to which an object's orbit differs from a perfect circle.

ecliptic — The plane of the Earth's orbit around the Sun.

ephemeris — A listing of the predicted positions of an object. The plural form is **ephemerides**.

flux — The amount of energy emitted by an object. Asteroids emit thermal energy (heat), in the form of infrared radiation. They also have a flux of energy at visible wavelengths, in the form of reflected sunlight.

igneous rock — Rock which has solidified from a molten state.

inclination — The degree to which an object's orbit is tilted, with respect to the ecliptic.

libration — An oscillation of the position of an object. The Moon 'librates', because we do not always see exactly the same face of the Moon. The Trojan asteroids librate, because they are not always exactly 60 degrees from Jupiter.

light curve — A graph of the brightness of an object versus time. If the brightness of an object changes as it rotates, we can determine its period of rotation and can obtain some idea of its shape.

magnitude — The brightness of an object, expressed on a logarithmic scale. Each magnitude step is a factor of 2.512 in brightness. The fainter an object is, the larger is its magnitude. Thus, a 5th-magnitude object is 2.512 times *brighter* than a 6th-magnitude object. This seemingly-backward system was devised by the ancient Greeks, and was made more quantitative in the nineteenth Century.

metamorphism — A change in the crystalline structure of a mineral, caused by heat or pressure.

occultation — An eclipse of a star by an asteroid. Basically, the asteroid casts its shadow on the Earth. Timing the duration of an occultation provides our most accurate means of measuring an asteroid's diameter.

opposition — The geometric situation in which a planet or asteroid is at the opposite point of the sky from the Sun.

perihelion — The point of an object's orbit at which the object is closest to the Sun.

phase — The degree of illumination of an object. For example, the phases of the moon.

photometer — An instrument for measuring the brightness of an object. A photometer can be photoelectric, photographic, or solid state.

photometry — Measuring the brightness of an object. Such measurements are usually made through differently colored filters, in order to determine the color, as well as the brightness, of the object.

photomultiplier — A vacuum tube which is sensitive to light, and which is used to measure the brightness of objects. Typically, photomultiplier tubes can amplify the energy which they 'see' by a factor of 100 000 to 1 000 000.

planetesimals — The small bodies in the early solar system which later coalesced into the planets and asteroids.

polarimetry — Measuring the degree of polarization of an object's light. For asteroids, such measurements give an indication of the surface texture of the object (whether dusty or rocky, for example).

radiometry — Measuring the infrared emission of an object, in order to determine its temperature, or other characteristics.

regolith — The fragmented, dusty or rocky surface of an object. A regolith may be a few millimeters, or a few kilometers, in depth.

semi-major axis — For asteroids and planets, this is the 'average' distance of the object from the Sun. Usually measured in astronomical units.

spectrophotometry — Measuring the brightness of an object at various wavelengths throughout its spectrum.

water of hydration — Water imbedded within the mineral structure of rock or other substance. Water of hydration gives clay its characteristic texture.

Bibliography

Recent results of asteroid research are published in technical journals such as *Icarus*, *The Astronomical Journal, Science,* and *Nature*. Less-technical information about asteroids is generally confined to single chapters in books about astronomy. Occasionally, articles about asteroids are published in magazines for amateur astronomers, and there are at least two small publications specifically for amateurs interested in asteroids and occultations.

Among the source materials used for this book, the following may be of general interest:

Technical

Physical Studies of Minor Planets. Tom Gehrels, ed. (1971). national Aeronautics and Space Administration, publication SP-267. US Government Printing Office, Washington, D.C.

Asteroids. Tom Gehrels, ed. (1979). University of Arizona Press, Tucson, Arizona.

The above two books are technical compilations of the state of asteroid science at the time they were published.They are essential reading for serious students of the asteroids.

History

Watchers of the Skies. Willy Ley (1963). The Viking Press, New York, N. Y.

Though long out of print, this is still my favorite book about the history of astronomy. You should be able to find it in large public libraries.

Space colonies

The High Frontier. Gerard O'Neill (1982). Anchor Press/Doubleday, Garden City, N.Y.

Several books about space colonization and industrialization have appeared, but this one is still the classic.

Celestial mechanics

Orbits for Amateurs. D. Tattersfield (1984–87). John Wiley, New York, N.Y. Two volumes.

Astronomical periodicals

Sky and Telescope. Sky Publishing, 49 Bay State Rd, Cambridge, Mass., 02238.

Astronomy. Astronomy Magazine 1027 N. Seventh St, Milwaukee, Wisconsin, 53233.

Tonight's Asteroids. Joseph F. Flowers, Route 4, Box 446, Wilson, North Carolina, 27893.

Occultation Newsletter. IOTA, 6 N 106 White Oak Lane, St Charles, Illinois, 60174.

Index

absolute magnitude, 27
Achilles, 13
achondrites, 42
Adonis, 15, 18
amateur astronomers, 8, 9, 89–91
Amor, 14
Amor asteroids, 15, 69–75
Apollo, 14, 18
Apollo asteroids, 15, 69–75, 8
asteroids
  brightness, 27–29
  captures, 59–61
  cataloging, 16–18
  classification, 37
  collisions, 15, 25, 52, 69–71
  density, 46
  distribution, 21, 23
  families, 22–24, 58
  mass, 46
  mining, 78
  naming, 17
  numbering, 17
  origin, 51
Aten, 15
Aten asteroids, 15, 69–75

Baade, Walter, 13, 65
Bode, Johann Elert, 11
Bode's Law, 11–12
breccia, 52
Budrosa family, 59

Callisto, 15
captured asteroids, 59–61
carbonaceous chondrites, 42
CCD, 37, 87
Central Bureau for Astronomical Telegrams, 18
Ceres, 8, 12, 56
Chaisson, Lola, 66
Chiron, 14, 65–68
chondrites, 42
chondrules, 42
collisions, among asteroids, 25, 52
collisions, with planets, 15, 69–71
colonies, space, 80, 81
Cook, A. F., 64
CRAF, 85

Delporte, E., 14, 15
differentiation, 53
dinosaurs, 16, 70

Eos family, 24, 59
Ephemerides of Minor Planets, 17
Eros, 13
Eucrite meteorites, 43, 58, 59

Flora region, 24

Galileo (space probe), 84
gamma-ray spectrometry, 86
Gauss, Carl Friedrich, 12
Gehrels, Tom, 15, 87
Glaser, Peter, 80
Gradie, J., 64

Harding, Karl, 12
Hartmann, W. K., 64
Hektor, 64
Helin, E. K., 15
Herget, Paul, 17
Hermes, 15, 18
Hidalgo, 13, 64
Hilda asteroids, 63
Hirayama families, 22–24, 58–59
Hubble Space Telescope, 83
Hungaria group, 24

Institute for Theoretical Astronomy, 16
International Astronomical Union Circulars, 17
Io, 15
IRAS (Infrared Astronomical Satellite), 26, 83

Kirkwood gaps, 21–23
Koronis family, 24, 59
Kowal, Charles, T., 14, 18, 65
Kuiper, Gerard, P., 25

Lagrange, Joseph Louis, 13, 63

Lagrangian points, 13, 63
Lebofsky, L., 56
liberation points (*see* Lagrangian points)
Liller, W., 66
Lyot, Bernard, 40

Malthus, Thomas, 77
Marsden, Brian, 17, 18
McCrosky, R. E., 18
McDonald Observatory, 25
meteorites, 41–44
mining, 78–79
Minor Planet Center, 16–18
Minor Planet Circulars, 17

NASA, 84, 85, 87
NEAR, 85

occultations, 44, 89–91
Olbers, Wilhelm, 12
Oljato, 73–74
O'Neill, Gerard, K., 80–81

Pallas, 12, 57
Palomar-Leiden Survey, 25
Phocaea group, 24, 59
photometry, 35, 90
Piazzi, Giuseppe, 12
planetesimals, 51
polarimetry, 40

radar, 47–48, 75
radiometry, 39
regoliths, 49, 52
Reinmuth, Karl, 14, 15
remote sensing, 86

Shao, C.-Y., 18
Shoemaker, Eugene, 15
SIRTF, 83
solar power satellites, 80
space colonies, 80–81
Space Telescope, 83
Spacewatch, 87
speckle interferometry, 44–46
spectrophotometry, 36
surveys, 25

Themis family, 24, 59
Thule, 63
Titius, Johannes, 11
Titius-Bode Law, 11–12
Trojan asteroids, 13, 63–64
Tunguska meteoroid, 42

van Houten, K., 25
van Houten-Groeneveld, I., 25
Vesta, 12, 58
Veverka, J., 64

water of hydration, 43, 50, 56
Witt, G., 13
Wolf, Max, 13

X-ray fluorescence, 86

1983TB, 74–75
1986DA, 75